AN INTRODUCTION TO
Thermal
Physics

Daniel V. Schroeder

Weber State University

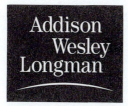

An imprint of Addison Wesley Longman

San Francisco, California • Reading, Massachusetts • New York • Harlow, England
Don Mills, Ontario • Sydney • Mexico City • Madrid • Amsterdam

Acquisitions Editor:	Sami Iwata
Publisher:	Robin J. Heyden
Marketing Manager:	Jennifer Schmidt
Production Coordination:	Joan Marsh
Cover Designer:	Mark Ong
Cover Printer:	Coral Graphics
Printer and Binder:	Maple-Vail Book Manufacturing Group

Library of Congress Cataloging-in-Publication Data

Schroeder, Daniel V.
      Introduction to thermal physics / Daniel V. Schroeder.
              p.   cm.
      Includes index.
      ISBN 0-201-38027-7
      1. Thermodynamics.   2. Statistical mechanics.   I. Title.
QC311.15.S32   1999
536'.7—dc21   99-31696
                                                 CIP

ISBN: 0-201-38027-7

25 24 23 22 21   —HPC—13 12 11

Contents

Part II: Thermodynamics

Part III: Statistical Mechanics

*   *   *

Preface

Thermal physics deals with collections of *large* numbers of particles—typically 10^{23} or so. Examples include the air in a balloon, the water in a lake, the electrons in a chunk of metal, and the photons (electromagnetic wave packets) given off by the sun. Anything big enough to see with our eyes (or even with a conventional microscope) has enough particles in it to qualify as a subject of thermal physics.

Consider a chunk of metal, containing perhaps 10^{23} ions and 10^{23} conduction electrons. We can't possibly follow every detail of the motions of all these particles, nor would we want to if we could. So instead, in thermal physics, we assume that the particles just jostle about randomly, and we use the laws of probability to predict how the chunk of metal as a whole ought to behave. Alternatively, we can measure the bulk properties of the metal (stiffness, conductivity, heat capacity, magnetization, and so on), and from these infer something about the particles it is made of.

Some of the properties of bulk matter don't really depend on the microscopic details of atomic physics. Heat always flows spontaneously from a hot object to a cold one, never the other way. Liquids always boil more readily at lower pressure. The maximum possible efficiency of an engine, working over a given temperature range, is the same whether the engine uses steam or air or anything else as its working substance. These kinds of results, and the principles that generalize them, comprise a subject called **thermodynamics**.

But to understand matter in more detail, we must also take into account both the quantum behavior of atoms and the laws of statistics that make the connection between one atom and 10^{23}. Then we can not only *predict* the properties of metals and other materials, but also explain *why* the principles of thermodynamics are what they are—why heat flows from hot to cold, for example. This underlying explanation of thermodynamics, and the many applications that come along with it, comprise a subject called **statistical mechanics**.

Physics instructors and textbook authors are in bitter disagreement over the proper content of a first course in thermal physics. Some prefer to cover only thermodynamics, it being less mathematically demanding and more readily applied to the everyday world. Others put a strong emphasis on statistical mechanics, with its spectacularly detailed predictions and concrete foundation in atomic physics. To some extent the choice depends on what application areas one has in mind: Thermodynamics is often sufficient in engineering or earth science, while statistical mechanics is essential in solid state physics or astrophysics.

In this book I have tried to do justice to *both* thermodynamics and statistical mechanics, without giving undue emphasis to either. The book is in three parts. Part I introduces the fundamental principles of thermal physics (the so-called first and second laws) in a unified way, going back and forth between the microscopic (statistical) and macroscopic (thermodynamic) viewpoints. This portion of the book also applies these principles to a few simple thermodynamic systems, chosen for their illustrative character. Parts II and III then develop more sophisticated techniques to treat further applications of thermodynamics and statistical mechanics, respectively. My hope is that this organizational plan will accomodate a variety of teaching philosophies in the middle of the thermo-to-statmech continuum. Instructors who are entrenched at one or the other extreme should look for a different book.

The thrill of thermal physics comes from using it to understand the world we live in. Indeed, thermal physics has so *many* applications that no single author can possibly be an expert on all of them. In writing this book I've tried to learn and include as many applications as possible, to such diverse areas as chemistry, biology, geology, meteorology, environmental science, engineering, low-temperature physics, solid state physics, astrophysics, and cosmology. I'm sure there are many fascinating applications that I've missed. But in my mind, a book like this one cannot have too many applications. Undergraduate physics students can and do go on to specialize in all of the subjects just named, so I consider it my duty to make you aware of some of the possibilities. Even if you choose a career entirely outside of the sciences, an understanding of thermal physics will enrich the experiences of every day of your life.

One of my goals in writing this book was to keep it short enough for a one-semester course. I have failed. Too many topics have made their way into the text, and it is now too long even for a very fast-paced semester. The book is still intended primarily *for* a one-semester course, however. Just be sure to omit several sections so you'll have time to cover what you *do* cover in some depth. In my own course I've been omitting Sections 1.7, 4.3, 4.4, 5.4 through 5.6, and all of Chapter 8. Many other portions of Parts II and III make equally good candidates for omission, depending on the emphasis of the course. If you're lucky enough to have *more* than one semester, then you can cover all of the main text and/or work some extra problems.

Listening to recordings won't teach you to play piano (though it can help), and reading a textbook won't teach you physics (though it too can help). To encourage you to learn actively while using this book, the publisher has provided ample margins for your notes, questions, and objections. I urge you to read with a pencil (not a highlighter). Even more important are the problems. All physics textbook authors tell their readers to work the problems, and I hereby do the same. In this book you'll encounter problems every few pages, at the end of almost every section. I've put them there (rather than at the ends of the chapters) to get your attention, to show you at every opportunity what you're now capable of doing. The problems come in all types: thought questions, short numerical calculations, order-of-magnitude estimates, derivations, extensions of the theory, new applications, and extended projects. The time required per problem varies by more than three orders of magnitude. Please work as many problems as you can, early and often. You won't have time to work all of them, but please *read* them all anyway, so you'll know what you're missing. Years later, when the mood strikes you, go back and work some of the problems you skipped the first time around.

Before reading this book you should have taken a year-long introductory physics course and a year of calculus. If your introductory course did not include any thermal physics you should spend some extra time studying Chapter 1. If your introductory course did not include any quantum physics you'll want to refer to Appendix A as necessary while reading Chapters 2, 6, and 7. Multivariable calculus is introduced in stages as the book goes on; a course in this subject would be a helpful, but not absolutely necessary, corequisite.

Some readers will be disappointed that this book does not cover certain topics, and covers others only superficially. As a partial remedy I have provided an annotated list of suggested further readings at the back of the book. A number of references on particular topics are given in the text as well. Except when I have borrowed some data or an illustration, I have not included any references merely to give credit to the originators of an idea. I am utterly unqualified to determine who deserves credit in any case. The occasional historical comments in the text are grossly oversimplified, intended to tell how things *could* have happened, not necessarily how they did happen.

No textbook is ever truly finished as it goes to press, and this one is no exception. Fortunately, the World-Wide Web gives authors a chance to continually provide updates. For the foreseeable future, the web site for this book will be at http://physics.weber.edu/thermal/. There you will find a variety of further information including a list of errors and corrections, platform-specific hints on solving problems requiring a computer, and additional references and links. You'll also find my e-mail address, to which you are welcome to send questions, comments, and suggestions.

Acknowledgments

It is a pleasure to thank the many people who have contributed to this project.

First there are the brilliant teachers who helped me learn thermal physics: Philip Wojak, Tom Moore, Bruce Thomas, and Michael Peskin. Tom and Michael have continued to teach me on a regular basis to this day, and I am sincerely grateful for these ongoing collaborations. In teaching thermal physics myself, I have especially depended on the insightful textbooks of Charles Kittel, Herbert Kroemer, and Keith Stowe.

As this manuscript developed, several brave colleagues helped by testing it in the classroom: Chuck Adler, Joel Cannon, Brad Carroll, Phil Fraundorf, Joseph Ganem, David Lowe, Juan Rodriguez, and Daniel Wilkins. I am indebted to each of them, and to their students, for enduring the many inconveniences of an unfinished textbook. I owe special thanks to my own students from seven years of teaching thermal physics at Grinnell College and Weber State University. I'm tempted to list all their names here, but instead let me choose just three to represent them all: Shannon Corona, Dan Dolan, and Mike Shay, whose questions pushed me to develop new approaches to important parts of the material.

Others who generously took the time to read and comment on early drafts of the manuscript were Elise Albert, W. Ariyasinghe, Charles Ebner, Alexander Fetter, Harvey Gould, Ying-Cheng Lai, Tom Moore, Robert Pelcovits, Michael Peskin, Andrew Rutenberg, Daniel Styer, and Larry Tankersley. Farhang Amiri, Lee Badger, and Adolph Yonkee provided essential feedback on individual chapters, while Colin Inglefield, Daniel Pierce, Spencer Seager, and John Sohl provided expert assistance with specific technical issues. Karen Thurber drew the magician and rabbit for Figures 1.15, 5.1, and 5.9. I am grateful to all of these individuals, and to the dozens of others who have answered questions, pointed to references, and given permission to reproduce their work.

I thank the faculty, staff, and administration of Weber State University, for providing support in a multitude of forms, and especially for creating an environment in which textbook writing is valued and encouraged.

It has been a pleasure to work with my editorial team at Addison Wesley Longman, especially Sami Iwata, whose confidence in this project has always exceeded my own, and Joan Marsh and Lisa Weber, whose expert advice has improved the appearance of every page.

In the space where most authors thank their immediate families, I would like to thank my family of friends, especially Deb, Jock, John, Lyall, Satoko, and Suzanne. Their encouragement and patience have been unlimited.

1 Energy in Thermal Physics

1.1 Thermal Equilibrium

The most familiar concept in thermodynamics is **temperature**. It's also one of the trickiest concepts—I won't be ready to tell you what temperature *really* is until Chapter 3. For now, however, let's start with a very naive definition:

> **Temperature** is what you measure with a thermometer.

If you want to measure the temperature of a pot of soup, you stick a thermometer (such as a mercury thermometer) into the soup, wait a while, then look at the reading on the thermometer's scale. This definition of temperature is what's called an **operational definition**, because it tells you how to *measure* the quantity in question.

Ok, but why does this procedure work? Well, the mercury in the thermometer expands or contracts, as its temperature goes up or down. Eventually the temperature of the mercury equals the temperature of the soup, and the volume occupied by the mercury tells us what that temperature is.

Notice that our thermometer (and any other thermometer) relies on the following fundamental fact: When you put two objects in contact with each other, and wait long enough, they tend to come to the same temperature. This property is so fundamental that we can even take it as an alternative *definition* of temperature:

> **Temperature** is the thing that's the same for two objects, after they've been in contact long enough.

I'll refer to this as the **theoretical definition** of temperature. But this definition is extremely vague: What kind of "contact" are we talking about here? How long is "long enough"? How do we actually ascribe a numerical value to the temperature? And what if there is more than one quantity that ends up being the same for both objects?

1

Before answering these questions, let me introduce some more terminology:

After two objects have been in contact long enough, we say that they are in **thermal equilibrium**.

The time required for a system to come to thermal equilibrium is called the **relaxation time**.

So when you stick the mercury thermometer into the soup, you have to wait for the relaxation time before the mercury and the soup come to the same temperature (so you get a good reading). After that, the mercury is in thermal equilibrium with the soup.

Now then, what do I mean by "contact"? A good enough definition for now is that "contact," in this sense, requires some means for the two objects to exchange energy spontaneously, in the form that we call "heat." Intimate mechanical contact (i.e., touching) usually works fine, but even if the objects are separated by empty space, they can "radiate" energy to each other in the form of electromagnetic waves. If you want to *prevent* two objects from coming to thermal equilibrium, you need to put some kind of thermal insulation in between, like spun fiberglass or the double wall of a thermos bottle. And even then, they'll eventually come to equilibrium; all you're really doing is increasing the relaxation time.

The concept of relaxation time is usually clear enough in particular examples. When you pour cold cream into hot coffee, the relaxation time for the contents of the cup is only a few seconds. However, the relaxation time for the coffee to come to thermal equilibrium with the surrounding room is many minutes.*

The cream-and-coffee example brings up another issue: Here the two substances not only end up at the same temperature, they also end up blended with each other. The blending is not necessary for *thermal* equilibrium, but constitutes a second type of equilibrium—**diffusive equilibrium**—in which the molecules of each substance (cream molecules and coffee molecules, in this case) are free to move around but no longer have any tendency to move one way or another. There is also **mechanical equilibrium**, when large-scale motions (such as the expansion of a balloon—see Figure 1.1) can take place but no longer do. For each type of equilibrium between two systems, there is a quantity that can be exchanged between the systems:

Exchanged quantity	Type of equilibrium
energy	thermal
volume	mechanical
particles	diffusive

Notice that for thermal equilibrium I'm claiming that the exchanged quantity is *energy*. We'll see some evidence for this in the following section.

When two objects are able to exchange energy, and energy tends to move spontaneously from one to the other, we say that the object that gives up energy is at

---

*Some authors define relaxation time more precisely as the time required for the temperature difference to decrease by a factor of $e \approx 2.7$. In this book all we'll need is a qualitative definition.

Figure 1.1. A hot-air balloon interacts thermally, mechanically, and diffusively with its environment—exchanging energy, volume, and particles. Not all of these interactions are at equilibrium, however.

a *higher* temperature, and the object that sucks in energy is at a *lower* temperature. With this convention in mind, let me now restate the theoretical definition of temperature:

> **Temperature** is a measure of the tendency of an object to spontaneously give up energy to its surroundings. When two objects are in thermal contact, the one that tends to spontaneously *lose* energy is at the *higher* temperature.

In Chapter 3 I'll return to this theoretical definition and make it much more precise, explaining, in the most fundamental terms, what temperature really *is*.

Meanwhile, I still need to make the *operational* definition of temperature (what you measure with a thermometer) more precise. How do you make a properly calibrated thermometer, to get a numerical *value* for temperature?

Most thermometers operate on the principle of thermal expansion: Materials tend to occupy more volume (at a given pressure) when they're hot. A mercury thermometer is just a convenient device for measuring the volume of a fixed amount of mercury. To define actual *units* for temperature, we pick two convenient temperatures, such as the freezing and boiling points of water, and assign them arbitrary numbers, such as 0 and 100. We then mark these two points on our mercury thermometer, measure off a hundred equally spaced intervals in between, and declare that this thermometer now measures temperature on the Celsius (or centigrade) scale, by definition!

Of course it doesn't have to be a mercury thermometer; we could instead exploit the thermal expansion of some other substance, such as a strip of metal, or a gas at fixed pressure. Or we could use an electrical property, such as the resistance, of some standard object. A few practical thermometers for various purposes are shown

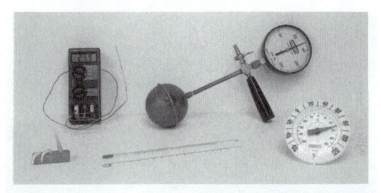

Figure 1.2. A selection of thermometers. In the center are two liquid-in-glass thermometers, which measure the expansion of mercury (for higher temperatures) and alcohol (for lower temperatures). The dial thermometer to the right measures the turning of a coil of metal, while the bulb apparatus behind it measures the pressure of a fixed volume of gas. The digital thermometer at left-rear uses a thermocouple—a junction of two metals—which generates a small temperature-dependent voltage. At left-front is a set of three potter's cones, which melt and droop at specified clay-firing temperatures.

in Figure 1.2. It's not obvious that the scales for various different thermometers would agree at all the intermediate temperatures between 0°C and 100°C. In fact, they generally won't, but in many cases the differences are quite small. If you ever have to measure temperatures with great precision you'll need to pay attention to these differences, but for our present purposes, there's no need to designate any one thermometer as the official standard.

A thermometer based on expansion of a gas is especially interesting, though, because if you extrapolate the scale down to very low temperatures, you are led to predict that for any low-density gas at constant pressure, the volume should go to *zero* at approximately −273°C. (In practice the gas will always liquefy first, but until then the trend is quite clear.) Alternatively, if you hold the volume of the gas fixed, then its *pressure* will approach zero as the temperature approaches −273°C (see Figure 1.3). This special temperature is called **absolute zero**, and defines the zero-point of the **absolute temperature scale**, first proposed by William Thomson in 1848. Thomson was later named Baron Kelvin of Largs, so the SI unit of absolute temperature is now called the **kelvin**.* A kelvin is the same size as a degree Celsius, but kelvin temperatures are measured up from absolute zero instead of from the freezing point of water. In round numbers, room temperature is approximately 300 K.

As we're about to see, many of the equations of thermodynamics are correct *only* when you measure temperature on the kelvin scale (or another absolute scale such as the Rankine scale defined in Problem 1.2). For this reason it's usually wise

---

*The Unit Police have decreed that it is impermissible to say "degree kelvin"—the name is simply "kelvin"—and also that the names of all Official SI Units shall not be capitalized.

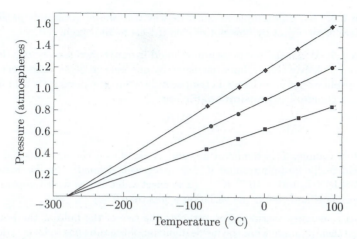

Figure 1.3. Data from a student experiment measuring the pressure of a fixed volume of gas at various temperatures (using the bulb apparatus shown in Figure 1.2). The three data sets are for three different amounts of gas (air) in the bulb. Regardless of the amount of gas, the pressure is a linear function of temperature that extrapolates to zero at approximately $-280°$C. (More precise measurements show that the zero-point does depend slightly on the amount of gas, but has a well-defined limit of $-273.15°$C as the density of the gas goes to zero.)

to convert temperatures to kelvins before plugging them into any formula. (Celsius is ok, though, when you're talking about the *difference* between two temperatures.)

Problem 1.1. The Fahrenheit temperature scale is defined so that ice melts at $32°$F and water boils at $212°$F.

(a) Derive the formulas for converting from Fahrenheit to Celsius and back.

(b) What is absolute zero on the Fahrenheit scale?

Problem 1.2. The Rankine temperature scale (abbreviated $°$R) uses the same size degrees as Fahrenheit, but measured up from absolute zero like kelvin (so Rankine is to Fahrenheit as kelvin is to Celsius). Find the conversion formula between Rankine and Fahrenheit, and also between Rankine and kelvin. What is room temperature on the Rankine scale?

Problem 1.3. Determine the kelvin temperature for each of the following:

(a) human body temperature;

(b) the boiling point of water (at the standard pressure of 1 atm);

(c) the coldest day you can remember;

(d) the boiling point of liquid nitrogen ($-196°$C);

(e) the melting point of lead ($327°$C).

Problem 1.4. Does it ever make sense to say that one object is "twice as hot" as another? Does it matter whether one is referring to Celsius or kelvin temperatures? Explain.

Problem 1.5. When you're sick with a fever and you take your temperature with a thermometer, approximately what is the relaxation time?

Problem 1.6. Give an example to illustrate why you *cannot* accurately judge the temperature of an object by how hot or cold it feels to the touch.

Problem 1.7. When the temperature of liquid mercury increases by one degree Celsius (or one kelvin), its volume increases by one part in 5500. The fractional increase in volume per unit change in temperature (when the pressure is held fixed) is called the **thermal expansion coefficient**, β:

$$\beta \equiv \frac{\Delta V/V}{\Delta T}$$

(where V is volume, T is temperature, and Δ signifies a change, which in this case should really be infinitesimal if β is to be well defined). So for mercury, $\beta = 1/5500 \text{ K}^{-1} = 1.81 \times 10^{-4} \text{ K}^{-1}$. (The exact value varies with temperature, but between 0°C and 200°C the variation is less than 1%.)

(a) Get a mercury thermometer, estimate the size of the bulb at the bottom, and then estimate what the inside diameter of the tube has to be in order for the thermometer to work as required. Assume that the thermal expansion of the glass is negligible.

(b) The thermal expansion coefficient of water varies significantly with temperature: It is $7.5 \times 10^{-4} \text{ K}^{-1}$ at 100°C, but decreases as the temperature is lowered until it becomes *zero* at 4°C. Below 4°C it is slightly *negative*, reaching a value of $-0.68 \times 10^{-4} \text{ K}^{-1}$ at 0°C. (This behavior is related to the fact that ice is less dense than water.) With this behavior in mind, imagine the process of a lake freezing over, and discuss in some detail how this process would be different if the thermal expansion coefficient of water were always positive.

Problem 1.8. For a solid, we also define the **linear thermal expansion coefficient**, α, as the fractional increase in length per degree:

$$\alpha \equiv \frac{\Delta L/L}{\Delta T}.$$

(a) For steel, α is $1.1 \times 10^{-5} \text{ K}^{-1}$. Estimate the total variation in length of a 1-km steel bridge between a cold winter night and a hot summer day.

(b) The dial thermometer in Figure 1.2 uses a coiled metal strip made of two different metals laminated together. Explain how this works.

(c) Prove that the volume thermal expansion coefficient of a solid is equal to the sum of its linear expansion coefficients in the three directions: $\beta = \alpha_x + \alpha_y + \alpha_z$. (So for an isotropic solid, which expands the same in all directions, $\beta = 3\alpha$.)

1.2 The Ideal Gas

Many of the properties of a low-density gas can be summarized in the famous **ideal gas law**,

$$PV = nRT, \tag{1.1}$$

where P = pressure, V = volume, n = number of moles of gas, R is a universal constant, and T is the temperature *in kelvins*. (If you were to plug a Celsius temperature into this equation you would get nonsense—it would say that the volume or pressure of a gas goes to zero at the freezing temperature of water and becomes negative at still lower temperatures.)

The constant R in the ideal gas law has the empirical value

$$R = 8.31 \; \frac{J}{mol \cdot K} \tag{1.2}$$

in SI units, that is, when you measure pressure in $N/m^2 = Pa$ (pascals) and volume in m^3. Chemists often measure pressure in atmospheres (1 atm = 1.013×10^5 Pa) or bars (1 bar = 10^5 Pa exactly) and volume in liters (1 liter = $(0.1 \; m)^3$), so be careful.

A **mole** of molecules is Avogadro's number of them,

$$N_A = 6.02 \times 10^{23}. \tag{1.3}$$

This is another "unit" that's more useful in chemistry than in physics. More often we will want to simply discuss the number of *molecules*, denoted by capital N:

$$N = n \times N_A. \tag{1.4}$$

If you plug in N/N_A for n in the ideal gas law, then group together the combination R/N_A and call it a new constant k, you get

$$PV = NkT. \tag{1.5}$$

This is the form of the ideal gas law that we'll usually use. The constant k is called **Boltzmann's constant**, and is tiny when expressed in SI units (since Avogadro's number is so huge):

$$k = \frac{R}{N_A} = 1.381 \times 10^{-23} \; J/K. \tag{1.6}$$

In order to remember how all the constants are related, I recommend memorizing

$$nR = Nk. \tag{1.7}$$

Units aside, though, the ideal gas law summarizes a number of important physical facts. For a given amount of gas at a given temperature, doubling the pressure squeezes the gas into exactly half as much space. Or, at a given volume, doubling the temperature causes the pressure to double. And so on. The problems below explore just a few of the implications of the ideal gas law.

Like nearly all the laws of physics, the ideal gas law is an *approximation*, never exactly true for a real gas in the real world. It is valid in the limit of low density, when the average space between gas molecules is much larger than the size of a molecule. For air (and other common gases) at room temperature and atmospheric pressure, the average distance between molecules is roughly ten times the size of a molecule, so the ideal gas law is accurate enough for most purposes.

Problem 1.9. What is the volume of one mole of air, at room temperature and 1 atm pressure?

Problem 1.10. Estimate the number of air molecules in an average-sized room.

Problem 1.11. Rooms A and B are the same size, and are connected by an open door. Room A, however, is warmer (perhaps because its windows face the sun). Which room contains the greater mass of air? Explain carefully.

Problem 1.12. Calculate the average volume per molecule for an ideal gas at room temperature and atmospheric pressure. Then take the cube root to get an estimate of the average distance between molecules. How does this distance compare to the size of a small molecule like N_2 or H_2O?

Problem 1.13. A mole is approximately the number of protons in a gram of protons. The mass of a neutron is about the same as the mass of a proton, while the mass of an electron is usually negligible in comparison, so if you know the total number of protons and neutrons in a molecule (i.e., its "atomic mass"), you know the approximate mass (in grams) of a mole of these molecules.* Referring to the periodic table at the back of this book, find the mass of a mole of each of the following: water, nitrogen (N_2), lead, quartz (SiO_2).

Problem 1.14. Calculate the mass of a mole of dry air, which is a mixture of N_2 (78% by volume), O_2 (21%), and argon (1%).

Problem 1.15. Estimate the average temperature of the air inside a hot-air balloon (see Figure 1.1). Assume that the total mass of the unfilled balloon and payload is 500 kg. What is the mass of the air inside the balloon?

Problem 1.16. The exponential atmosphere.

(a) Consider a horizontal slab of air whose thickness (height) is dz. If this slab is at rest, the pressure holding it up from below must balance both the pressure from above and the weight of the slab. Use this fact to find an expression for dP/dz, the variation of pressure with altitude, in terms of the density of air.

(b) Use the ideal gas law to write the density of air in terms of pressure, temperature, and the average mass m of the air molecules. (The information needed to calculate m is given in Problem 1.14.) Show, then, that the pressure obeys the differential equation

$$\frac{dP}{dz} = -\frac{mg}{kT}P,$$

called the **barometric equation**.

(c) Assuming that the temperature of the atmosphere is independent of height (not a great assumption but not terrible either), solve the barometric equation to obtain the pressure as a function of height: $P(z) = P(0)e^{-mgz/kT}$. Show also that the density obeys a similar equation.

---

*The precise definition of a mole is the number of carbon-12 atoms in 12 grams of carbon-12. The **atomic mass** of a substance is then the mass, in grams, of exactly one mole of that substance. Masses of *individual* atoms and molecules are often given in **atomic mass units**, abbreviated "u", where 1 u is defined as exactly 1/12 the mass of a carbon-12 atom. The mass of an isolated proton is actually slightly greater than 1 u, while the mass of an isolated neutron is slightly greater still. But in this problem, as in most thermal physics calculations, it's fine to round atomic masses to the nearest integer, which amounts to counting the total number of protons and neutrons.

(d) Estimate the pressure, in atmospheres, at the following locations: Ogden, Utah (4700 ft or 1430 m above sea level); Leadville, Colorado (10,150 ft, 3090 m); Mt. Whitney, California (14,500 ft, 4420 m); Mt. Everest, Nepal/Tibet (29,000 ft, 8850 m). (Assume that the pressure at sea level is 1 atm.)

Problem 1.17. Even at low density, real gases don't quite obey the ideal gas law. A systematic way to account for deviations from ideal behavior is the **virial expansion**,

$$PV = nRT\left(1 + \frac{B(T)}{(V/n)} + \frac{C(T)}{(V/n)^2} + \cdots\right),$$

where the functions $B(T)$, $C(T)$, and so on are called the **virial coefficients**. When the density of the gas is fairly low, so that the volume per mole is large, each term in the series is much smaller than the one before. In many situations it's sufficient to omit the third term and concentrate on the second, whose coefficient $B(T)$ is called the second virial coefficient (the first coefficient being 1). Here are some measured values of the second virial coefficient for nitrogen (N_2):

T (K)	B (cm^3/mol)
100	-160
200	-35
300	-4.2
400	9.0
500	16.9
600	21.3

(a) For each temperature in the table, compute the second term in the virial equation, $B(T)/(V/n)$, for nitrogen at atmospheric pressure. Discuss the validity of the ideal gas law under these conditions.

(b) Think about the forces between molecules, and explain why we might expect $B(T)$ to be negative at low temperatures but positive at high temperatures.

(c) Any proposed relation between P, V, and T, like the ideal gas law or the virial equation, is called an **equation of state**. Another famous equation of state, which is qualitatively accurate even for dense fluids, is the **van der Waals equation**,

$$\left(P + \frac{an^2}{V^2}\right)(V - nb) = nRT,$$

where a and b are constants that depend on the type of gas. Calculate the second and third virial coefficients (B and C) for a gas obeying the van der Waals equation, in terms of a and b. (Hint: The binomial expansion says that $(1+x)^p \approx 1 + px + \frac{1}{2}p(p-1)x^2$, provided that $|px| \ll 1$. Apply this approximation to the quantity $[1 - (nb/V)]^{-1}$.)

(d) Plot a graph of the van der Waals prediction for $B(T)$, choosing a and b so as to approximately match the data given above for nitrogen. Discuss the accuracy of the van der Waals equation over this range of conditions. (The van der Waals equation is discussed much further in Section 5.3.)

Microscopic Model of an Ideal Gas

In Section 1.1 I defined the concepts of "temperature" and "thermal equilibrium," and briefly noted that thermal equilibrium arises through the exchange of *energy* between two systems. But how, exactly, is temperature related to energy? The answer to this question is not simple in general, but it *is* simple for an ideal gas, as I'll now attempt to demonstrate.

I'm going to construct a mental "model" of a container full of gas.* The model will not be accurate in all respects, but I hope to preserve some of the most important aspects of the behavior of real low-density gases. To start with, I'll make the model as simple as possible: Imagine a cylinder containing just *one* gas molecule, as shown in Figure 1.4. The length of the cylinder is L, the area of the piston is A, and therefore the volume inside is $V = LA$. At the moment, the molecule has a velocity vector $\vec{v}$, with horizontal component v_x. As time passes, the molecule bounces off the walls of the cylinder, so its velocity changes. I'll assume, however, that these collisions are always elastic, so the molecule doesn't lose any kinetic energy; its *speed* never changes. I'll also assume that the surfaces of the cylinder and piston are perfectly smooth, so the molecule's path as it bounces is symmetrical about a line normal to the surface, just like light bouncing off a mirror.[†]

Here's my plan. I want to know how the *temperature* of a gas is related to the kinetic *energy* of the molecules it contains. But the only thing I know about temperature so far is the ideal gas law,

$$PV = NkT \tag{1.8}$$

(where P is pressure). So what I'll first try to do is figure out how the *pressure* is related to the kinetic energy; then I'll invoke the ideal gas law to relate pressure to temperature.

Well, what is the pressure of my simplified gas? Pressure means force per unit area, exerted in this case on the piston (and the other walls of the cylinder). What

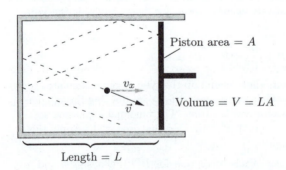

Piston area = A

v_x

$\vec{v}$

Volume = $V = LA$

Length = L

Figure 1.4. A greatly simplified model of an ideal gas, with just one molecule bouncing around elastically.

---

*This model dates back to a 1738 treatise by Daniel Bernoulli, although many of its implications were not worked out until the 1840s.

[†]These assumptions are actually valid only for the *average* behavior of molecules bouncing off surfaces; in any *particular* collision a molecule might gain or lose energy, and can leave the surface at almost any angle.

is the pressure exerted on the piston by the molecule? Usually it's zero, since the molecule isn't even touching the piston. But periodically the molecule crashes into the piston and bounces off, exerting a relatively large force on the piston for a brief moment. What I really want to know is the *average* pressure exerted on the piston over long time periods. I'll use an overbar to denote an average taken over some long time period, like this: $\overline{P}$. I can calculate the average pressure as follows:

$$\overline{P} = \frac{\overline{F}_{x,\text{ on piston}}}{A} = \frac{-\overline{F}_{x,\text{ on molecule}}}{A} = -\frac{m\left(\overline{\frac{\Delta v_x}{\Delta t}}\right)}{A}. \tag{1.9}$$

In the first step I've written the pressure in terms of the x component of the force exerted by the molecule on the piston. In the second step I've used Newton's third law to write this in terms of the force exerted by the piston on the molecule. Finally, in the third step, I've used Newton's second law to replace this force by the mass m of the molecule times its acceleration, $\Delta v_x/\Delta t$. I'm still supposed to average over some long time period; I can do this simply by taking Δt to be fairly large. However, I should include only those accelerations that are caused by the piston, not those caused by the wall on the opposite side. The best way to accomplish this is to take Δt to be exactly the time it takes for the molecule to undergo one round-trip from the left to the right and back again:

$$\Delta t = 2L/v_x. \tag{1.10}$$

(Collisions with the perpendicular walls will not affect the molecule's motion in the x direction.) During this time interval, the molecule undergoes exactly one collision with the piston, and the change in its x velocity is

$$\Delta v_x = (v_{x,\text{ final}}) - (v_{x,\text{ initial}}) = (-v_x) - (v_x) = -2v_x. \tag{1.11}$$

Putting these expressions into equation 1.9, I find for the average pressure on the piston

$$\overline{P} = -\frac{m}{A}\frac{(-2v_x)}{(2L/v_x)} = \frac{mv_x^2}{AL} = \frac{mv_x^2}{V}. \tag{1.12}$$

It's interesting to think about why there are *two* factors of v_x in this equation. One of them came from Δv_x: If the molecule is moving faster, each collision is more violent and exerts more pressure. The other one came from Δt: If the molecule is moving faster, collisions occur more frequently.

Now imagine that the cylinder contains not just one molecule, but some large number, N, of identical molecules, with random* positions and directions of motion. I'll pretend that the molecules don't collide or interact with each other—just with

---

*What, exactly, does the word *random* mean? Philosophers have filled thousands of pages with attempts to answer this question. Fortunately, we won't be needing much more than an everyday understanding of the word. Here I simply mean that the distribution of molecular positions and velocity vectors is more or less uniform; there's no obvious tendency toward any particular direction.

the walls. Since each molecule periodically collides with the piston, the average pressure is now given by a sum of terms of the form of equation 1.12:

$$\overline{P}V = mv_{1x}^2 + mv_{2x}^2 + mv_{3x}^2 + \cdots. \qquad (1.13)$$

If the number of molecules is large, the collisions will be so frequent that the pressure is essentially continuous, and we can forget the overbar on the P. On the other hand, the sum of v_x^2 for all N molecules is just N times the *average* of their v_x^2 values. Using the same overbar to denote this average over all molecules, equation 1.13 then becomes

$$PV = Nm\overline{v_x^2}. \qquad (1.14)$$

So far I've just been exploring the consequences of my model, without bringing in any facts about the real world (other than Newton's laws). But now let me invoke the ideal gas law (1.8), treating it as an experimental fact. This allows me to substitute NkT for PV on the left-hand side of equation 1.14. Canceling the N's, we're left with

$$kT = m\overline{v_x^2} \qquad \text{or} \qquad \tfrac{1}{2}m\overline{v_x^2} = \tfrac{1}{2}kT. \qquad (1.15)$$

I wrote this equation the second way because the left-hand side is almost equal to the average translational **kinetic energy** of the molecules. The only problem is the x subscript, which we can get rid of by realizing that the same equation must also hold for y and z:

$$\tfrac{1}{2}m\overline{v_y^2} = \tfrac{1}{2}m\overline{v_z^2} = \tfrac{1}{2}kT. \qquad (1.16)$$

The average translational kinetic energy is then

$$\overline{K}_{\text{trans}} = \overline{\tfrac{1}{2}mv^2} = \tfrac{1}{2}m(\overline{v_x^2} + \overline{v_y^2} + \overline{v_z^2}) = \tfrac{1}{2}kT + \tfrac{1}{2}kT + \tfrac{1}{2}kT = \tfrac{3}{2}kT. \qquad (1.17)$$

(Note that the average of a sum is the sum of the averages.)

This is a good place to pause and think about what just happened. I started with a naive model of a gas as a bunch of molecules bouncing around inside a cylinder. I also invoked the ideal gas law as an experimental fact. Conclusion: The average translational kinetic energy of the molecules in a gas is given by a simple constant times the temperature. So if this model is accurate, the temperature of a gas is a direct measure of the average translational kinetic energy of its molecules.

This result gives us a nice interpretation of Boltzmann's constant, k. Recall that k has just the right units, J/K, to convert a temperature into an energy. Indeed, we now see that k is essentially a *conversion factor* between temperature and molecular energy, at least for this simple system. Think about the numbers, though: For an air molecule at room temperature (300 K), the quantity kT is

$$(1.38 \times 10^{-23} \text{ J/K})(300 \text{ K}) = 4.14 \times 10^{-21} \text{ J}, \qquad (1.18)$$

and the average translational energy is 3/2 times as much. Of course, since molecules are so small, we would expect their kinetic energies to be tiny. The joule, though, is not a very convenient unit for dealing with such small energies. Instead we often use the **electron-volt** (eV), which is the kinetic energy of an electron that has been accelerated through a voltage difference of one volt: $1 \text{ eV} = 1.6 \times 10^{-19}$ J. Boltzmann's constant is 8.62×10^{-5} eV/K, so at room temperature,

$$kT = (8.62 \times 10^{-5} \text{ eV/K})(300 \text{ K}) = 0.026 \text{ eV} \approx \frac{1}{40} \text{ eV}. \qquad (1.19)$$

Even in electron-volts, molecular energies at room temperature are rather small.

If you want to know the average *speed* of the molecules in a gas, you can *almost* get it from equation 1.17, but not quite. Solving for $\overline{v^2}$ gives

$$\overline{v^2} = \frac{3kT}{m}, \qquad (1.20)$$

but if you take the square root of both sides, you get not the average speed, but rather the square root of the average of the squares of the speeds (root-mean-square, or rms for short):

$$v_{\text{rms}} \equiv \sqrt{\overline{v^2}} = \sqrt{\frac{3kT}{m}}. \qquad (1.21)$$

We'll see in Section 6.4 that v_{rms} is only slightly larger than $\overline{v}$, so if you're not too concerned about accuracy, v_{rms} is a fine estimate of the average speed. According to equation 1.21, light molecules tend to move faster than heavy ones, at a given temperature. If you plug in some numbers, you'll find that small molecules at ordinary temperatures are bouncing around at *hundreds* of meters per second.

Getting back to our main result, equation 1.17, you may be wondering whether it's really true for real gases, given all the simplifying assumptions I made in deriving it. Strictly speaking, my derivation breaks down if molecules exert forces on each other, or if collisions with the walls are inelastic, or if the ideal gas law itself fails. Brief interactions between molecules are generally no big deal, since such collisions won't change the average velocities of the molecules. The only serious problem is when the gas becomes so dense that the space occupied by the molecules themselves becomes a substantial fraction of the total volume of the container. Then the basic picture of molecules flying in straight lines through empty space no longer applies. In this case, however, the ideal gas law also breaks down, in such a way as to precisely preserve equation 1.17. Consequently, this equation is still true, not only for dense gases but also for most liquids and sometimes even solids! I'll prove it in Section 6.3.

Problem 1.18. Calculate the rms speed of a nitrogen molecule at room temperature.

Problem 1.19. Suppose you have a gas containing hydrogen molecules and oxygen molecules, in thermal equilibrium. Which molecules are moving faster, on average? By what factor?

Problem 1.20. Uranium has two common isotopes, with atomic masses of 238 and 235. One way to separate these isotopes is to combine the uranium with fluorine to make uranium hexafluoride gas, UF_6, then exploit the difference in the average thermal speeds of molecules containing the different isotopes. Calculate the rms speed of each type of molecule at room temperature, and compare them.

Problem 1.21. During a hailstorm, hailstones with an average mass of 2 g and a speed of 15 m/s strike a window pane at a 45° angle. The area of the window is 0.5 m^2 and the hailstones hit it at a rate of 30 per second. What average pressure do they exert on the window? How does this compare to the pressure of the atmosphere?

Problem 1.22. If you poke a hole in a container full of gas, the gas will start leaking out. In this problem you will make a rough estimate of the rate at which gas escapes through a hole. (This process is called **effusion**, at least when the hole is sufficiently small.)

(a) Consider a small portion (area = A) of the inside wall of a container full of gas. Show that the number of molecules colliding with this surface in a time interval Δt is $P A \Delta t / (2m\overline{v_x})$, where P is the pressure, m is the average molecular mass, and $\overline{v_x}$ is the average x velocity of those molecules that collide with the wall.

(b) It's not easy to calculate $\overline{v_x}$, but a good enough approximation is $(\overline{v_x^2})^{1/2}$, where the bar now represents an average over all molecules in the gas. Show that $(\overline{v_x^2})^{1/2} = \sqrt{kT/m}$.

(c) If we now take away this small part of the wall of the container, the molecules that *would* have collided with it will instead escape through the hole. Assuming that nothing *enters* through the hole, show that the number N of molecules inside the container as a function of time is governed by the differential equation

$$\frac{dN}{dt} = -\frac{A}{2V}\sqrt{\frac{kT}{m}}\,N.$$

Solve this equation (assuming constant temperature) to obtain a formula of the form $N(t) = N(0)e^{-t/\tau}$, where τ is the "characteristic time" for N (and P) to drop by a factor of e.

(d) Calculate the characteristic time for air at room temperature to escape from a 1-liter container punctured by a 1-mm^2 hole.

(e) Your bicycle tire has a slow leak, so that it goes flat within about an hour after being inflated. Roughly how big is the hole? (Use any reasonable estimate for the volume of the tire.)

(f) In Jules Verne's *Round the Moon*, the space travelers dispose of a dog's corpse by quickly opening a window, tossing it out, and closing the window. Do you think they can do this quickly enough to prevent a significant amount of air from escaping? Justify your answer with some rough estimates and calculations.

1.3 Equipartition of Energy

Equation 1.17 is a special case of a much more general result, called the **equipartition theorem**. This theorem concerns not just translational kinetic energy but *all* forms of energy for which the formula is a quadratic function of a coordinate or velocity component. Each such form of energy is called a **degree of freedom**. So far, the only degrees of freedom I've talked about are translational motion in the x, y, and z directions. Other degrees of freedom might include rotational motion, vibrational motion, and elastic potential energy (as stored in a spring). Look at the similarities of the formulas for all these types of energy:

$$\tfrac{1}{2}mv_x^2, \quad \tfrac{1}{2}mv_y^2, \quad \tfrac{1}{2}mv_z^2, \quad \tfrac{1}{2}I\omega_x^2, \quad \tfrac{1}{2}I\omega_y^2, \quad \tfrac{1}{2}k_sx^2, \quad \text{etc.} \tag{1.22}$$

The fourth and fifth expressions are for rotational kinetic energy, a function of the moment of inertia I and the angular velocity ω. The sixth expression is for elastic potential energy, a function of the spring constant k_s and the amount of displacement from equilibrium, x. The equipartition theorem simply says that for each degree of freedom, the average energy will be $\tfrac{1}{2}kT$:

> **Equipartition theorem:** At temperature T, the average energy of any quadratic degree of freedom is $\tfrac{1}{2}kT$.

If a system contains N molecules, each with f degrees of freedom, and there are no other (non-quadratic) temperature-dependent forms of energy, then its *total* thermal energy is

$$U_{\text{thermal}} = N \cdot f \cdot \tfrac{1}{2}kT. \tag{1.23}$$

Technically this is just the *average* total thermal energy, but if N is large, fluctuations away from the average will be negligible.

I'll prove the equipartition theorem in Section 6.3. For now, though, it's important to understand exactly what it says. First of all, the quantity U_{thermal} is almost never the *total* energy of a system; there's also "static" energy that doesn't change as you change the temperature, such as energy stored in chemical bonds or the rest energies (mc^2) of all the particles in the system. So it's safest to apply the equipartition theorem only to *changes* in energy when the temperature is raised or lowered, and to avoid phase transformations and other reactions in which bonds between particles may be broken.

Another difficulty with the equipartition theorem is in counting how many degrees of freedom a system has. This is a skill best learned through examples. In a gas of monatomic molecules like helium or argon, only translational motion counts, so each molecule has three degrees of freedom, that is, $f = 3$. In a diatomic gas like oxygen (O_2) or nitrogen (N_2), each molecule can also *rotate* about two different axes (see Figure 1.5). Rotation about the axis running down the length of the molecule doesn't count, for reasons having to do with quantum mechanics. The

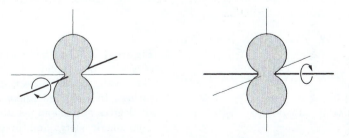

Figure 1.5. A diatomic molecule can rotate about two independent axes, perpendicular to each other. Rotation about the third axis, down the length of the molecule, is not allowed.

same is true for carbon dioxide (CO_2), since it also has an axis of symmetry down its length. However, most polyatomic molecules can rotate about all three axes.

It's not obvious why a rotational degree of freedom should have exactly the same average energy as a translational degree of freedom. However, if you imagine gas molecules knocking around inside a container, colliding with each other and with the walls, you can see how the average rotational energy should eventually reach some equilibrium value that is larger if the molecules are moving fast (high temperature) and smaller if the molecules are moving slow (low temperature). In any particular collision, rotational energy might be converted to translational energy or vice versa, but on average these processes should balance out.

A diatomic molecule can also *vibrate*, as if the two atoms were held together by a spring. This vibration should count as *two* degrees of freedom, one for the vibrational kinetic energy and one for the potential energy. (You may recall from classical mechanics that the average kinetic and potential energies of a simple harmonic oscillator are equal—a result that is consistent with the equipartition theorem.) More complicated molecules can vibrate in a variety of ways: stretching, flexing, twisting. Each "mode" of vibration counts as two degrees of freedom.

However, at room temperature many vibrational degrees of freedom do *not* contribute to a molecule's thermal energy. Again, the explanation lies in quantum mechanics, as we will see in Chapter 3. So air molecules (N_2 and O_2), for instance, have only five degrees of freedom, not seven, at room temperature. At higher temperatures, the vibrational modes *do* eventually contribute. We say that these modes are "frozen out" at room temperature; evidently, collisions with other molecules are sufficiently violent to make an air molecule rotate, but hardly ever violent enough to make it vibrate.

In a solid, each atom can vibrate in three perpendicular directions, so for each atom there are six degrees of freedom (three for kinetic energy and three for potential energy). A simple model of a crystalline solid is shown in Figure 1.6. If we let N stand for the number of *atoms* and f stand for the number of degrees of freedom *per atom*, then we can use equation 1.23 with $f = 6$ for a solid. Again, however, some of the degrees of freedom may be "frozen out" at room temperature.

Liquids are more complicated than either gases or solids. You can generally use the formula $\frac{3}{2}kT$ to find the average translational kinetic energy of molecules in a

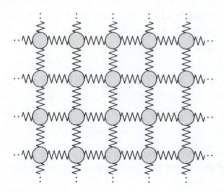

Figure 1.6. The "bed-spring" model of a crystalline solid. Each atom is like a ball, joined to its neighbors by springs. In three dimensions, there are six degrees of freedom per atom: three from kinetic energy and three from potential energy stored in the springs.

liquid, but the equipartition theorem doesn't work for the rest of the thermal energy, because the intermolecular potential energies are not nice quadratic functions.

You might be wondering what practical consequences the equipartition theorem has: How can we *test* it, experimentally? In brief, we would have to add some energy to a system, measure how much its temperature changes, and compare to equation 1.23. I'll discuss this procedure in more detail, and show some experimental results, in Section 1.6.

> **Problem 1.23.** Calculate the total thermal energy in a liter of helium at room temperature and atmospheric pressure. Then repeat the calculation for a liter of air.

> **Problem 1.24.** Calculate the total thermal energy in a gram of lead at room temperature, assuming that none of the degrees of freedom are "frozen out" (this happens to be a good assumption in this case).

> **Problem 1.25.** List all the degrees of freedom, or as many as you can, for a molecule of water vapor. (Think carefully about the various ways in which the molecule can vibrate.)

1.4 Heat and Work

Much of thermodynamics deals with three closely related concepts: **temperature**, **energy**, and **heat**. Much of students' difficulty with thermodynamics comes from confusing these three concepts with each other. Let me remind you that temperature, fundamentally, is a measure of an object's tendency to spontaneously give up energy. We have just seen that in many cases, when the energy content of a system increases, so does its temperature. But please don't think of this as the *definition* of temperature—it's merely a statement *about* temperature that happens to be true.

To further clarify matters, I really should give you a precise definition of **energy**. Unfortunately, I can't do this. Energy is the most fundamental dynamical concept in all of physics, and for this reason, I can't tell you what it is in terms of something more fundamental. I can, however, list the various *forms* of energy—kinetic, electrostatic, gravitational, chemical, nuclear—and add the statement that, while energy can often be converted from one form to another, the *total* amount of energy in the universe never changes. This is the famous law of **conservation of energy**. I sometimes picture energy as a perfectly indestructible (and unmakable) fluid, which moves about from place to place but whose total amount never changes. (This image is convenient but *wrong*—there simply isn't any such fluid.)

Suppose, for instance, that you have a container full of gas or some other thermodynamic system. If you notice that the energy of the system increases, you can conclude that some energy came in from outside; it can't have been manufactured on the spot, since this would violate the law of conservation of energy. Similarly, if the energy of your system decreases, then some energy must have escaped and gone elsewhere. There are all sorts of mechanisms by which energy can be put into or taken out of a system. However, in thermodynamics, we usually classify these mechanisms under two categories: **heat** and **work**.

Heat is defined as any spontaneous flow of energy from one object to another, caused by a difference in temperature between the objects. We say that "heat" flows from a warm radiator into a cold room, from hot water into a cold ice cube, and from the hot sun to the cool earth. The *mechanism* may be different in each case, but in each of these processes the energy transferred is called "heat."

Work, in thermodynamics, is defined as any other transfer of energy into or out of a system. You do work on a system whenever you push on a piston, stir a cup of coffee, or run current through a resistor. In each case, the system's energy will increase, and usually its temperature will too. But we don't say that the system is being "heated," because the flow of energy is not a spontaneous one caused by a difference in temperature. Usually, with work, we can identify some "agent" (possibly an inanimate object) that is "actively" putting energy into the system; it wouldn't happen "automatically."

The definitions of heat and work are not easy to internalize, because both of these words have very different meanings in everyday language. It is strange to think that there is no "heat" entering your hands when you rub them together to warm them up, or entering a cup of tea that you are warming in the microwave. Nevertheless, both of these processes are classified as work, not heat.

Notice that both heat and work refer to energy *in transit*. You can talk about the total *energy* inside a system, but it would be meaningless to ask how much heat, or how much work, is *in* a system. We can only discuss how much heat *entered* a system, or how much work *was done on* a system.

I'll use the symbol U for the total energy inside a system. The symbols Q and W will represent the amounts of energy that enter a system as heat and work, respectively, during any time period of interest. (Either one could be negative, if energy *leaves* the system.) The sum $Q + W$ is then the total energy that enters the system, and, by conservation of energy, this is the amount by which the system's energy changes (see Figure 1.7). Written as an equation, this statement is

$$\Delta U = Q + W, \tag{1.24}$$

the change in energy equals the heat added plus the work done.* This equation is

---

*Many physics and engineering texts define W to be positive when work-energy *leaves* the system rather than enters. Then equation 1.24 instead reads $\Delta U = Q - W$. This sign convention is convenient when dealing with heat engines, but I find it confusing in other situations. My sign convention is consistently followed by chemists, and seems to be catching on among physicists.

Another notational issue concerns the fact that we'll often want ΔU, Q, and W to be infinitesimal. In such cases I'll usually write dU instead of ΔU, but I'll leave the symbols Q and W alone. Elsewhere you may see "dQ" and "dW" used to represent infinitesimal amounts of heat and work. Whatever you do, *don't* read these as the "changes" in Q and W—that would be meaningless. To caution you not to commit this crime, many authors put a little bar through the d, writing $đQ$ and $đW$. To me, though, that $đ$ still looks like it should be pronounced "change." So I prefer to do away with the d entirely and just remember when Q and W are infinitesimal and when they're not.

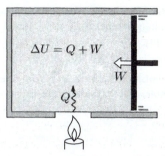

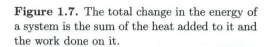

Figure 1.7. The total change in the energy of a system is the sum of the heat added to it and the work done on it.

really just a statement of the law of conservation of energy. However, it dates from a time when this law was just being discovered, and the relation between energy and heat was still controversial. So the equation was given a more mysterious name, which is still in use: the **first law of thermodynamics**.

The official SI unit of energy is the **joule**, defined as $1 \text{ kg·m}^2/\text{s}^2$. (So a 1-kg object traveling at 1 m/s has $\frac{1}{2}$ J of kinetic energy, $\frac{1}{2}mv^2$.) Traditionally, however, heat has been measured in **calories**, where 1 cal was defined as the amount of heat needed to raise the temperature of a gram of water by 1°C (while no work is being done on it). It was James Joule (among others*) who demonstrated that the same temperature increase could be accomplished by doing mechanical work (for instance, by vigorously stirring the water) instead of adding heat. In modern units, Joule showed that 1 cal equals approximately 4.2 J. Today the calorie is *defined* to equal exactly 4.186 J, and many people still use this unit when dealing with thermal or chemical energy. The well-known food calorie (sometimes spelled with a capital C) is actually a *kilo*calorie, or 4186 J.

Processes of heat transfer are further classified into three categories, according to the mechanism involved. **Conduction** is the transfer of heat by molecular contact: Fast-moving molecules bump into slow-moving molecules, giving up some of their energy in the process. **Convection** is the bulk motion of a gas or liquid, usually driven by the tendency of warmer material to expand and rise in a gravitational field. **Radiation** is the emission of electromagnetic waves, mostly infrared for objects at room temperature but including visible light for hotter objects like the filament of a lightbulb or the surface of the sun.

Problem 1.26. A battery is connected in series to a resistor, which is immersed in water (to prepare a nice hot cup of tea). Would you classify the flow of energy from the battery to the resistor as "heat" or "work"? What about the flow of energy from the resistor to the water?

Problem 1.27. Give an example of a process in which no heat is added to a system, but its temperature increases. Then give an example of the opposite: a process in which heat is added to a system but its temperature does not change.

---

*Among the many others who helped establish the first law were Benjamin Thompson (Count Rumford), Robert Mayer, William Thomson, and Hermann von Helmholtz.

Problem 1.28. Estimate how long it should take to bring a cup of water to boiling temperature in a typical 600-watt microwave oven, assuming that all the energy ends up in the water. (Assume any reasonable initial temperature for the water.) Explain why no heat is involved in this process.

Problem 1.29. A cup containing 200 g of water is sitting on your dining room table. After carefully measuring its temperature to be $20°C$, you leave the room. Returning ten minutes later, you measure its temperature again and find that it is now $25°C$. What can you conclude about the amount of heat added to the water? (Hint: This is a trick question.)

Problem 1.30. Put a few spoonfuls of water into a bottle with a tight lid. Make sure everything is at room temperature, measuring the temperature of the water with a thermometer to make sure. Now close the bottle and shake it as hard as you can for several minutes. When you're exhausted and ready to drop, shake it for several minutes more. Then measure the temperature again. Make a rough calculation of the expected temperature change, and compare.

1.5 Compression Work

We'll deal with more than one type of work in this book, but the most important type is work done on a system (often a gas) by *compressing* it, as when you push on a piston. You may recall from classical mechanics that in such a case the amount of work done is equal to the force you exert dotted into the displacement:

$$W = \vec{F} \cdot \vec{dr}. \tag{1.25}$$

(There is some ambiguity in this formula when the system is more complicated than a point particle: Does $\vec{dr}$ refer to the displacement of the center of mass, or the point of contact (if any), or what? In thermodynamics, it is always the point of contact, and we won't deal with work done by long-range forces such as gravity. In this case the work-energy theorem tells us that the total energy of the system increases by W.*)

For a gas, though, it's much more convenient to express the work done in terms of the pressure and volume. For definiteness, consider the typical cylinder-piston arrangement shown in Figure 1.8. The force is parallel to the displacement, so we can forget about dot products and just write

$$W = F \, \Delta x. \tag{1.26}$$

(I'm taking Δx to be positive when the piston moves inward.)

What I want to do next is replace F by PA, the pressure of the gas times the area of the piston. But in order to make this replacement, I need to assume that as the gas is compressed it always remains in internal equilibrium, so that its pressure is uniform from place to place (and hence well defined). For this to be the case, the

---

*For a detailed discussion of different definitions of "work," see A. John Mallinckrodt and Harvey S. Leff, "All About Work," *American Journal of Physics* **60**, 356–365 (1992).

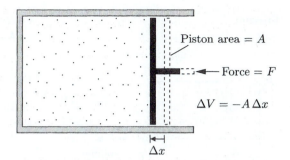

Figure 1.8. When the piston moves inward, the volume of the gas changes by ΔV (a negative amount) and the work done on the gas (assuming quasistatic compression) is $-P\Delta V$.

piston's motion must be reasonably slow, so that the gas has time to continually equilibrate to the changing conditions. The technical term for a volume change that is slow in this sense is **quasistatic**. Although perfectly quasistatic compression is an idealization, it is usually a good approximation in practice. To compress the gas non-quasistatically you would have to slam the piston very hard, so it moves faster than the gas can "respond" (the speed must be at least comparable to the speed of sound in the gas).

For quasistatic compression, then, the force exerted on the gas equals the pressure of the gas times the area of the piston.* Thus,

$$W = PA\,\Delta x \qquad \text{(for quasistatic compression).} \qquad (1.27)$$

But the product $A\,\Delta x$ is just minus the change in the volume of the gas (minus because the volume *decreases* when the piston moves in), so

$$W = -P\,\Delta V \qquad \text{(quasistatic).} \qquad (1.28)$$

For example, if you have a tank of air at atmospheric pressure (10^5 N/m^2) and you wish to reduce its volume by one liter (10^{-3} m^3), you must perform 100 J of work. You can easily convince yourself that the same formula holds if the gas *expands*; then ΔV is positive, so the work done on the gas is negative, as required.

There is one possible flaw in the derivation of this formula. Usually the pressure will *change* during the compression. In that case, what pressure should you use—initial, final, average, or what? There's no difficulty for very small ("infinitesimal") changes in volume, since then any change in the pressure will be negligible. Ah—but we can always think of a large change as a bunch of small changes, one after another. So when the pressure *does* change significantly during the compression, we need to mentally divide the process into many tiny steps, apply equation 1.28 to each step, and add up all the little works to get the total work.

---

*Even for quasistatic compression, friction between the piston and the cylinder walls could upset the balance between the force exerted from outside and the backward force exerted on the piston by the gas. If W represents the work done on the gas by the piston, this isn't a problem. But if it represents the work *you* do when pushing on the piston, then I'll need to assume that friction is negligible in what follows.

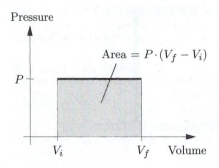
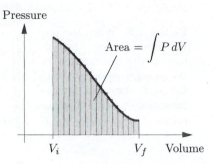

Figure 1.9. When the volume of a gas changes and its pressure is constant, the work done on the gas is minus the area under the graph of pressure vs. volume. The same is true even when the pressure is not constant.

This procedure is easier to understand graphically. If the pressure is *constant*, then the work done is just minus the *area* under a graph of pressure vs. volume (see Figure 1.9). If the pressure is *not* constant, we divide the process into a bunch of tiny steps, compute the area under the graph for each step, then add up all the areas to get the total work. That is, the work is *still* minus the total area under the graph of P vs. V.

If you happen to know a formula for the pressure as a function of volume, $P(V)$, then you can compute the total work as an integral:

$$W = - \int_{V_i}^{V_f} P(V)\, dV \qquad \text{(quasistatic)}. \qquad (1.29)$$

This is a good formula, since it is valid whether the pressure changes during the process or not. It isn't always easy, however, to carry out the integral and get a simple formula for W.

It's important to remember that compression-expansion work is not the *only* type of work that can be done on thermodynamic systems. For instance, the chemical reactions in a battery cause *electrical* work to be done on the circuit it is connected to. We'll see plenty of examples in this book where compression-expansion work *is* the only kind of relevant work, and plenty of examples where it isn't.

Problem 1.31. Imagine some helium in a cylinder with an initial volume of 1 liter and an initial pressure of 1 atm. Somehow the helium is made to expand to a final volume of 3 liters, in such a way that its pressure rises in direct proportion to its volume.

(a) Sketch a graph of pressure vs. volume for this process.

(b) Calculate the work done on the gas during this process, assuming that there are no "other" types of work being done.

(c) Calculate the change in the helium's energy content during this process.

(d) Calculate the amount of heat added to or removed from the helium during this process.

(e) Describe what you might do to *cause* the pressure to rise as the helium expands.

Problem 1.32. By applying a pressure of 200 atm, you can compress water to 99% of its usual volume. Sketch this process (not necessarily to scale) on a PV diagram, and estimate the work required to compress a liter of water by this amount. Does the result surprise you?

Problem 1.33. An ideal gas is made to undergo the cyclic process shown in Figure 1.10(a). For each of the steps A, B, and C, determine whether each of the following is positive, negative, or zero: (a) the work done on the gas; (b) the change in the energy content of the gas; (c) the heat added to the gas. Then determine the sign of each of these three quantities for the whole cycle. What does this process accomplish?

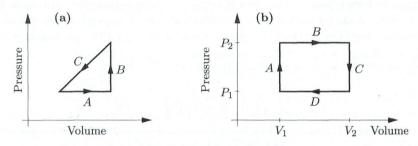

Figure 1.10. PV diagrams for Problems 1.33 and 1.34.

Problem 1.34. An ideal diatomic gas, in a cylinder with a movable piston, undergoes the rectangular cyclic process shown in Figure 1.10(b). Assume that the temperature is always such that rotational degrees of freedom are active, but vibrational modes are "frozen out." Also assume that the only type of work done on the gas is quasistatic compression-expansion work.

(a) For each of the four steps A through D, compute the work done on the gas, the heat added to the gas, and the change in the energy content of the gas. Express all answers in terms of P_1, P_2, V_1, and V_2. (Hint: Compute ΔU before Q, using the ideal gas law and the equipartition theorem.)

(b) Describe in words what is physically being done during each of the four steps; for example, during step A, heat is added to the gas (from an external flame or something) while the piston is held fixed.

(c) Compute the net work done on the gas, the net heat added to the gas, and the net change in the energy of the gas during the entire cycle. Are the results as you expected? Explain briefly.

Compression of an Ideal Gas

To get a feel for some of the preceding formulas, I'd like to apply them to the compression of an ideal gas. Since most familiar gases (such as air) are fairly close to ideal, the results we obtain will actually be quite useful.

When you compress a container full of gas, you're doing work on it, that is, adding energy. Generally this causes the temperature of the gas to increase, as you know if you've ever pumped up a bicycle tire. However, if you compress the gas very slowly, or if the container is in good thermal contact with its environment, heat will escape as the gas is compressed and its temperature won't rise very much.[*] The difference between fast compression and slow compression is therefore very important in thermodynamics.

In this section I'll consider two idealized ways of compressing an ideal gas: **isothermal compression**, which is so slow that the temperature of the gas doesn't rise at all; and **adiabatic compression**, which is so fast that no heat escapes from the gas during the process. Most *real* compression processes will be somewhere between these extremes, usually closer to the adiabatic approximation. I'll start with the isothermal case, though, since it's simpler.

Suppose, then, that you compress an ideal gas isothermally, that is, without changing its temperature. This almost certainly implies that the process is quasistatic, so I can use formula 1.29 to calculate the work done, with P determined by the ideal gas law. On a PV diagram, the formula $P = NkT/V$, for constant T, is a concave-up hyperbola (called an **isotherm**), as shown in Figure 1.11. The work done is minus the area under the graph:

$$W = -\int_{V_i}^{V_f} P\,dV = -NkT \int_{V_i}^{V_f} \frac{1}{V}\,dV$$

$$= -NkT\left(\ln V_f - \ln V_i\right) = NkT \ln\frac{V_i}{V_f}. \tag{1.30}$$

Notice that the work done is positive if $V_i > V_f$, that is, if the gas is being *compressed*. If the gas *expands* isothermally, the same equation applies but with $V_i < V_f$, that is, the work done *on* the gas is negative.

As the gas is compressed isothermally, heat must be flowing out, into the environment. To calculate how much, we can use the first law of thermodynamics and the fact that for an ideal gas U is proportional to T:

$$Q = \Delta U - W = \Delta(\tfrac{1}{2}NfkT) - W = 0 - W = NkT \ln\frac{V_f}{V_i}. \tag{1.31}$$

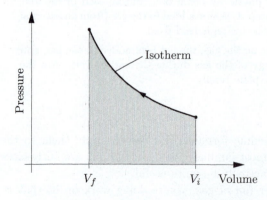

Figure 1.11. For isothermal compression of an ideal gas, the PV graph is a concave-up hyperbola, called an **isotherm**. As always, the work done is minus the area under the graph.

---

[*]Scuba tanks are usually held under water as they are filled, to prevent the compressed air inside from getting too hot.

Thus the heat input is just minus the work done. For compression, Q is negative because heat *leaves* the gas; for isothermal expansion, heat must *enter* the gas so Q is positive.

Now let's consider adiabatic compression, which is so fast that *no* heat flows out of (or into) the gas. I'll still assume, however, that the compression is quasistatic. In practice this usually isn't a bad approximation.

If you do work on a gas but don't let any heat escape, the internal energy of the gas will increase:

$$\Delta U = Q + W = W. \tag{1.32}$$

If it's an *ideal* gas, U is proportional to T so the temperature increases as well. The curve describing this process on a PV diagram must connect a low-temperature isotherm to a high-temperature isotherm, and therefore must be steeper than either of the isotherms (see Figure 1.12).

To find an equation describing the exact shape of this curve, let me first use the equipartition theorem to write

$$U = \frac{f}{2}NkT, \tag{1.33}$$

where f is the number of degrees of freedom per molecule—3 for a monatomic gas, 5 for a diatomic gas near room temperature, etc. Then the energy change along any infinitesimal segment of the curve is

$$dU = \frac{f}{2}Nk\,dT. \tag{1.34}$$

Meanwhile, the work done during quasistatic compression is $-P\,dV$, so equation 1.32, applied to an infinitesimal part of the process, becomes

$$\frac{f}{2}Nk\,dT = -P\,dV. \tag{1.35}$$

This differential equation relates the changes in temperature and volume during the compression process. To solve the equation, however, we need to write the pressure P in terms of the variables T and V. The needed relation is just the ideal gas law; plugging in NkT/V for P and canceling the Nk gives

$$\frac{f}{2}\frac{dT}{T} = -\frac{dV}{V}. \tag{1.36}$$

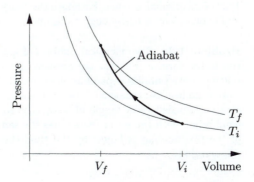

Figure 1.12. The PV curve for adiabatic compression (called an **adiabat**) begins on a lower-temperature isotherm and ends on a higher-temperature isotherm.

Now we can integrate both sides from the initial values (V_i and T_i) to the final values (V_f and T_f):

$$\frac{f}{2} \ln \frac{T_f}{T_i} = -\ln \frac{V_f}{V_i}. \tag{1.37}$$

To simplify this equation, exponentiate both sides and gather the i's and f's. After a couple of lines of algebra you'll find

$$V_f \, T_f^{f/2} = V_i \, T_i^{f/2}, \tag{1.38}$$

or more compactly,

$$VT^{f/2} = \text{constant}. \tag{1.39}$$

Given any starting point and any final volume, you can now calculate the final temperature. To find the final pressure you can use the ideal gas law to eliminate T on both sides of equation 1.38. The result can be written

$$V^\gamma P = \text{constant}, \tag{1.40}$$

where γ, called the **adiabatic exponent**, is an abbreviation for $(f + 2)/f$.

Problem 1.35. Derive equation 1.40 from equation 1.39.

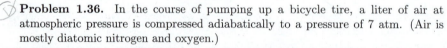 **Problem 1.36.** In the course of pumping up a bicycle tire, a liter of air at atmospheric pressure is compressed adiabatically to a pressure of 7 atm. (Air is mostly diatomic nitrogen and oxygen.)

    (a) What is the final volume of this air after compression?

    (b) How much work is done in compressing the air?

    (c) If the temperature of the air is initially 300 K, what is the temperature after compression?

Problem 1.37. In a Diesel engine, atmospheric air is quickly compressed to about 1/20 of its original volume. Estimate the temperature of the air after compression, and explain why a Diesel engine does not require spark plugs.

Problem 1.38. Two identical bubbles of gas form at the bottom of a lake, then rise to the surface. Because the pressure is much lower at the surface than at the bottom, both bubbles expand as they rise. However, bubble A rises very quickly, so that no heat is exchanged between it and the water. Meanwhile, bubble B rises slowly (impeded by a tangle of seaweed), so that it always remains in thermal equilibrium with the water (which has the same temperature everywhere). Which of the two bubbles is larger by the time they reach the surface? Explain your reasoning fully.

Problem 1.39. By applying Newton's laws to the oscillations of a continuous medium, one can show that the speed of a sound wave is given by

$$c_s = \sqrt{\frac{B}{\rho}},$$

where ρ is the density of the medium (mass per unit volume) and B is the **bulk modulus**, a measure of the medium's stiffness. More precisely, if we imagine applying an increase in pressure ΔP to a chunk of the material, and this increase results in a (negative) change in volume ΔV, then B is defined as the change in pressure divided by the magnitude of the fractional change in volume:

$$B \equiv \frac{\Delta P}{-\Delta V/V}.$$

This definition is *still* ambiguous, however, because I haven't said whether the compression is to take place isothermally or adiabatically (or in some other way).

(a) Compute the bulk modulus of an ideal gas, in terms of its pressure P, for both isothermal and adiabatic compressions.

(b) Argue that for purposes of computing the speed of a sound wave, the *adiabatic* B is the one we should use.

(c) Derive an expression for the speed of sound in an ideal gas, in terms of its temperature and average molecular mass. Compare your result to the formula for the rms speed of the molecules in the gas. Evaluate the speed of sound numerically for air at room temperature.

(d) When Scotland's Battlefield Band played in Utah, one musician remarked that the high altitude threw their bagpipes out of tune. Would you expect altitude to affect the speed of sound (and hence the frequencies of the standing waves in the pipes)? If so, in which direction? If not, why not?

Problem 1.40. In Problem 1.16 you calculated the pressure of earth's atmosphere as a function of altitude, assuming constant temperature. Ordinarily, however, the temperature of the bottommost 10–15 km of the atmosphere (called the **troposphere**) decreases with increasing altitude, due to heating from the ground (which is warmed by sunlight). If the temperature gradient $|dT/dz|$ exceeds a certain critical value, convection will occur: Warm, low-density air will rise, while cool, high-density air sinks. The decrease of pressure with altitude causes a rising air mass to expand adiabatically and thus to cool. The condition for convection to occur is that the rising air mass must *remain* warmer than the surrounding air despite this adiabatic cooling.

(a) Show that when an ideal gas expands adiabatically, the temperature and pressure are related by the differential equation

$$\frac{dT}{dP} = \frac{2}{f+2}\frac{T}{P}.$$

(b) Assume that dT/dz is just at the critical value for convection to begin, so that the vertical forces on a convecting air mass are always approximately in balance. Use the result of Problem 1.16(b) to find a formula for dT/dz in this case. The result should be a constant, independent of temperature and pressure, which evaluates to approximately $-10°C/km$. This fundamental meteorological quantity is known as the **dry adiabatic lapse rate**.

1.6 Heat Capacities

The **heat capacity** of an object is the amount of heat needed to raise its temperature, per degree temperature increase:

$$C \equiv \frac{Q}{\Delta T}. \tag{1.41}$$

(The symbol for heat capacity is a capital C.) Of course, the more of a substance you have, the larger its heat capacity will be. A more fundamental quantity is the **specific heat capacity**, defined as the heat capacity per unit mass:

$$c \equiv \frac{C}{m}. \tag{1.42}$$

(The symbol for specific heat capacity is a lowercase c.)

The most important thing to know about the definition (1.41) of heat capacity is that it is *ambiguous*. The amount of heat needed to raise an object's temperature by one degree *depends on the circumstances*, specifically, on whether you are also doing *work* on the object (and if so, how much). To see this, just plug the first law of thermodynamics into equation 1.41:

$$C = \frac{Q}{\Delta T} = \frac{\Delta U - W}{\Delta T}. \tag{1.43}$$

Even if the energy of an object is a well-defined function of its temperature alone (which is sometimes but not always the case), the work W done on the object can be anything, so C can be anything, too.

In practice, there are two types of circumstances (and choices for W) that are most likely to occur. Perhaps the most obvious choice is $W = 0$, when there is *no* work being done on the system. Usually this means that the system's volume isn't changing, since if it were, there would be compression work equal to $-P\Delta V$. So the heat capacity, for the particular case where $W = 0$ and V is constant, is called the **heat capacity at constant volume**, denoted C_V. From equation 1.43,

$$C_V = \left(\frac{\Delta U}{\Delta T}\right)_V = \left(\frac{\partial U}{\partial T}\right)_V. \tag{1.44}$$

(The subscript V indicates that the changes are understood to occur with the volume held fixed. The symbol ∂ indicates a *partial* derivative, in this case treating U as a function of T and V, with only T, not V, varying as the derivative is taken.) A better name for this quantity would be "energy capacity," since it is the *energy* needed to raise the object's temperature, per degree, regardless of whether the energy actually enters as heat. For a gram of water, C_V is 1 cal/°C or about 4.2 J/°C.

In everyday life, however, objects often *expand* as they are heated. In this case they do work on their surroundings, so W is negative, so C is *larger* than C_V: you need to add additional heat to compensate for the energy lost as work. If the *pressure* surrounding your object happens to be constant, then the total heat needed is unambiguous, and we refer to the heat needed per degree as C_P, the **heat capacity at constant pressure**. Plugging the formula for compression-expansion work into equation 1.43 gives

$$C_P = \left(\frac{\Delta U - (-P\Delta V)}{\Delta T}\right)_P = \left(\frac{\partial U}{\partial T}\right)_P + P\left(\frac{\partial V}{\partial T}\right)_P. \qquad (1.45)$$

The last term on the right is the additional heat needed to compensate for the energy lost as work. Notice that the more the volume increases, the larger this term is. For solids and liquids, $\partial V/\partial T$ is usually small and can often be neglected. For gases, however, the second term is quite significant. (The first term, $(\partial U/\partial T)_P$, is not quite the same as C_V, since it is P, not V, that is held fixed in the partial derivative.)

Equations 1.41 through 1.45 are essentially definitions, so they apply to any object whatsoever. To determine the heat capacity of some *particular* object, you generally have three choices: measure it (see Problem 1.41); look it up in a reference work where measured values are tabulated; or try to predict it theoretically. The last is the most fun, as we'll see repeatedly throughout this book. For some objects we already know enough to predict the heat capacity.

Suppose that our system stores thermal energy only in quadratic "degrees of freedom," as described in Section 1.3. Then the equipartition theorem says $U = \frac{1}{2}NfkT$ (neglecting any "static" energy, which doesn't depend on temperature), so

$$C_V = \frac{\partial U}{\partial T} = \frac{\partial}{\partial T}\left(\frac{NfkT}{2}\right) = \frac{Nfk}{2}, \qquad (1.46)$$

assuming that f is independent of temperature. (Note that in this case it doesn't matter whether V or P is held fixed in the derivative $\partial U/\partial T$.) This result gives us a direct method of measuring the number of degrees of freedom in an object, or, if we know this number, of testing the equipartition theorem. For instance, in a monatomic gas like helium, $f = 3$, so we expect $C_V = \frac{3}{2}Nk = \frac{3}{2}nR$; that is, the heat capacity *per mole* should be $\frac{3}{2}R = 12.5$ J/K. For diatomic and polyatomic molecules the heat capacity should be larger, in proportion to the number of degrees of freedom per molecule. Figure 1.13 (see the following page) shows a graph of C_V vs. temperature for a mole of hydrogen (H_2) gas, showing how the vibrational and rotational degrees of freedom freeze out at low temperatures. For a solid, there are six degrees of freedom per atom, so the heat capacity per mole should be $\frac{6}{2}R = 3R$; this general result is called the **rule of Dulong and Petit**. In this case, though, *all* of the degrees of freedom freeze out at low temperature, so the heat capacity approaches zero as $T \rightarrow 0$. What qualifies as "low" temperature depends on the material, as shown in Figure 1.14.

What about heat capacities of gases at constant pressure? For an *ideal* gas, the derivative $\partial U/\partial T$ is the same with P fixed as with V fixed, and we can compute the second term in equation 1.45 using the ideal gas law. At constant pressure,

$$\left(\frac{\partial V}{\partial T}\right)_P = \frac{\partial}{\partial T}\left(\frac{NkT}{P}\right) = \frac{Nk}{P} \qquad \text{(ideal gas)}. \qquad (1.47)$$

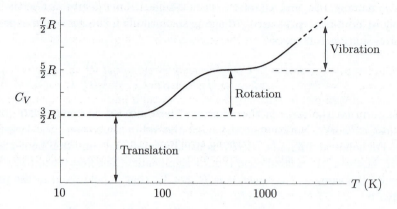

Figure 1.13. Heat capacity at constant volume of one mole of hydrogen (H$_2$) gas. Note that the temperature scale is logarithmic. Below about 100 K only the three translational degrees of freedom are active. Around room temperature the two rotational degrees of freedom are active as well. Above 1000 K the two vibrational degrees of freedom also become active. At atmospheric pressure, hydrogen liquefies at 20 K and begins to dissociate at about 2000 K. Data from Woolley et al. (1948).

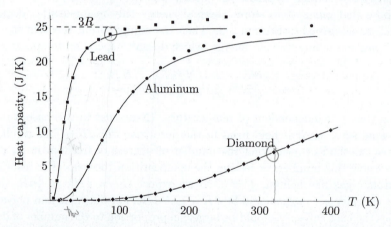

Figure 1.14. Measured heat capacities at constant pressure (data points) for one mole each of three different elemental solids. The solid curves show the heat capacity at constant *volume* predicted by the model used in Section 7.5, with the horizontal scale chosen to best fit the data for each substance. At sufficiently high temperatures, C_V for each material approaches the value $3R$ predicted by the equipartition theorem. The discrepancies between the data and the solid curves at high T are mostly due to the differences between C_P and C_V. At $T = 0$ all degrees of freedom are frozen out, so both C_P and C_V go to zero. Data from Y. S. Touloukian, ed., *Thermophysical Properties of Matter* (Plenum, New York, 1970).

Therefore,

$$C_P = C_V + Nk = C_V + nR \qquad \text{(ideal gas)}. \qquad (1.48)$$

In other words, for each mole of an ideal gas, the heat capacity at constant pressure exceeds the heat capacity at constant volume by R, the gas constant. Oddly, the additional term in the heat capacity doesn't depend on *what* the pressure is, so long as it is constant. Apparently, if the pressure is high the gas expands less, in such a way that the work done on the environment is independent of P.

Problem 1.41. To measure the heat capacity of an object, all you usually have to do is put it in thermal contact with another object whose heat capacity you know. As an example, suppose that a chunk of metal is immersed in boiling water (100°C), then is quickly transferred into a Styrofoam cup containing 250 g of water at 20°C. After a minute or so, the temperature of the contents of the cup is 24°C. Assume that during this time no significant energy is transferred between the contents of the cup and the surroundings. The heat capacity of the cup itself is negligible.

(a) How much heat is gained by the water?

(b) How much heat is lost by the metal?

(c) What is the heat capacity of this chunk of metal?

(d) If the mass of the chunk of metal is 100 g, what is its specific heat capacity?

Problem 1.42. The specific heat capacity of Albertson's *Rotini Tricolore* is approximately 1.8 J/g·°C. Suppose you toss 340 g of this pasta (at 25°C) into 1.5 liters of boiling water. What effect does this have on the temperature of the water (before there is time for the stove to provide more heat)?

Problem 1.43. Calculate the heat capacity of liquid water *per molecule*, in terms of k. Suppose (incorrectly) that all the thermal energy of water is stored in quadratic degrees of freedom. How many degrees of freedom would each molecule have to have?

Problem 1.44. At the back of this book is a table of thermodynamic data for selected substances at room temperature. Browse through the C_P values in this table, and check that you can account for most of them (approximately) using the equipartition theorem. Which values seem anomalous?

Problem 1.45. As an illustration of why it matters which variables you hold fixed when taking partial derivatives, consider the following mathematical example. Let $w = xy$ and $x = yz$.

(a) Write w purely in terms of x and z, and then purely in terms of y and z.

(b) Compute the partial derivatives

$$\left(\frac{\partial w}{\partial x}\right)_y \quad \text{and} \quad \left(\frac{\partial w}{\partial x}\right)_z,$$

and show that they are not equal. (Hint: To compute $(\partial w/\partial x)_y$, use a formula for w in terms of x and y, not z. Similarly, compute $(\partial w/\partial x)_z$ from a formula for w in terms of only x and z.)

(c) Compute the other four partial derivatives of w (two each with respect to y and z), and show that it matters which variable is held fixed.

Problem 1.46. Measured heat capacities of solids and liquids are almost always at constant pressure, not constant volume. To see why, estimate the pressure needed to keep V fixed as T increases, as follows.

(a) First imagine slightly increasing the temperature of a material at constant pressure. Write the change in volume, dV_1, in terms of dT and the thermal expansion coefficient β introduced in Problem 1.7.

(b) Now imagine slightly compressing the material, holding its temperature fixed. Write the change in volume for *this* process, dV_2, in terms of dP and the **isothermal compressibility** κ_T, defined as

$$\kappa_T \equiv -\frac{1}{V}\left(\frac{\partial V}{\partial P}\right)_T.$$

(This is the reciprocal of the isothermal bulk modulus defined in Problem 1.39.)

(c) Finally, imagine that you compress the material just enough in part (b) to offset the expansion in part (a). Then the ratio of dP to dT is equal to $(\partial P/\partial T)_V$, since there is no net change in volume. Express this partial derivative in terms of β and κ_T. Then express it more abstractly in terms of the partial derivatives used to define β and κ_T. For the second expression you should obtain

$$\left(\frac{\partial P}{\partial T}\right)_V = -\frac{(\partial V/\partial T)_P}{(\partial V/\partial P)_T}.$$

This result is actually a purely mathematical relation, true for any three quantities that are related in such a way that any two determine the third.

(d) Compute β, κ_T, and $(\partial P/\partial T)_V$ for an ideal gas, and check that the three expressions satisfy the identity you found in part (c).

(e) For water at 25°C, $\beta = 2.57 \times 10^{-4}$ K^{-1} and $\kappa_T = 4.52 \times 10^{-10}$ Pa^{-1}. Suppose you increase the temperature of some water from 20°C to 30°C. How much pressure must you apply to prevent it from expanding? Repeat the calculation for mercury, for which (at 25°C) $\beta = 1.81 \times 10^{-4}$ K^{-1} and $\kappa_T = 4.04 \times 10^{-11}$ Pa^{-1}. Given the choice, would you rather measure the heat capacities of these substances at constant V or at constant P?

Latent Heat

In some situations you can put heat into a system without increasing its temperature at *all*. This normally happens at a **phase transformation**, such as melting ice or boiling water. Technically, the heat capacity is then *infinite*:

$$C = \frac{Q}{\Delta T} = \frac{Q}{0} = \infty \qquad \text{(during a phase transformation).} \qquad (1.49)$$

However, you still might want to know how much heat is required to melt or boil the substance completely. This amount, divided by the mass of the substance, is called the **latent heat** of the transformation, and denoted L:

$$L \equiv \frac{Q}{m} \qquad \text{to accomplish the transformation.} \qquad (1.50)$$

Like the definition of heat capacity, this definition is ambiguous, since any amount of work could also be done during the process. By convention, however, we assume that the pressure is constant (usually 1 atm), and that no other work is done besides the usual constant-pressure expansion or compression. The latent heat for melting ice is 333 J/g, or 80 cal/g. The latent heat for boiling water is 2260 J/g, or 540 cal/g. (To get a feel for these numbers, recall that raising the temperature of water from 0°C to 100°C requires 100 cal/g.)

Problem 1.47. Your 200-g cup of tea is boiling-hot. About how much ice should you add to bring it down to a comfortable sipping temperature of 65°C? (Assume that the ice is initially at −15°C. The specific heat capacity of ice is 0.5 cal/g·°C.)

Problem 1.48. When spring finally arrives in the mountains, the snow pack may be two meters deep, composed of 50% ice and 50% air. Direct sunlight provides about 1000 watts/m^2 to earth's surface, but the snow might reflect 90% of this energy. Estimate how many weeks the snow pack should last, if direct solar radiation is the only source of energy.

Enthalpy

Constant-pressure processes occur quite often, both in the natural world and in the laboratory. Keeping track of the compression-expansion work done during these processes gets to be a pain after a while, but there is a convenient trick that makes it a bit easier. Instead of always talking about the *energy* content of a system, we can agree to always add in the work needed to make room for it (under a constant pressure, usually 1 atm). This work is PV, the pressure of the environment times the total volume of the system (that is, the total space you would need to clear out to make room for it). Adding PV onto the energy gives a quantity called the **enthalpy**, denoted H:

$$H \equiv U + PV. \tag{1.51}$$

This is the *total* energy you would have to come up with, to create the system out of nothing and put it into this environment (see Figure 1.15). Or, put another way, if you could somehow annihilate the system, the energy you could extract is not just U, but also the work (PV) done by the atmosphere as it collapses to fill the vacuum left behind.

Figure 1.15. To create a rabbit out of nothing and place it on the table, the magician must summon up not only the energy U of the rabbit, but also some additional energy, equal to PV, to push the atmosphere out of the way to make room. The *total* energy required is the **enthalpy**, $H = U + PV$.

To see the usefulness of enthalpy, suppose that some change takes place in the system—you add some heat, or chemicals react, or whatever—while the pressure is always held constant. The energy, volume, and enthalpy can all change, by amounts that I'll call ΔV, ΔU, and ΔH. The new enthalpy is

$$
\begin{aligned}
H + \Delta H &= (U + \Delta U) + P(V + \Delta V) \\
&= (U + PV) + (\Delta U + P\,\Delta V) \\
&= H + (\Delta U + P\,\Delta V),
\end{aligned} \tag{1.52}
$$

so the change in enthalpy during a constant-pressure process is

$$
\Delta H = \Delta U + P\,\Delta V \qquad (\text{constant } P). \tag{1.53}
$$

This says that enthalpy can increase for two reasons: either because the energy increases, or because the system expands and work is done on the atmosphere to make room for it.

Now recall the first law of thermodynamics: The change in energy equals the heat added to the system, plus the compression-expansion work done on it, plus any other work (e.g., electrical) done on it:

$$
\Delta U = Q + (-P\,\Delta V) + W_{\text{other}}. \tag{1.54}
$$

Combining this law with equation 1.53, we obtain

$$
\Delta H = Q + W_{\text{other}} \qquad (\text{constant } P), \tag{1.55}
$$

that is, the change in enthalpy is caused *only* by heat and other forms of work, *not* by compression-expansion work (during constant-pressure processes). In other words, you can forget all about compression-expansion work if you deal with enthalpy instead of energy. If no "other" types of work are being done, the change in enthalpy tells you *directly* how much heat has been added to the system. (That's why we use the symbol H.)

For the simple case of raising an object's temperature, the change in enthalpy per degree, at constant pressure, is the same as the heat capacity at constant pressure, C_P:

$$
C_P = \left(\frac{\partial H}{\partial T} \right)_P. \tag{1.56}
$$

This formula is really the best way to define C_P, though you can easily see that it is equivalent to equation 1.45. Just as C_V should really be called "energy capacity," C_P should really be called "enthalpy capacity." And as with C_V, there doesn't have to be any heat involved at all, since the enthalpy could just as well enter as "other" work, as in a microwave oven.

Chemistry books are full of tables of ΔH values for more dramatic processes: phase transformations, chemical reactions, ionization, dissolution in solvents, and so on. For instance, standard tables say that the change in enthalpy when you boil one mole of water at 1 atm is 40,660 J. Since a mole of water is about 18 grams

(16 for the oxygen and 2 for the hydrogen), this means that the change in enthalpy when you boil one *gram* of water should be $(40{,}660 \text{ J})/18 = 2260$ J, precisely the number I quoted earlier for the latent heat. However, not all of this energy ends up in the vaporized water. The volume of one mole of water vapor, according to the ideal gas law, is RT/P (while the initial volume of the liquid is negligible), so the work needed to push the atmosphere away is

$$PV = RT = (8.31 \text{ J/K})(373 \text{ K}) = 3100 \text{ J}. \tag{1.57}$$

This is only 8% of the 40,660 J of energy put in, but sometimes it's necessary to keep track of such things.

As another example, consider the chemical reaction in which hydrogen and oxygen gas combine to form liquid water:

$$H_2 + \tfrac{1}{2}O_2 \longrightarrow H_2O. \tag{1.58}$$

For each mole of water produced, ΔH for this reaction is -286 kJ; in tables this quantity is referred to as the **enthalpy of formation** of water, because it's being "formed" out of elemental constituents in their most stable states. (The numerical value assumes that both the reactants and the product are at room temperature and atmospheric pressure. This number and others like it are tabulated in the data section at the back of this book.) If you simply burn a mole of hydrogen, then 286 kJ is the amount of heat you get out. Nearly all of this energy comes from the thermal and chemical energy of the molecules themselves, but a small amount comes from work done by the atmosphere as it collapses to fill the space left behind by the consumed gases.

You might wonder, though, whether some of the 286 kJ can't be extracted as *work* (perhaps electrical work) rather than as heat. Certainly this would be a good thing, since electricity is so much more useful and versatile than heat. In general the answer is that much of the energy from a chemical reaction *can* be extracted as work, but there are limits, as we'll see in Chapter 5.

Problem 1.49. Consider the combustion of one mole of H_2 with 1/2 mole of O_2 under standard conditions, as discussed in the text. How much of the heat energy produced comes from a decrease in the internal energy of the system, and how much comes from work done by the collapsing atmosphere? (Treat the volume of the liquid water as negligible.)

Problem 1.50. Consider the combustion of one mole of methane gas:

$$CH_4(\text{gas}) + 2O_2(\text{gas}) \longrightarrow CO_2(\text{gas}) + 2H_2O(\text{gas}).$$

The system is at standard temperature (298 K) and pressure (10^5 Pa) both before and after the reaction.

(a) First imagine the process of converting a mole of methane into its elemental consituents (graphite and hydrogen gas). Use the data at the back of this book to find ΔH for this process.

(b) Now imagine forming a mole of CO_2 and two moles of water vapor from their elemental constituénts. Determine ΔH for this process.

(c) What is ΔH for the actual reaction in which methane and oxygen form carbon dioxide and water vapor directly? Explain.

(d) How much heat is given off during this reaction, assuming that no "other" forms of work are done?

(e) What is the change in the system's *energy* during this reaction? How would your answer differ if the H_2O ended up as liquid water instead of vapor?

(f) The sun has a mass of 2×10^{30} kg and gives off energy at a rate of 3.9×10^{26} watts. If the source of the sun's energy were ordinary combustion of a chemical fuel such as methane, about how long could it last?

Problem 1.51. Use the data at the back of this book to determine ΔH for the combustion of a mole of glucose,

$$C_6H_{12}O_6 + 6O_2 \longrightarrow 6CO_2 + 6H_2O.$$

This is the (net) reaction that provides most of the energy needs in our bodies.

Problem 1.52. The enthalpy of combustion of a gallon (3.8 liters) of gasoline is about 31,000 kcal. The enthalpy of combustion of an ounce (28 g) of corn flakes is about 100 kcal. Compare the cost of gasoline to the cost of corn flakes, per calorie.

Problem 1.53. Look up the enthalpy of formation of atomic hydrogen in the back of this book. This is the enthalpy change when a mole of atomic hydrogen is formed by dissociating 1/2 mole of molecular hydrogen (the more stable state of the element). From this number, determine the energy needed to dissociate a single H_2 molecule, in electron-volts.

Problem 1.54. A 60-kg hiker wishes to climb to the summit of Mt. Ogden, an ascent of 5000 vertical feet (1500 m).

(a) Assuming that she is 25% efficient at converting chemical energy from food into mechanical work, and that essentially all the mechanical work is used to climb vertically, roughly how many bowls of corn flakes (standard serving size 1 ounce, 100 kilocalories) should the hiker eat before setting out?

(b) As the hiker climbs the mountain, three-quarters of the energy from the corn flakes is converted to thermal energy. If there were no way to dissipate this energy, by how many degrees would her body temperature increase?

(c) In fact, the extra energy does not warm the hiker's body significantly; instead, it goes (mostly) into evaporating water from her skin. How many liters of water should she drink during the hike to replace the lost fluids? (At 25°C, a reasonable temperature to assume, the latent heat of vaporization of water is 580 cal/g, 8% more than at 100°C.)

Problem 1.55. Heat capacities are normally positive, but there is an important class of exceptions: systems of particles held together by gravity, such as stars and star clusters.

(a) Consider a system of just two particles, with identical masses, orbiting in circles about their center of mass. Show that the gravitational potential energy of this system is -2 times the total kinetic energy.

(b) The conclusion of part (a) turns out to be true, at least on average, for *any* system of particles held together by mutual gravitational attraction:

$$\overline{U}_{\text{potential}} = -2\overline{U}_{\text{kinetic}}.$$

Here each $\overline{U}$ refers to the total energy (of that type) for the entire system, averaged over some sufficiently long time period. This result is known as the **virial theorem**. (For a proof, see Carroll and Ostlie (1996), Section 2.4.) Suppose, then, that you add some energy to such a system and then wait for the system to equilibrate. Does the average total kinetic energy increase or decrease? Explain.

(c) A star can be modeled as a gas of particles that interact with each other only gravitationally. According to the equipartition theorem, the average kinetic energy of the particles in such a star should be $\frac{3}{2}kT$, where T is the average temperature. Express the total energy of a star in terms of its average temperature, and calculate the heat capacity. Note the sign.

(d) Use dimensional analysis to argue that a star of mass M and radius R should have a total potential energy of $-GM^2/R$, times some constant of order 1.

(e) Estimate the average temperature of the sun, whose mass is 2×10^{30} kg and whose radius is 7×10^8 m. Assume, for simplicity, that the sun is made entirely of protons and electrons.

1.7 Rates of Processes

Usually, to determine *what* the equilibrium state of a system is, we need not worry about *how long* the system takes to reach equilibrium. Thermodynamics, by many people's definitions, includes only the study of equilibrium states themselves. Questions about time and rates of processes are then considered a separate (though related) subject, sometimes called **transport theory** or **kinetics**.

In this book I won't say much about rates of processes, because these kinds of questions are often quite difficult and require somewhat different tools. But transport theory is important enough that I should say *something* about it, at least the simpler aspects. That is the purpose of this section.*

Heat Conduction

At what rate does heat flow from a hot object to a cold object? The answer depends on many factors, particularly on what *mechanisms* of heat transfer are possible under the circumstances.

If the objects are separated by empty space (like the sun and the earth, or the inner and outer walls of a thermos bottle) then the only possible heat transfer mechanism is radiation. I'll derive a formula for the rate of radiation in Chapter 7.

If a fluid (gas or liquid) can mediate the heat transfer, then convection—bulk motion of the fluid—is often the dominant mechanism. Convection rates depend on all sorts of factors, including the heat capacity of the fluid and the many possible forces acting on it. I won't try to calculate any convection rates in this book.

That leaves conduction: heat transfer by direct contact at the molecular level. Conduction can happen through a solid, liquid, or gas. In a liquid or a gas the

---

*This section is somewhat outside the main development of the book. No other sections depend on it, so you may omit or postpone it if you wish.

energy is transferred through molecular collisions: When a fast molecule hits a slow molecule, energy is usually transferred from the former to the latter. In solids, heat is conducted via lattice vibrations and, in metals, via conduction electrons. Good electrical conductors tend to be good heat conductors as well, because the same conduction electrons can carry both electric current and energy, while lattice vibrations are much less efficient than electrons at conducting heat.

Regardless of these details, the rate of heat conduction obeys a mathematical law that is not hard to guess. For definiteness, imagine a glass window separating the warm interior of a building from the cold outdoors (see Figure 1.16). We would expect the amount of heat Q that passes through the window to be directly proportional to the window's total area A, and to the amount of time that passes, Δt. We would probably expect Q to be *inversely* proportional to the thickness of the window, Δx. Finally, we would expect Q to depend on the indoor and outdoor temperatures, in such a way that $Q = 0$ if these temperatures are the same. The simplest guess is that Q is directly proportional to the temperature difference, $\Delta T = T_2 - T_1$; this guess turns out to be correct for any heat transfer by conduction (though not for radiation). Summarizing these proportionalities, we can write

$$Q \propto \frac{A\,\Delta T\,\Delta t}{\Delta x}, \qquad \text{or} \qquad \frac{Q}{\Delta t} \propto A\,\frac{dT}{dx}. \tag{1.59}$$

The constant of proportionality depends on the material through which the heat is being conducted (in this case, glass). This constant is called the **thermal conductivity** of the material. The usual symbol for thermal conductivity is k, but to distinguish it from Boltzmann's constant I'll called it k_t. I'll also put a minus sign into the equation to remind us that if T increases from left to right, Q flows from right to left. The law of heat conduction is then

$$\frac{Q}{\Delta t} = -k_t A\,\frac{dT}{dx}. \tag{1.60}$$

This equation is known as the **Fourier heat conduction law**, after the same J. B. J. Fourier who invented Fourier analysis.

To *derive* the Fourier heat conduction law, and to predict the value of k_t for a particular material, we would have to invoke a detailed molecular model of what happens during heat conduction. I'll do this for the easiest case, an ideal gas, in the

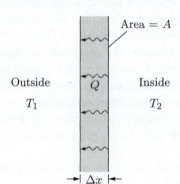

Figure 1.16. The rate of heat conduction through a pane of glass is proportional to its area A and inversely proportional to its thickness Δx.

following subsection. For now, though, let's just take Fourier's law as an empirical fact and treat k_t as a property that you need to measure for any material of interest.

Thermal conductivities of common materials vary by more than four orders of magnitude. In SI units (watts per meter per kelvin), a few representative values are: air, 0.026; wood, 0.08; water, 0.6; glass, 0.8; iron, 80; copper, 400. Again, good electrical conductors tend to be good thermal conductors. Note that the values for air and water apply to conduction only, even though convection can often be extremely important.

Back to our window, suppose it has an area of one square meter and a thickness of 3.2 mm (1/8 inch). Then if the temperature just inside the window is 20°C and the temperature just outside is 0°C, the rate of heat flow through it is

$$\frac{Q}{\Delta t} = \frac{(0.8 \text{ W/m·K})(1 \text{ m}^2)(293 \text{ K} - 273 \text{ K})}{0.0032 \text{ m}} = 5000 \text{ watts}. \qquad (1.61)$$

If this number seems absurdly high to you, you're right. My assumption of such a large temperature difference between "just inside" and "just outside" the window is unrealistic, because there is always a thin layer of still air on each side of the glass. The two air layers can provide many times more thermal insulation than the glass itself, bringing the heat loss down into the range of a few hundred watts (see Problem 1.57).

Problem 1.56. Calculate the rate of heat conduction through a layer of still air that is 1 mm thick, with an area of 1 m^2, for a temperature difference of 20°C.

Problem 1.57. Home owners and builders discuss thermal conductivities in terms of the **R value** (R for *resistance*) of a material, defined as the thickness divided by the thermal conductivity:

$$R \equiv \frac{\Delta x}{k_t}.$$

(a) Calculate the R value of a 1/8-inch (3.2 mm) piece of plate glass, and then of a 1 mm layer of still air. Express both answers in SI units.

(b) In the United States, R values of building materials are normally given in English units, °F·ft^2·hr/Btu. A Btu, or British thermal unit, is the energy needed to raise the temperature of a pound of water by 1°F. Work out the conversion factor between the SI and English units for R values. Convert your answers from part (a) to English units.

(c) Prove that for a compound layer of two different materials sandwiched together (such as air and glass, or brick and wood), the effective total R value is the sum of the individual R values.

(d) Calculate the effective R value of a single piece of plate glass with a 1.0-mm layer of still air on each side. (The effective thickness of the air layer will depend on how much wind is blowing; 1 mm is of the right order of magnitude under most conditions.) Using this effective R value, make a revised estimate of the heat loss through a 1-m^2 single-pane window when the temperature in the room is 20°C higher than the outdoor temperature.

Problem 1.58. According to a standard reference table, the R value of a 3.5-inch-thick vertical air space (within a wall) is 1.0 (in English units), while the R value of a 3.5-inch thickness of fiberglass batting is 10.9. Calculate the R value of a 3.5-inch thickness of *still* air, then discuss whether these two numbers are reasonable. (Hint: These reference values include the effects of convection.)

Problem 1.59. Make a rough estimate of the total rate of conductive heat loss through the windows, walls, floor, and roof of a typical house in a cold climate. Then estimate the cost of replacing this lost energy over the course of a month. If possible, compare your estimate to a real utility bill. (Utility companies measure electricity by the **kilowatt-hour**, a unit equal to 3.6 MJ. In the United States, natural gas is billed in **therms**, where 1 therm = 10^5 Btu. Utility rates vary by region; I currently pay about 7 cents per kilowatt-hour for electricity and 50 cents per therm for natural gas.)

Problem 1.60. A frying pan is quickly heated on the stovetop to $200°$C. It has an iron handle that is 20 cm long. Estimate how much time should pass before the end of the handle is too hot to grab with your bare hand. (Hint: The cross-sectional area of the handle doesn't matter. The density of iron is about 7.9 g/cm^3 and its specific heat is 0.45 J/g·$°$C).

Problem 1.61. Geologists measure conductive heat flow out of the earth by drilling holes (a few hundred meters deep) and measuring the temperature as a function of depth. Suppose that in a certain location the temperature increases by $20°$C per kilometer of depth and the thermal conductivity of the rock is 2.5 W/m·K. What is the rate of heat conduction per square meter in this location? Assuming that this value is typical of other locations over all of earth's surface, at approximately what rate is the earth losing heat via conduction? (The radius of the earth is 6400 km.)

Problem 1.62. Consider a uniform rod of material whose temperature varies only along its length, in the x direction. By considering the heat flowing from both directions into a small segment of length Δx, derive the **heat equation**,

$$\frac{\partial T}{\partial t} = K \frac{\partial^2 T}{\partial x^2},$$

where $K = k_t/c\rho$, c is the specific heat of the material, and ρ is its density. (Assume that the only motion of energy is heat conduction within the rod; no energy enters or leaves along the sides.) Assuming that K is independent of temperature, show that a solution of the heat equation is

$$T(x,t) = T_0 + \frac{A}{\sqrt{t}} e^{-x^2/4Kt},$$

where T_0 is a constant background temperature and A is any constant. Sketch (or use a computer to plot) this solution as a function of x, for several values of t. Interpret this solution physically, and discuss in some detail how energy spreads through the rod as time passes.

Conductivity of an Ideal Gas

In a gas, the rate of heat conduction is limited by how *far* a molecule can travel before it collides with another molecule. The average distance traveled between collisions is called the **mean free path**. In a dilute gas the mean free path is many times larger than the average distance between molecules, because a molecule can pass by many of its neighbors before actually hitting one of them. Let me now make a rough estimate of the mean free path in a dilute gas.

For simplicity, imagine that all the molecules in a gas except one are frozen in place. How far does the remaining molecule travel between collisions? Well, a collision happens when the center of our molecule comes within one molecular diameter ($2r$, where r is the radius of a molecule) of the center of some other molecule (see Figure 1.17). Collisions would occur just as often if our molecule were twice as wide and all the others were points; let's therefore pretend that this is the case. Then, as our molecule travels along, it sweeps out an imaginary cylinder of space whose radius is $2r$. When the volume of this cylinder equals the average volume per molecule in the gas, we're likely to get a collision. The mean free path, ℓ, is roughly the length of the cylinder when this condition is met:

$$\text{volume of cylinder} = \text{average volume per molecule}$$

$$\Rightarrow \quad \pi(2r)^2\ell \approx \frac{V}{N}$$

$$\Rightarrow \quad \ell \approx \frac{1}{4\pi r^2}\frac{V}{N}. \tag{1.62}$$

The $\approx$ symbol indicates that this formula is only a rough approximation for ℓ, because I've neglected the motion of the other molecules as well as the variation in path lengths between collisions. The actual mean free path will differ by a numerical factor that shouldn't be too different from 1. But there's not much point in being more precise, because r itself is not well defined: Molecules don't have sharp edges,

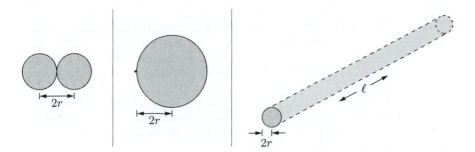

Figure 1.17. A collision between molecules occurs when their centers are separated by twice the molecular radius r. The same would be true if one molecule had radius $2r$ and the other were a point. When a sphere of radius $2r$ moves in a straight line of length ℓ, it sweeps out a cylinder whose volume is $4\pi r^2\ell$.

and most of them aren't even spherical.*

The effective radius of a nitrogen or oxygen molecule should be one or two ångstroms; let's say $r = 1.5$ Å $= 1.5 \times 10^{-10}$ m. Treating air as an ideal gas, the volume per particle is $V/N = kT/P = 4 \times 10^{-26}$ m^3 at room temperature and atmospheric pressure. With these numbers, equation 1.62 predicts a mean free path of 150 nm, about 40 times greater than the average separation between air molecules. We can also estimate the average *time* between collisions:

$$\overline{\Delta t} = \frac{\ell}{\overline{v}} \approx \frac{\ell}{v_{\mathrm{rms}}} \approx \frac{1.5 \times 10^{-7} \text{ m}}{500 \text{ m/s}} = 3 \times 10^{-10} \text{ s}. \qquad (1.63)$$

Now back to heat conduction. Consider a small region within a gas where the temperature increases in the x direction (see Figure 1.18). The heavy dotted line in the figure represents a plane perpendicular to the x direction; my intent is to estimate the amount of heat that flows across this plane. Let Δt be the average time between collisions, so that each molecule travels a distance of roughly one mean free path during this time. Then, during this time, the molecules that cross the dotted line from the left will have started from somewhere within box 1 (whose thickness is ℓ), while the molecules that cross the dotted line from the right will have started from somewhere within box 2 (whose thickness is also ℓ). Both of these boxes have the same area A in the yz plane. If the total energy of all the molecules in box 1 is U_1, then the energy crossing the dotted line from the left is roughly $U_1/2$, since only half of the molecules will have positive x velocities at this moment. Similarly, the energy crossing the line from the right is half the total energy in box 2, or $U_2/2$. The net heat flow across the line is therefore

$$Q = \frac{1}{2}(U_1 - U_2) = -\frac{1}{2}(U_2 - U_1) = -\frac{1}{2}C_V(T_2 - T_1) = -\frac{1}{2}C_V\ell\frac{dT}{dx}, \qquad (1.64)$$

where C_V is the heat capacity of all the gas in either box and T_1 and T_2 are the average temperatures in the two boxes. (In the last step I've used the fact that the distance between the centers of the two boxes is ℓ.)

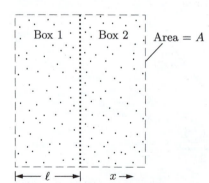

Figure 1.18. Heat conduction across the dotted line occurs because the molecules moving from box 1 to box 2 have a different average energy than the molecules moving from box 2 to box 1. For free motion between these boxes, each should have a width of roughly one mean free path.

---

*For that matter, I haven't even given a precise definition of what constitutes a collision. After all, even when molecules pass at a distance, they attract and deflect each other somewhat. For a more careful treatment of transport processes in gases, see Reif (1965).

Equation 1.64 confirms Fourier's law, that the rate of heat conduction is directly proportional to the difference in temperatures. Furthermore, comparison to equation 1.60 yields an explicit prediction for the thermal conductivity:

$$k_t = \frac{1}{2}\frac{C_V \ell}{A\,\Delta t} = \frac{1}{2}\frac{C_V}{A\ell}\frac{\ell^2}{\Delta t} = \frac{1}{2}\frac{C_V}{V}\ell\,\bar{v},\tag{1.65}$$

where $\bar{v}$ is the average speed of the molecules. The quantity C_V/V is the heat capacity of the gas per unit volume, which can be evaluated as

$$\frac{C_V}{V} = \frac{\frac{f}{2}Nk}{V} = \frac{f}{2}\frac{P}{T},\tag{1.66}$$

where f is the number of degrees of freedom per molecule. Recall, however, that ℓ for a gas is proportional to V/N. Therefore the thermal conductivity of a given gas should depend only on its temperature, through $\bar{v} \propto \sqrt{T}$ and possibly through f. Over limited ranges of temperature the number of degrees of freedom is fairly constant, so k_t should be proportional to the square root of the absolute temperature. Experiments on a wide variety of gases have confirmed this prediction (see Figure 1.19).

For air at room temperature and atmospheric pressure, $f = 5$ so $C_V/V = \frac{5}{2}(10^5 \text{ N/m}^2)/(300 \text{ K}) \approx 800 \text{ J/m}^3\cdot\text{K}$. Equation 1.65 therefore predicts a thermal conductivity of

$$k_t \approx \tfrac{1}{2}(800 \text{ J/m}^3\cdot\text{K})(1.5 \times 10^{-7} \text{ m})(500 \text{ m/s}) = 0.031 \text{ W/m}\cdot\text{K},\tag{1.67}$$

only a little higher than the measured value of 0.026. Not bad, considering all the crude approximations I've made in this section.

The preceding analysis of the thermal conductivities of gases is an example of what's called **kinetic theory**, an approach to thermal physics based on actual molecular motions. Another example was the microscopic model of an ideal gas presented in Section 1.2. While kinetic theory is the most direct and concrete approach to thermal physics, it is also the most difficult. Fortunately, there are

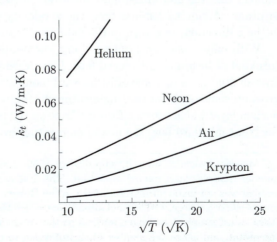

Figure 1.19. Thermal conductivities of selected gases, plotted vs. the square root of the absolute temperature. The curves are approximately linear, as predicted by equation 1.65. Data from Lide (1994).

much easier methods for predicting most of the *equilibrium* properties of materials, without having to know the details of how molecules move. To predict the *rates* of processes, however, we usually have to resort to kinetic theory.

Problem 1.63. At about what pressure would the mean free path of an air molecule at room temperature equal 10 cm, the size of a typical laboratory apparatus?

Problem 1.64. Make a rough estimate of the thermal conductivity of helium at room temperature. Discuss your result, explaining why it differs from the value for air.

Problem 1.65. Pretend that you live in the 19th century and don't know the value of Avogadro's number* (or of Boltzmann's constant or of the mass or size of any molecule). Show how you could make a rough estimate of Avogadro's number from a measurement of the thermal conductivity of a gas, together with other measurements that are relatively easy.

Viscosity

Energy isn't the only thing that can spread through a fluid at the molecular level; another is *momentum*.

Consider the situation shown in Figure 1.20: two parallel solid surfaces moving past one another, separated by a small gap containing a liquid or gas. Let's work in the reference frame where the bottom surface is at rest and the top surface is moving in the $+x$ direction. What about the motion of the fluid? At normal temperatures the fluid molecules will be jostling with thermal velocities of hundreds of meters per second, but let's ignore this motion for the moment and instead ask about the average motion at the macroscopic scale. Taking a macroscopic view, it's natural to guess that just above the bottom surface the fluid should be at rest; a thin layer of fluid "sticks" to the surface. For the same reason (since reference frames are arbitrary), a thin layer "sticks" to the top surface and moves along with it. In between the motion of the fluid could be turbulent and chaotic, but let's assume that this is not the case: The motion is slow enough, or the gap is narrow enough, that the flow of the fluid is entirely horizontal. Then the flow is said to be **laminar**. Assuming laminar flow, the x velocity of the fluid will increase steadily in the z direction, as shown in the figure.

With only a few exceptions at very low temperatures, all fluids tend to resist this kind of shearing, differential flow. This resistance is called **viscosity**. The top layer of fluid gives up some of its forward momentum to the next layer down, which gives up some of its forward momentum to the next layer, and so on down to the bottom layer which exerts a forward force on the bottom surface. At the same time (by Newton's third law) the loss of momentum by the top layer causes it to exert a

---

*Amedeo Avogadro himself, who died in 1856, never knew the numerical value of the number that was later named after him. The first *accurate* determination of Avogadro's number was not made until around 1913, when Robert Millikan measured the fundamental unit of electric charge. Others had already measured the charge-to-mass ratio of the proton (then called simply a hydrogen ion), so at that point it was easy to calculate the mass of the proton and hence the number of them needed to make a gram.

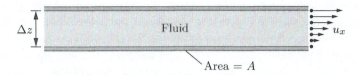

Figure 1.20. The simplest arrangement for demonstrating viscosity: two parallel surfaces sliding past each other, separated by a narrow gap containing a fluid. If the motion is slow enough and the gap narrow enough, the fluid flow is **laminar**: At the macroscopic scale the fluid moves only horizontally, with no turbulence.

backward force on the top surface. The more "viscous" the fluid, the more efficient the momentum transfer and the greater these forces will be. Air isn't very viscous; corn syrup is.

As with thermal conductivity, it isn't hard to guess how the viscous drag force depends on the geometry of the situation. The simplest guess (which turns out to be correct) is that the force is proportional to the common area of the surfaces, inversely proportional to the width of the gap, and directly proportional to the difference in velocity between the two surfaces. In the notation of Figure 1.20 (using u_x for the macroscopic velocity to distinguish it from the much faster thermal velocities),

$$F_x \propto \frac{A \cdot (u_{x,\text{top}} - u_{x,\text{bottom}})}{\Delta z} \qquad \text{or} \qquad \frac{F_x}{A} \propto \frac{\Delta u_x}{\Delta z}. \qquad (1.68)$$

The constant of proportionality is called the **coefficient of viscosity** or simply the **viscosity** of the fluid; the standard symbol for this coefficient is η, the Greek letter eta. Our formula for the force is then

$$\frac{|F_x|}{A} = \eta \frac{du_x}{dz}, \qquad (1.69)$$

where I've put absolute value bars around F_x because it could represent the force on either plate, these two forces being equal in magnitude but opposite in direction. The force per unit area has units of pressure (Pa or N/m^2), but please don't *call* it a pressure because it's exerted parallel to the surface, not perpendicular. The correct term for such a force per unit area is **shear stress**.

From equation 1.69 you can see that the coefficient of viscosity has units of pascal-seconds in the SI system. (Sometimes you'll still see viscosities given in a unit called the **poise**; this is the cgs unit, equal to a dyne-second per cm^2, which turns out to be 10 times smaller than the SI unit.) Viscosities vary enormously from one fluid to another and also vary considerably with temperature. The viscosity of water is 0.0018 Pa·s at 0°C but only 0.00028 Pa·s at 100°C. Low-viscosity motor oil (SAE 10) has a room-temperature viscosity of about 0.25 Pa·s. Gases have much lower viscosities, for example, 19 μPa·s for air at room temperature. Surprisingly, the viscosity of an ideal gas is *independent* of its pressure and *increases* as a function of temperature. This strange behavior requires some explanation.

Recall from the previous subsection that the thermal conductivity of an ideal gas behaves in a similar way: It is independent of pressure and increases with temperature in proportion to $\sqrt{T}$. Although the *amount* of energy carried by a parcel of gas is proportional to the density of particles N/V, this dependence cancels in k_t because the mean free path, which controls *how far* the energy can travel at once, is proportional to V/N. The temperature dependence of k_t comes from the remaining factor of $\bar{v}$, the average thermal speed of the gas molecules (see equation 1.65).

In exactly the same way, the transfer of horizontal momentum vertically through a gas depends on three factors: the momentum density in the gas, the mean free path, and the average thermal speed. The first two factors depend on the particle density, but this dependence cancels: Although a dense gas carries *more* momentum, random thermal motions transport that momentum through *less distance* at a time. The molecules do move *faster* at high temperature, however. According to this picture the viscosity of a gas should be proportional to $\sqrt{T}$ just like the thermal conductivity, and experiments confirm this prediction.

Why, then, does the viscosity of a liquid *decrease* as its temperature increases? In a liquid the density and the mean free path are essentially independent of temperature and pressure, but another factor comes into play: When the temperature is low and the thermal motions are slow, the molecules can better latch onto each other as they collide. This binding allows a very efficient transfer of momentum from one molecule to another. In the extreme case of a solid, the molecules are more or less permanently bonded together and the viscosity is almost infinite; solids *can* flow like fluids, but only on geological time scales.

> **Problem 1.66.** In analogy with the thermal conductivity, derive an approximate formula for the viscosity of an ideal gas in terms of its density, mean free path, and average thermal speed. Show explicitly that the viscosity is independent of pressure and proportional to the square root of the temperature. Evaluate your formula numerically for air at room temperature and compare to the experimental value quoted in the text.

Diffusion

Heat conduction is the transport of *energy* by random thermal motions. Viscosity results from the transport of *momentum*, which in gases is accomplished mainly by random thermal motions. A third entity that can be transported by random thermal motions is *particles*, which tend to spread from areas of high concentration to areas of low concentration. For example, if you drop a drop of food coloring into a cup of still water, you'll see the dye gradually spreading out in all directions. This spreading out of particles is called **diffusion**.[*]

Like the flow of energy and momentum, the flow of particles by diffusion obeys an equation that is fairly easy to guess. Just as heat conduction is caused by a temperature difference and viscous drag is caused by a velocity difference, diffusion is caused by a difference in the *concentration* of particles, that is, the number of particles per unit volume, N/V. In this section (and only in this section) I'll use the

---

[*]Problem 1.22 treats the simpler process of a gas escaping through a hole into a vacuum, called **effusion**.

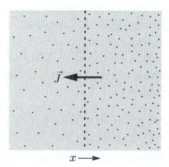

Figure 1.21. When the concentration of a certain type of molecule increases from left to right, there will be **diffusion**, a net flow of molecules, from right to left.

symbol n for particle concentration. To keep the geometry simple, imagine a region where n for a certain type of particle increases uniformly in the x direction (see Figure 1.21). The **flux** of these particles across any surface is the net number that cross it per unit area per unit time; the symbol for particle flux is $\vec{J}$. Then, in analogy with equations 1.60 and 1.69, we would probably guess that $|\vec{J}|$ is proportional to dn/dx. Again, this guess turns out to be correct under most circumstances. Using the symbol D for the constant of proportionality, we can write

$$J_x = -D\frac{dn}{dx}. \tag{1.70}$$

The minus sign indicates that if dn/dx is positive, the flux is in the negative x direction. This equation is known as **Fick's law**, after the 19th century German physiologist Adolf Eugen Fick.

The constant D is called the **diffusion coefficient**; it depends both on the type of molecule that is diffusing and on what it is diffusing through. In SI units (m^2/s), diffusion coefficients in water near room temperature range from 9×10^{-9} for H^+ ions to 5×10^{-10} for sucrose to a few times 10^{-11} for very large molecules such as proteins. Diffusion in gases is faster: For CO molecules diffusing through air at room temperature and atmospheric pressure, $D = 2 \times 10^{-5}$ m^2/s. Other small molecules diffusing through air have similar D values. As you would probably expect, diffusion coefficients generally increase with increasing temperature.

Although diffusion is extremely important on the small scales of biological cells, cloud droplets, and semiconductor fabrication, the small D values quoted above indicate that it is not an efficient mechanism for large-scale mixing. As a quick example, consider a drop of food coloring added to a glass of water. Imagine that the dye has already spread uniformly through half of the glass. How long would it take to diffuse into the other half? According to Fick's law, I can write very roughly

$$\frac{N}{A\,\Delta t} = D\frac{N/V}{\Delta x}, \tag{1.71}$$

where N is the total number of dye molecules, Δx is about 0.1 m and $V \approx A \cdot \Delta x$. I've written the particle flux in terms of the same N to indicate that I want Δt to be the time for approximately all (that is, half) of the molecules to cross from one side of the glass to the other. I don't know how big a molecule of food coloring is, but it can't be too different in size from sucrose so I'll guess $D = 10^{-9}$ m^2/s. Solving for Δt then gives 10^7 seconds, or almost four months. If you actually perform an experiment with water and food coloring, you'll probably find that they mix *much* faster than this, due to bulk motion of the water—convection. You *can* see the diffusion, though, if you look very closely at the interface between the colored and clear water.

Problem 1.67. Make a rough estimate of how far food coloring (or sugar) will diffuse through water in one minute.

Problem 1.68. Suppose you open a bottle of perfume at one end of a room. Very roughly, how much time would pass before a person at the other end of the room could smell the perfume, *if* diffusion were the only transport mechanism? Do you think diffusion *is* the dominant transport mechanism in this situation?

Problem 1.69. Imagine a narrow pipe, filled with fluid, in which the concentration of a certain type of molecule varies only along the length of the pipe (in the x direction). By considering the flux of these particles from both directions into a short segment Δx, derive **Fick's second law**,

$$\frac{\partial n}{\partial t} = D \frac{\partial^2 n}{\partial x^2}.$$

Noting the similarity to the heat equation derived in Problem 1.62, discuss the implications of this equation in some detail.

Problem 1.70. In analogy with the thermal conductivity, derive an approximate formula for the diffusion coefficient of an ideal gas in terms of the mean free path and the average thermal speed. Evaluate your formula numerically for air at room temperature and atmospheric pressure, and compare to the experimental value quoted in the text. How does D depend on T, at fixed pressure?

Humans are to a large degree sensitive to energy fluxes rather than temperatures, which you can verify for yourself on a cold, dark morning in the outhouse of a mountain cabin equipped with wooden and metal toilet seats. Both seats are at the same temperature, but your backside, which is not a very good thermometer, is nevertheless very effective at telling you which is which.

—Craig F. Bohren and Bruce A. Albrecht,
Atmospheric Thermodynamics (Oxford
University Press, New York, 1998).

2 The Second Law

The previous chapter explored the law of energy conservation as it applies to thermodynamic systems. It also introduced the concepts of heat, work, and temperature. However, some very fundamental questions remain unanswered: What is temperature, *really*, and why does heat flow spontaneously from a hotter object to a cooler object, never the other way? More generally, why do so many thermodynamic processes happen in one direction but never the reverse? This is the Big Question of thermal physics, which we now set out to answer.

In brief, the answer is this: Irreversible processes are not *inevitable*, they are just overwhelmingly *probable*. For instance, when heat flows from a hot object to a cooler object, the energy is just moving around more or less randomly. After we wait a while, the chances are overwhelming that we will find the energy distributed more "uniformly" (in a sense that I will make precise later) among all the parts of a system. "Temperature" is a way of quantifying the tendency of energy to enter or leave an object during the course of these random rearrangements.

To make these ideas precise, we need to study *how* systems store energy, and learn to count all the ways that the energy might be arranged. The mathematics of counting ways of arranging things is called **combinatorics**, and this chapter begins with a brief introduction to this subject.

2.1 Two-State Systems

Suppose that I flip three coins: a penny, a nickel, and a dime. How many possible outcomes are there? Not very many, so I've listed them all explicitly in Table 2.1. By this brute-force method, I count *eight* possible outcomes. If the coins are fair, each outcome is equally probable, so the probability of getting three heads or three tails is one in eight. There are three different ways of getting two heads and a tail, so the probability of getting exactly two heads is 3/8, as is the probability of

Penny	Nickel	Dime
H	H	H
H	H	T
H	T	H
T	H	H
H	T	T
T	H	T
T	T	H
T	T	T

Table 2.1. A list of all possible "microstates" of a set of three coins (where H is for heads and T is for tails).

getting exactly one head and two tails.

Now let me introduce some fancy terminology. Each of the eight different outcomes is called a **microstate**. In general, to specify the microstate of a system, we must specify the state of each individual particle, in this case the state of each coin. If we specify the state more generally, by merely saying how *many* heads or tails there are, we call it a **macrostate**. Of course, if you know the microstate of the system (say HHT), then you also know its macrostate (in this case, two heads). But the reverse is not true: Knowing that there are exactly two heads does *not* tell you the state of each coin, since there are three microstates corresponding to this macrostate. The number of microstates corresponding to a given macrostate is called the **multiplicity** of that macrostate, in this case 3.

The symbol I'll use for multiplicity is the Greek letter capital omega, Ω. In the example of the three coins, $\Omega(3 \text{ heads}) = 1$, $\Omega(2 \text{ heads}) = 3$, $\Omega(1 \text{ head}) = 3$, and $\Omega(0 \text{ heads}) = 1$. Note that the total multiplicity of all four macrostates is $1 + 3 + 3 + 1 = 8$, the total number of microstates. I'll call this quantity $\Omega(\text{all})$. Then the *probability* of any particular macrostate can be written

$$\text{probability of } n \text{ heads} = \frac{\Omega(n)}{\Omega(\text{all})}. \tag{2.1}$$

For instance, the probability of getting 2 heads is $\Omega(2)/\Omega(\text{all}) = 3/8$. Again, I'm assuming here that the coins are fair, so that all 8 microstates are equally probable.

To make things a little more interesting, suppose now that there are not just three coins but 100. The total number of *micro*states is now very large: 2^{100}, since each of the 100 coins has two possible states. The number of *macro*states, however, is only 101: 0 heads, 1 head, ... up to 100 heads. What about the multiplicities of these macrostates?

Let's start with the 0-heads macrostate. If there are zero heads, then every coin faces tails-up, so the exact microstate has been specified, that is, $\Omega(0) = 1$.

What if there is exactly one head? Well, the heads-up coin could be the first one, or the second one, etc., up to the 100th one; that is, there are exactly 100 possible microstates: $\Omega(1) = 100$. If you imagine all the coins starting heads-down, then $\Omega(1)$ is the number of ways of *choosing* one of them to turn over.

To find $\Omega(2)$, consider the number of ways of choosing two coins to turn heads-up. You have 100 choices for the first coin, and for each of these choices you have

99 remaining choices for the second coin. But you could choose any pair in either order, so the number of *distinct* pairs is

$$\Omega(2) = \frac{100 \cdot 99}{2}. \tag{2.2}$$

If you're going to turn three coins heads-up, you have 100 choices for the first, 99 for the second, and 98 for the third. But any triplet could be chosen in several ways: 3 choices for which one to flip first, and for each of these, 2 choices for which to flip second. Thus, the number of distinct triplets is

$$\Omega(3) = \frac{100 \cdot 99 \cdot 98}{3 \cdot 2}. \tag{2.3}$$

Perhaps you can now see the pattern. To find $\Omega(n)$, we write the product of n factors, starting with 100 and counting down, in the numerator. Then we divide by the product of n factors, starting with n and counting down to 1:

$$\Omega(n) = \frac{100 \cdot 99 \cdots (100 - n + 1)}{n \cdots 2 \cdot 1}. \tag{2.4}$$

The denominator is just n-factorial, denoted "$n!$". We can also write the numerator in terms of factorials, as $100!/(100-n)!$. (Imagine writing the product of all integers from 100 down to 1, then canceling all but the first n of them.) Thus the general formula can be written

$$\Omega(n) = \frac{100!}{n! \cdot (100 - n)!} \equiv \binom{100}{n}. \tag{2.5}$$

The last expression is just a standard abbreviation for this quantity, sometimes spoken "100 choose n"—the number of different ways of choosing n items out of 100, or the number of "combinations" of n items chosen from 100.

If instead there are N coins, the multiplicity of the macrostate with n heads is

$$\Omega(N, n) = \frac{N!}{n! \cdot (N - n)!} = \binom{N}{n}, \tag{2.6}$$

the number of ways of choosing n objects out of N.

Problem 2.1. Suppose you flip four fair coins.
   (a) Make a list of all the possible outcomes, as in Table 2.1.
   (b) Make a list of all the different "macrostates" and their probabilities.
   (c) Compute the multiplicity of each macrostate using the combinatorial formula 2.6, and check that these results agree with what you got by brute-force counting.

Problem 2.2. Suppose you flip 20 fair coins.
   (a) How many possible outcomes (microstates) are there?
   (b) What is the probability of getting the sequence HTHHTTTHTHHHTHH-HTHT (in exactly that order)?
   (c) What is the probability of getting 12 heads and 8 tails (in any order)?

Problem 2.3. Suppose you flip 50 fair coins.

(a) How many possible outcomes (microstates) are there?

(b) How many ways are there of getting exactly 25 heads and 25 tails?

(c) What is the probability of getting exactly 25 heads and 25 tails?

(d) What is the probability of getting exactly 30 heads and 20 tails?

(e) What is the probability of getting exactly 40 heads and 10 tails?

(f) What is the probability of getting 50 heads and no tails?

(g) Plot a graph of the probability of getting n heads, as a function of n.

Problem 2.4. Calculate the number of possible five-card poker hands, dealt from a deck of 52 cards. (The order of cards in a hand does not matter.) A royal flush consists of the five highest-ranking cards (ace, king, queen, jack, 10) of any one of the four suits. What is the probability of being dealt a royal flush (on the first deal)?

The Two-State Paramagnet

You may be wondering what this silly coin-flipping example has to do with physics. Not much yet, but actually there are important physical systems whose combinatorics are exactly the same. Perhaps the most important of these is a **two-state paramagnet**.

All materials will respond in some way to a magnetic field, because of the electrical nature of electrons and atomic nuclei. A **paramagnet** is a material in which the constituent particles act like tiny compass needles that tend to align *parallel* to any externally applied magnetic field. (If the particles interact strongly enough with each other, the material can magnetize even *without* any externally applied field. We then call it a **ferromagnet**, after the most famous example, iron. Paramagnetism, in constrast, is a magnetic alignment that lasts only as long as an external field is applied.)

I'll refer to the individual magnetic particles as **dipoles**, because each has its own magnetic dipole moment vector. In practice each dipole could be an individual electron, a group of electrons in an atom, or an atomic nucleus. For any such microscopic dipole, quantum mechanics allows the component of the dipole moment vector along any given axis to take on only certain discrete values—intermediate values are not allowed. In the simplest case only *two* values are allowed, one positive and the other negative. We then have a **two-state paramagnet**, in which each elementary compass needle can have only two possible orientations, either parallel or antiparallel to the applied field. I'll draw this system as a bunch of little arrows, each pointing either up or down, as in Figure 2.1.*

Now for the combinatorics. Let's define $N_\uparrow$ to be the number of elementary dipoles that point up (at some particular time), and $N_\downarrow$ to be the number of dipoles that point down. The total number of dipoles is then $N = N_\uparrow + N_\downarrow$, and we'll

---

*A particle's dipole moment vector is proportional to its angular momentum vector; the simple two-state case occurs for particles with "spin 1/2." For a more complete discussion of quantum mechanics and angular momentum, see Appendix A.

Figure 2.1. A symbolic representation of a two-state paramagnet, in which each elementary dipole can point either parallel or antiparallel to the externally applied magnetic field.

consider this number to be fixed. This system has one macrostate for each possible value of $N_\uparrow$, from 0 to N. The multiplicity of any macrostate is given by the same formula as in the coin-tossing example:

$$\Omega(N_\uparrow) = \binom{N}{N_\uparrow} = \frac{N!}{N_\uparrow! \, N_\downarrow!}. \tag{2.7}$$

The external magnetic field exerts a torque on each little dipole, trying to twist it to point parallel to the field. If the external field points up, then an up-dipole has *less* energy than a down-dipole, since you would have add energy to twist it from up to down. The total energy of the system (neglecting any interactions *between* dipoles) is determined by the total numbers of up- and down-dipoles, so specifying which macrostate this system is in is the same as specifying its total energy. In fact, in nearly all physical examples, the macrostate of a system is characterized, at least in part, by its total energy.

2.2   The Einstein Model of a Solid

Now let's move on to a system that's a bit more complicated, but also more representative of the systems typically encountered in physics. Consider a collection of microscopic systems that can each store any number of energy "units," all of the same size. Equal-size energy units occur for any quantum-mechanical *harmonic oscillator*, whose potential energy function has the form $\frac{1}{2}k_s x^2$ (where k_s is the "spring constant"). The size of the energy units is then hf,* where h is **Planck's constant** $(6.63 \times 10^{-34}$ J·s$)$ and f is the natural frequency of the oscillator $(\frac{1}{2\pi}\sqrt{k_s/m})$. An abstract way of picturing a collection of many such oscillators is shown in Figure 2.2.

---

*As explained in Appendix A, the lowest possible energy of a quantum harmonic oscillator is actually $\frac{1}{2}hf$, not zero. But this "zero-point" energy never moves around, so it plays no role in thermal interactions. The excited-state energies are $\frac{3}{2}hf$, $\frac{5}{2}hf$, and so on, each with an additional energy "unit" of hf. For our purposes, it's fine to measure all energies relative to the ground state; then the allowed energies are 0, hf, $2hf$, etc.

Elsewhere you may see the energy unit of a quantum oscillator written as $\hbar\omega$, where $\hbar = h/2\pi$ and $\omega = 2\pi f$. The difference between $\hbar\omega$ and hf is nothing but a matter of where to put the factors of 2π.

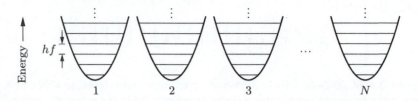

Figure 2.2. In quantum mechanics, any system with a quadratic potential energy function has evenly spaced energy levels separated in energy by hf, where f is the classical oscillation frequency. An Einstein solid is a collection of N such oscillators, all with the same frequency.

Examples of quantum oscillators include the vibrational motions of diatomic and polyatomic gas molecules. But an even more common example is the oscillation of atoms in a solid (see Figure 1.6). In a three-dimensional solid, each atom can oscillate in three independent directions, so if there are N oscillators, there are only $N/3$ atoms. The model of a solid as a collection of identical oscillators with quantized energy units was first proposed by Albert Einstein in 1907, so I will refer to this system as an **Einstein solid**.

Let's start with a very small Einstein solid, containing only three oscillators: $N = 3$. Table 2.2 lists the various microstates that this system could have, in order of increasing total energy; each row in the table corresponds to a different microstate. There is just one microstate with total energy 0, while there are three microstates with one unit of energy, six with two units, and ten with three units. That is,

$$\Omega(0) = 1, \qquad \Omega(1) = 3, \qquad \Omega(2) = 6, \qquad \Omega(3) = 10. \qquad (2.8)$$

The general formula for the multiplicity of an Einstein solid with N oscillators

Oscillator:	#1	#2	#3
Energy:	0	0	0
	1	0	0
	0	1	0
	0	0	1
	2	0	0
	0	2	0
	0	0	2
	1	1	0
	1	0	1
	0	1	1

Oscillator:	#1	#2	#3
Energy:	3	0	0
	0	3	0
	0	0	3
	2	1	0
	2	0	1
	1	2	0
	0	2	1
	1	0	2
	0	1	2
	1	1	1

Table 2.2. Microstates of a small Einstein solid consisting of only three oscillators, containing a total of zero, one, two, or three units of energy.

and q energy units is

$$\Omega(N, q) = \binom{q + N - 1}{q} = \frac{(q + N - 1)!}{q! \, (N - 1)!}. \tag{2.9}$$

Please check this formula for the examples just given. To *prove* this formula, let me adopt the following graphical representation of the microstate of an Einstein solid: I'll use a dot to represent each energy unit, and a vertical line to represent a partition between one oscillator and the next. So in a solid with four oscillators, the sequence

$$\bullet \; | \; \bullet \; \bullet \; \bullet \; | \; | \; \bullet \; \bullet \; \bullet \; \bullet$$

represents the microstate in which the first oscillator has one unit of energy, the second oscillator has three, the third oscillator has none, and the fourth oscillator has four. Notice that any microstate can be represented uniquely in this way, and that every possible sequence of dots and lines corresponds to a microstate. There are always q dots and $N - 1$ lines, for a total of $q + N - 1$ symbols. Given q and N, the number of possible arrangements is

of the symbols to be dots, that is, $\binom{q + N - 1}{q}$.

Problem 2.5. For an Einstein solid with
list all of the possible microstates, count th

   (a) $N = 3, q = 4$

   (b) $N = 3, q = 5$

   (c) $N = 3, q = 6$

   (d) $N = 4, q = 2$

   (e) $N = 4, q = 3$

   (f) $N = 1, q =$ anything

   (g) $N =$ anything, $q = 1$

Problem 2.6. Calculate the multiplicity
and 30 units of energy. (Do not attempt t

Problem 2.7. For an Einstein solid with
represent each possible microstate as a se
the text to prove equation 2.9.

Compare # of microstates w/ multiplicity in 2.9

You know, the most amazing thing happened to me tonight. I was coming here, on the way to the lecture, and I came in through the parking lot. And you won't believe what happened. I saw a car with the license plate ARW 357! Can you imagine? Of all the millions of license plates in the state, what was the chance that I would see that particular one tonight? Amazing!

                     —Richard Feynman, quoted by David
                     Goodstein, *Physics Today* **42**, 73
                     (February, 1989).

2.3 Interacting Systems

We now know how to count the microstates of an Einstein solid. To understand heat flow and irreversible processes, however, we need to consider a system of *two* Einstein solids that can share energy back and forth.* I'll call the two solids A and B (see Figure 2.3).

First I should be clear about what is meant by the "macrostate" of such a composite system. For simplicity, I'll assume that the two solids are **weakly coupled**, so that the exchange of energy between them is much slower than the exchange of energy among atoms within each solid. Then the individual energies of the solids, U_A and U_B, will change only slowly; over sufficiently short time scales they are essentially fixed. I will use the word "macrostate" to refer to the state of the combined system, as specified by the (temporarily) constrained values of U_A and U_B. For any such macrostate we can compute the multiplicity, as we shall soon see. However, on longer time scales the values of U_A and U_B *will* change, so I'll also talk about the total multiplicity for *all* allowed values of U_A and U_B, counting all possible microstates with only the sum $U_{\text{total}} = U_A + U_B$ held fixed.

Let's start with a very small system, in which each of the "solids" contains only three harmonic oscillators and they contain a total of six units of energy:

$$N_A = N_B = 3; \qquad q_{\text{total}} = q_A + q_B = 6. \tag{2.10}$$

(Again I'm using q to denote the number of units of energy. The actual value of the energy is $U = qhf$.) Given these parameters, I must still specify the individual value of q_A or q_B to describe the macrostate of the system. There are seven possible macrostates, with $q_A = 0, 1, \ldots, 6$, as listed in Figure 2.4. I've used the standard formula $\binom{q+N-1}{q}$ to compute the individual multiplicities Ω_A and Ω_B for each macrostate. (I also computed some of them in the previous section by explicitly counting the microstates.) The total multiplicity of any macrostate, Ω_{total}, is just the product of the individual multiplicities, since the systems are independent of each other: For *each* of the Ω_A microstates available to solid A, there are Ω_B microstates available to solid B. The total multiplicity is also plotted in the bar

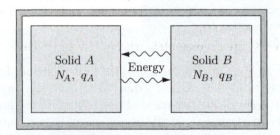

Figure 2.3. Two Einstein solids that can exchange energy with each other, isolated from the rest of the universe.

---

*This section and parts of Sections 3.1 and 3.3 are based on an article by T. A. Moore and D. V. Schroeder, *American Journal of Physics* **65**, 26–36 (1997).

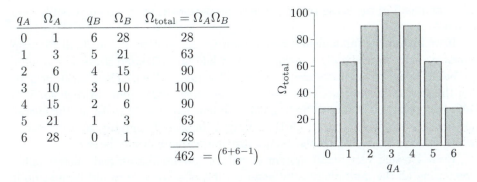

q_A	Ω_A	q_B	Ω_B	$\Omega_{\text{total}} = \Omega_A \Omega_B$
0	1	6	28	28
1	3	5	21	63
2	6	4	15	90
3	10	3	10	100
4	15	2	6	90
5	21	1	3	63
6	28	0	1	28

$$462 = \binom{6+6-1}{6}$$

Figure 2.4. Macrostates and multiplicities of a system of two Einstein solids, each containing three oscillators, sharing a total of six units of energy.

graph. Over long time scales, the number of microstates accessible to the system is 462, the sum of the last column in the table. This number can also be checked by applying the standard formula to the entire system of six oscillators and six energy units.

Now let me introduce a big assumption: Let's assume that, over long time scales, the energy gets passed around randomly* in such a way that *all 462 microstates are equally probable.* So if you look at the system at any instant, you are equally likely to find it in any of the 462 microstates. This assumption is called the **fundamental assumption of statistical mechanics**:

> In an isolated system in thermal equilibrium, all accessible microstates are equally probable.

I can't prove this assumption, though it should seem plausible. At the microscopic level, we expect that any process that would take the system from state X to state Y is reversible, so that the system can just as easily go from state Y to state X.[†] In that case, the system should have no preference for one state over another. Still, it's not obvious that all of the supposedly "accessible" microstates can actually be reached within a reasonable amount of time. In fact, we'll soon see that for a large system, the number of "accessible" microstates is usually so huge that only a miniscule fraction of them could possibly occur within a lifetime. What we're assuming is that the microstates that *do* occur, over "long" but not unthinkably long time scales, constitute a representative sample. We assume that the transitions

---

*Exchange of energy requires some kind of interaction among the oscillators. Fortunately, the precise nature of this interaction doesn't really matter. There is a danger, though, that interactions among oscillators could affect the energy levels of each particular oscillator. This would spoil our assumption that the energy levels of each oscillator are evenly spaced. Let us therefore assume that the interactions among oscillators are strong enough to allow the exchange of energy, but too weak to have much effect on the energy levels themselves. This assumption is not fundamental to statistical mechanics, but it makes explicit calculations a whole lot easier.

[†]This idea is called the **principle of detailed balance**.

are "random," in the sense that they have no pattern that we could possibly care about.*

If we invoke the fundamental assumption for our system of two small Einstein solids, we can immediately conclude that, while all 462 *microstates* are equally probable, some *macrostates* are more probable than others. The chance of finding the system in the fourth macrostate (with three energy units in each solid) is 100/462, while the chance of finding it in the first macrostate (with all the energy in solid B) is only 28/462. If all the energy is in solid B initially, and we wait a while, chances are we'll find the energy distributed more evenly later on.

Even for this very small system of only a few oscillators and energy units, computing all the multiplicities by hand is a bit of a chore. I would hate to do it for a system of a hundred oscillators and energy units. Fortunately, it's not hard to instruct a *computer* to do the arithmetic. Using a computer spreadsheet program, or comparable software, or perhaps even a graphing calculator, you should be able to reproduce the table and graph in Figure 2.4 without too much difficulty (see Problem 2.9).

Figure 2.5 shows a computer-generated table and graph for a system of two Einstein solids with

$$N_A = 300, \qquad N_B = 200, \qquad q_\text{total} = 100. \qquad (2.11)$$

Now there are 101 possible macrostates, of which only a few are shown in the table. Look at the multiplicities: Even the *least* likely macrostate, with all the energy in solid B, has a multiplicity of 3×10^{81}. The *most* likely macrostate, with $q_A = 60$, has a multiplicity of 7×10^{114}. But what is important about these numbers is not that they are large, but that their *ratio* is large: The most likely macrostate is more than 10^{33} times more probable than the least likely macrostate.

Let's look at this example in a little more detail. The total number of microstates for all the macrostates is 9×10^{115}, so the probability of finding the system in its *most* likely macrostate is not particularly large: about 7%. There are several other macrostates, with q_A slightly smaller or larger than 60, whose probabilities are nearly as large. But as q_A gets farther away from 60, on either side, the probability drops off very sharply. The probability of finding q_A to be less than 30 or greater than 90 is less than one in a million, and the probability of finding $q_A < 10$ is less than 10^{-20}. The age of the universe is less than 10^{18} seconds, so you would need to check this system a hundred times each second over the entire age of the universe before you had a decent chance of ever finding it with $q_A < 10$. Even then, you would never find it with $q_A = 0$.

---

*There can be whole classes of states that are not accessible at all, perhaps because they have the wrong total energy. There can also be classes of states that are accessible only over time scales that are much longer than we are willing to wait. The concept of "accessible," like that of "macrostate," depends on the time scale under consideration. In the case of the Einstein solids, I'm assuming that all microstates with a given energy are accessible.

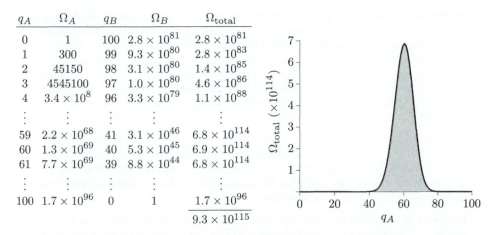

q_A	Ω_A	q_B	Ω_B	Ω_{total}
0	1	100	2.8×10^{81}	2.8×10^{81}
1	300	99	9.3×10^{80}	2.8×10^{83}
2	45150	98	3.1×10^{80}	1.4×10^{85}
3	4545100	97	1.0×10^{80}	4.6×10^{86}
4	3.4×10^{8}	96	3.3×10^{79}	1.1×10^{88}
$\vdots$	$\vdots$	$\vdots$	$\vdots$	$\vdots$
59	2.2×10^{68}	41	3.1×10^{46}	6.8×10^{114}
60	1.3×10^{69}	40	5.3×10^{45}	6.9×10^{114}
61	7.7×10^{69}	39	8.8×10^{44}	6.8×10^{114}
$\vdots$	$\vdots$	$\vdots$	$\vdots$	$\vdots$
100	1.7×10^{96}	0	1	1.7×10^{96}
				9.3×10^{115}

Figure 2.5. Macrostates and multiplicities of a system of two Einstein solids, with 300 and 200 oscillators respectively, sharing a total of 100 units of energy.

Suppose, however, that this system is *initially* in a state with q_A much less than 60; perhaps all the energy starts out in solid B. If you now wait a while for the energy to rearrange itself, then check again, you are more or less *certain* to find that energy has flowed from B to A. This system exhibits *irreversible* behavior: Energy flows spontaneously from B to A, but never (aside from small fluctuations around $q_A = 60$) from A to B. Apparently, we have discovered the physical explanation of *heat*: It is a *probabilistic* phenomenon, not absolutely certain, but extremely likely.

We have also stumbled upon a new law of physics: The spontaneous flow of energy *stops* when a system is at, or very near, its *most likely macrostate*, that is, the macrostate with the greatest multiplicity. This "law of increase of multiplicity" is one version of the famous **second law of thermodynamics**. Notice, though, that it's not a *fundamental* law at all—it's just a very strong statement about probabilities.

To make the statement stronger, and to be more realistic in general, we really should consider systems with not just a few hundred particles, but more like 10^{23}. Unfortunately, even a computer cannot calculate the number of ways of arranging 10^{23} units of energy among 10^{23} oscillators. Fortunately, there are some nice approximations we can make, to tackle this problem analytically. That is the subject of the following section.

Problem 2.8. Consider a system of two Einstein solids, A and B, each containing 10 oscillators, sharing a total of 20 units of energy. Assume that the solids are weakly coupled, and that the total energy is fixed.

(a) How many different *macrostates* are available to this system?

(b) How many different *microstates* are available to this system?

(c) Assuming that this system is in thermal equilibrium, what is the probability of finding all the energy in solid A?

(d) What is the probability of finding exactly half of the energy in solid A?

(e) Under what circumstances would this system exhibit irreversible behavior?

Problem 2.9. Use a computer to reproduce the table and graph in Figure 2.4: two Einstein solids, each containing three harmonic oscillators, with a total of six units of energy. Then modify the table and graph to show the case where one Einstein solid contains six harmonic oscillators and the other contains four harmonic oscillators (with the total number of energy units still equal to six). Assuming that all microstates are equally likely, what is the most probable macrostate, and what is its probability? What is the least probable macrostate, and what is its probability?

Problem 2.10. Use a computer to produce a table and graph, like those in this section, for the case where one Einstein solid contains 200 oscillators, the other contains 100 oscillators, and there are 100 units of energy in total. What is the most probable macrostate, and what is its probability? What is the least probable macrostate, and what is its probability?

Problem 2.11. Use a computer to produce a table and graph, like those in this section, for two interacting two-state paramagnets, each containing 100 elementary magnetic dipoles. Take a "unit" of energy to be the amount needed to flip a single dipole from the "up" state (parallel to the external field) to the "down" state (antiparallel). Suppose that the total number of units of energy, relative to the state with all dipoles pointing up, is 80; this energy can be shared in any way between the two paramagnets. What is the most probable macrostate, and what is its probability? What is the least probable macrostate, and what is its probability?

2.4 Large Systems

In the previous section we saw that, for a system of two interacting Einstein solids, each with a hundred or so oscillators, certain macrostates are *much* more probable than others. However, a significant fraction of the macrostates, roughly 20%, were still fairly probable. Next we'll look at what happens when the system is much larger, so that each solid contains, say, 10^{20} or more oscillators. My goal, by the end of this section, is to show you that out of all the macrostates, only a tiny fraction are reasonably probable. In other words, the multiplicity function becomes very sharp (see Figure 2.6). To analyze such large systems, however, we must first make a detour into the mathematics of very large numbers.

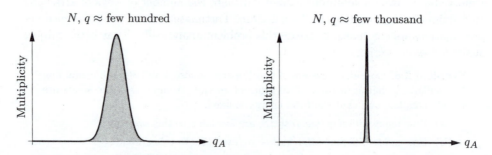

Figure 2.6. Typical multiplicity graphs for two interacting Einstein solids, containing a few hundred oscillators and energy units (left) and a few thousand (right). As the size of the system increases, the peak becomes very narrow relative to the full horizontal scale. For $N \approx q \approx 10^{20}$, the peak is much too sharp to draw.

Very Large Numbers

There are three kinds of numbers that commonly occur in statistical mechanics: small numbers, large numbers, and very large numbers.

  Small numbers are small numbers, like 6, 23, and 42. You already know how to manipulate small numbers.

  Large numbers are much larger than small numbers, and are frequently made by exponentiating small numbers. The most important large number in statistical mechanics is Avogadro's number, which is of order 10^{23}. The most important property of large numbers is that you can *add* a small number to a large number without changing it. For example,

$$10^{23} + 23 = 10^{23}. \tag{2.12}$$

(The only exception to this rule is when you plan to eventually subtract off the same large number: $10^{23} + 42 - 10^{23} = 42$.)

  Very large numbers are even larger than large numbers, and can be made by exponentiating large numbers. An example would be* $10^{10^{23}}$. Very large numbers have the amazing property that you can *multiply* them by large numbers without changing them. For instance,

$$10^{10^{23}} \times 10^{23} = 10^{(10^{23}+23)} = 10^{10^{23}}, \tag{2.13}$$

by virtue of equation 2.12. This property takes some getting used to, but can be extremely convenient when manipulating very large numbers. (Again, there is an exception: When you plan to eventually *divide* by the same very large number, you need to keep track of any leftover factors.)

  One common trick for manipulating very large numbers is to take the logarithm. This operation turns a *very* large number into an ordinary *large* number, which is much more familiar and can be manipulated more straightforwardly. Then at the end you can exponentiate to get back the very large number. I'll use this trick later in this section.

  Problem 2.12. The natural logarithm function, ln, is defined so that $e^{\ln x} = x$ for any positive number x.

  (a) Sketch a graph of the natural logarithm function.

  (b) Prove the identities

$$\ln ab = \ln a + \ln b \qquad \text{and} \qquad \ln a^b = b \ln a.$$

  (c) Prove that $\dfrac{d}{dx} \ln x = \dfrac{1}{x}$.

  (d) Derive the useful approximation

$$\ln(1 + x) \approx x,$$

  which is valid when $|x| \ll 1$. Use a calculator to check the accuracy of this approximation for $x = 0.1$ and $x = 0.01$.

---

*Note that x^{y^z} means $x^{(y^z)}$, not $(x^y)^z$.

Problem 2.13. Fun with logarithms.

(a) Simplify the expression $e^{a \ln b}$. (That is, write it in a way that doesn't involve logarithms.)

(b) Assuming that $b \ll a$, prove that $\ln(a+b) \approx (\ln a) + (b/a)$. (Hint: Factor out the a from the argument of the logarithm, so that you can apply the approximation of part (d) of the previous problem.)

Problem 2.14. Write $e^{10^{23}}$ in the form 10^x, for some x.

Stirling's Approximation

Our formulas for multiplicities involve "combinations," which involve factorials. To apply these formulas to *large* systems, we need a trick for evaluating factorials of large numbers. The trick is called **Stirling's approximation**:

$$N! \approx N^N e^{-N} \sqrt{2\pi N}. \tag{2.14}$$

This approximation is accurate in the limit where $N \gg 1$. Let me try to explain why.

The quantity $N!$ is the product of N factors, from 1 up to N. A very crude approximation would be to replace each of the N factors in the factorial by N, so $N! \approx N^N$. This is a gross *over*estimate, since nearly all of the N factors in $N!$ are actually smaller than N. It turns out that, on average, each factor is effectively smaller by a factor of e:

$$N! \approx \left(\frac{N}{e}\right)^N = N^N e^{-N}. \tag{2.15}$$

This is still off by a large factor, roughly $\sqrt{2\pi N}$. But if N is a large number, then $N!$ is a *very* large number, and often this correction factor (which is only a *large* number) can be omitted.

If all you care about is the *logarithm* of $N!$, then equation 2.15 is usually good enough. Another way to write it is

$$\ln N! \approx N \ln N - N. \tag{2.16}$$

It's fun to test Stirling's approximation on some not-very-large numbers, using a calculator or a computer. Table 2.3 shows a sampling of results. As you can see, N does not have to be particularly large before Stirling's approximation becomes useful. Equation 2.14 is quite accurate even for $N = 10$, while equation 2.16 is quite accurate for $N = 100$ (if all you care about is the logarithm).

For a *derivation* of Stirling's approximation, see Appendix B.

N	$N!$	$N^N e^{-N}\sqrt{2\pi N}$	Error	$\ln N!$	$N \ln N - N$	Error
1	1	.922	7.7%	0	-1	∞
10	3628800	3598696	.83%	15.1	13.0	13.8%
100	9×10^{157}	9×10^{157}	.083%	364	360	.89%

Table 2.3. Comparison of Stirling's approximation (equations 2.14 and 2.16) to exact values for $N = 1$, 10, and 100.

Problem 2.15. Use a pocket calculator to check the accuracy of Stirling's approximation for $N = 50$. Also check the accuracy of equation 2.16 for $\ln N!$.

Problem 2.16. Suppose you flip 1000 coins.

(a) What is the probability of getting *exactly* 500 heads and 500 tails? (Hint: First write down a formula for the total number of possible outcomes. Then, to determine the "multiplicity" of the 500-500 "macrostate," use Stirling's approximation. If you have a fancy calculator that makes Stirling's approximation unnecessary, multiply all the numbers in this problem by 10, or 100, or 1000, until Stirling's approximation becomes necessary.)

(b) What is the probability of getting exactly 600 heads and 400 tails?

Multiplicity of a Large Einstein Solid

Armed with Stirling's approximation, let me now estimate the multiplicity of an Einstein solid containing a *large* number of oscillators and energy units. Rather than working it out in complete generality, I'll consider only the case $q \gg N$, when there are many more energy units than oscillators. (This is the "high-temperature" limit.)

I'll start with the exact formula:

$$\Omega(N, q) = \binom{q + N - 1}{q} = \frac{(q + N - 1)!}{q!\,(N - 1)!} \approx \frac{(q + N)!}{q!\,N!}. \qquad (2.17)$$

I'm making the last approximation because the ratio of $N!$ to $(N - 1)!$ is only a large factor (N), which is insignificant in a *very* large number like Ω. Next I'll take the natural logarithm and apply Stirling's approximation in the form 2.16:

$$
\begin{aligned}
\ln \Omega &= \ln\left(\frac{(q + N)!}{q!\,N!}\right) \\
&= \ln(q + N)! - \ln q! - \ln N! \qquad (2.18) \\
&\approx (q + N)\ln(q + N) - (q + N) - q\ln q + q - N\ln N + N \\
&= (q + N)\ln(q + N) - q\ln q - N\ln N.
\end{aligned}
$$

So far I haven't assumed that $q \gg N$—only that both q and N are large. But now let me manipulate the first logarithm as in Problem 2.13:

$$
\begin{aligned}
\ln(q + N) &= \ln\left[q\left(1 + \frac{N}{q}\right)\right] \\
&= \ln q + \ln\left(1 + \frac{N}{q}\right) \qquad (2.19) \\
&\approx \ln q + \frac{N}{q}.
\end{aligned}
$$

The last step follows from the Taylor expansion of the logarithm, $\ln(1 + x) \approx x$ for $|x| \ll 1$. Plugging this result into equation 2.18 and canceling the $q\ln q$ terms, we obtain

$$\ln \Omega \approx N \ln \frac{q}{N} + N + \frac{N^2}{q}. \qquad (2.20)$$

The last term becomes negligible compared to the others in the limit $q \gg N$. Exponentiating the first two terms gives

$$\Omega(N, q) \approx e^{N \ln(q/N)} e^N = \left(\frac{eq}{N}\right)^N \qquad \text{(when } q \gg N\text{)}. \qquad (2.21)$$

This formula is nice and simple, but it's bizarre. The exponent is a *large* number, so Ω is a *very* large number, as we already knew. Furthermore, if you increase either N or q by just a little bit, Ω will increase by a *lot*, due to the large exponent N.

Problem 2.17. Use the methods of this section to derive a formula, similar to equation 2.21, for the multiplicity of an Einstein solid in the "low-temperature" limit, $q \ll N$. assume $q \gg 1$, $N \gg 1$

Problem 2.18. Use Stirling's approximation to show that the multiplicity of an Einstein solid, for any large values of N and q, is approximately

$$\Omega(N, q) \approx \frac{\left(\frac{q+N}{q}\right)^q \left(\frac{q+N}{N}\right)^N}{\sqrt{2\pi q(q+N)/N}}.$$

The square root in the denominator is merely large, and can often be neglected. However, it is needed in Problem 2.22. (Hint: First show that $\Omega = \frac{N}{q+N} \frac{(q+N)!}{q! \, N!}$. Do not neglect the $\sqrt{2\pi N}$ in Stirling's approximation.)

Problem 2.19. Use Stirling's approximation to find an approximate formula for the multiplicity of a two-state paramagnet. Simplify this formula in the limit $N_\downarrow \ll N$ to obtain $\Omega \approx (Ne/N_\downarrow)^{N_\downarrow}$. This result should look very similar to your answer to Problem 2.17; explain why these two systems, in the limits considered, are essentially the same.

Sharpness of the Multiplicity Function

Finally we're ready to return to the issue raised at the beginning of this section: For a system of *two* large, interacting Einstein solids, just how skinny is the peak in the multiplicity function?

For simplicity, let me assume that each solid has N oscillators. I'll call the total number of energy units simply q (instead of q_{total}) for brevity, and I'll assume that this is much larger than N, so we can use formula 2.21. Then the multiplicity of the combined system, for any given macrostate, is

$$\Omega = \left(\frac{eq_A}{N}\right)^N \left(\frac{eq_B}{N}\right)^N = \left(\frac{e}{N}\right)^{2N} (q_A q_B)^N, \qquad (2.22)$$

where q_A and q_B are the numbers of energy units in solids A and B. (Note that $q_A + q_B$ must equal q.)

If you graph equation 2.22 as a function of q_A, it will have a very sharp peak at $q_A = q/2$, where the energy is distributed equally between the solids. The height of this peak is a very large number:

$$\Omega_{\text{max}} = \left(\frac{e}{N}\right)^{2N} \left(\frac{q}{2}\right)^{2N}. \qquad (2.23)$$

I want to know what the graph looks like *near* this peak, so let me set

$$q_A = \frac{q}{2} + x, \qquad q_B = \frac{q}{2} - x, \tag{2.24}$$

where x can be any number that is much smaller than q (but still possibly quite large). Plugging these expressions into equation 2.22 gives

$$\Omega = \left(\frac{e}{N}\right)^{2N} \left[\left(\frac{q}{2}\right)^2 - x^2\right]^N. \tag{2.25}$$

To simplify the second factor, I'll take its logarithm and manipulate it as I did in equation 2.19:

$$
\begin{aligned}
\ln\left[\left(\frac{q}{2}\right)^2 - x^2\right]^N &= N \ln\left[\left(\frac{q}{2}\right)^2 - x^2\right] \\
&= N \ln\left[\left(\frac{q}{2}\right)^2 \left(1 - \left(\frac{2x}{q}\right)^2\right)\right] \\
&= N\left[\ln\left(\frac{q}{2}\right)^2 + \ln\left(1 - \left(\frac{2x}{q}\right)^2\right)\right] \\
&\approx N\left[\ln\left(\frac{q}{2}\right)^2 - \left(\frac{2x}{q}\right)^2\right].
\end{aligned}
\tag{2.26}
$$

Now I can exponentiate the last expression and plug this back into equation 2.25:

$$\Omega = \left(\frac{e}{N}\right)^{2N} e^{N\ln(q/2)^2} e^{-N(2x/q)^2} = \Omega_{\max} \cdot e^{-N(2x/q)^2}. \tag{2.27}$$

A function of this form is called a **Gaussian**; it has a peak at $x = 0$ and a sharp fall-off on either side, as shown in Figure 2.7. The multiplicity falls off to $1/e$ of its maximum value when

$$N\left(\frac{2x}{q}\right)^2 = 1 \qquad \text{or} \qquad x = \frac{q}{2\sqrt{N}}. \tag{2.28}$$

This is actually a rather large number. But if $N = 10^{20}$, it's only one part in ten billion of the entire scale of the graph! On the scale used in the figure, where the

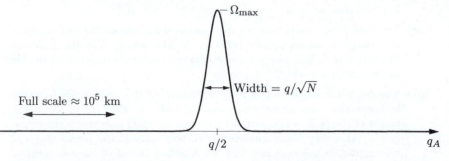

Figure 2.7. Multiplicity of a system of two large Einstein solids with many energy units per oscillator (high-temperature limit). Only a tiny fraction of the full horizontal scale is shown.

width of the peak is about 1 cm, the full scale of the graph would have to stretch 10^{10} cm, or 100,000 km—more than twice around the earth. And near the edge of the page, where x is only ten times larger than $q/2\sqrt{N}$, the multiplicity is less than its maximum value by a factor of $e^{-100} \approx 10^{-44}$.

This result tells us that, when two large Einstein solids are in thermal equilibrium with each other, any random fluctuations away from the most likely macrostate will be *utterly unmeasurable*. To measure such fluctuations we would have to measure the energy to an accuracy of ten significant figures. Once the system has had time to come to thermal equilibrium, so that all *microstates* are equally probable, we might as well assume that it is in its *most* likely macrostate. The limit where a system becomes infinitely large, so that measurable fluctuations away from the most likely macrostate never occur, is called the **thermodynamic limit**.

Problem 2.20. Suppose you were to shrink Figure 2.7 until the entire horizontal scale fits on the page. How wide would the peak be?

Problem 2.21. Use a computer to plot formula 2.22 directly, as follows. Define $z = q_A/q$, so that $(1-z) = q_B/q$. Then, aside from an overall constant that we'll ignore, the multiplicity function is $[4z(1-z)]^N$, where z ranges from 0 to 1 and the factor of 4 ensures that the height of the peak is equal to 1 for any N. Plot this function for $N = 1, 10, 100, 1000,$ and $10,000$. Observe how the width of the peak decreases as N increases.

Problem 2.22. This problem gives an alternative approach to estimating the width of the peak of the multiplicity function for a system of two large Einstein solids.

(a) Consider two identical Einstein solids, each with N oscillators, in thermal contact with each other. Suppose that the total number of energy units in the combined system is exactly $2N$. How many different macrostates (that is, possible values for the total energy in the first solid) are there for this combined system?

(b) Use the result of Problem 2.18 to find an approximate expression for the total number of microstates for the combined system. (Hint: Treat the combined system as a single Einstein solid. Do *not* throw away factors of "large" numbers, since you will eventually be dividing two "very large" numbers that are nearly equal. *Answer:* $2^{4N}/\sqrt{8\pi N}$.)

(c) The most likely macrostate for this system is (of course) the one in which the energy is shared equally between the two solids. Use the result of Problem 2.18 to find an approximate expression for the multiplicity of this macrostate. (*Answer:* $2^{4N}/(4\pi N)$.)

(d) You can get a rough idea of the "sharpness" of the multiplicity function by comparing your answers to parts (b) and (c). Part (c) tells you the height of the peak, while part (b) tells you the total area under the entire graph. As a very crude approximation, pretend that the peak's shape is rectangular. In this case, how wide would it be? Out of all the macrostates, what fraction have reasonably large probabilities? Evaluate this fraction numerically for the case $N = 10^{23}$.

Problem 2.23. Consider a two-state paramagnet with 10^{23} elementary dipoles, with the total energy fixed at zero so that exactly half the dipoles point up and half point down.

(a) How many microstates are "accessible" to this system?

(b) Suppose that the microstate of this system changes a billion times per second. How many microstates will it explore in ten billion years (the age of the universe)?

(c) Is it correct to say that, if you wait long enough, a system will eventually be found in every "accessible" microstate? Explain your answer, and discuss the meaning of the word "accessible."

Problem 2.24. For a single *large* two-state paramagnet, the multiplicity function is very sharply peaked about $N_\uparrow = N/2$.

(a) Use Stirling's approximation to estimate the height of the peak in the multiplicity function.

(b) Use the methods of this section to derive a formula for the multiplicity function in the vicinity of the peak, in terms of $x \equiv N_\uparrow - (N/2)$. Check that your formula agrees with your answer to part (a) when $x = 0$.

(c) How wide is the peak in the multiplicity function?

(d) Suppose you flip 1,000,000 coins. Would you be surprised to obtain 501,000 heads and 499,000 tails? Would you be surprised to obtain 510,000 heads and 490,000 tails? Explain.

Problem 2.25. The mathematics of the previous problem can also be applied to a one-dimensional **random walk**: a journey consisting of N steps, all the same size, each chosen randomly to be either forward or backward. (The usual mental image is that of a drunk stumbling along an alley.)

(a) Where are you *most* likely to find yourself, after the end of a long random walk?

(b) Suppose you take a random walk of 10,000 steps (say each a yard long). About how far from your starting point would you expect to be at the end?

(c) A good example of a random walk in nature is the **diffusion** of a molecule through a gas; the average step length is then the mean free path, as computed in Section 1.7. Using this model, and neglecting any small numerical factors that might arise from the varying step size and the multidimensional nature of the path, estimate the expected net displacement of an air molecule (or perhaps a carbon monoxide molecule traveling through air) in one second, at room temperature and atmospheric pressure. Discuss how your estimate would differ if the elapsed time or the temperature were different. Check that your estimate is consistent with the treatment of diffusion in Section 1.7.

(d) Sketch plot of $P_t(y)$ for 3 values of time $t_1 < t_2 < t_3$
(e) How does width Δy of probability distribution $P_t(y)$ depend on time t?

It all works because Avogadro's number is closer to infinity than to 10.

—Ralph Baierlein, *American Journal of Physics* **46**, 1045 (1978). Copyright 1978, American Association of Physics Teachers. Reprinted with permission.

2.5 The Ideal Gas

The conclusion of the previous section—that only a *tiny* fraction of the macrostates of a large interacting system have reasonably large probabilities—applies to many other systems besides Einstein solids. In fact, it is true for essentially *any* pair of interacting objects, provided that the number of particles and the number of energy units are both "large." In this section I'll argue that it is true for ideal gases.

An ideal gas is more complicated than an Einstein solid, because its multiplicity depends on its volume as well as its total energy and number of particles. Furthermore, when two gases interact, they can often expand and contract, and even exchange molecules, in addition to exchanging energy. We will still find, however, that the multiplicity function for two interacting gases is very sharply peaked around a relatively small subset of macrostates.

Multiplicity of a Monatomic Ideal Gas

For simplicity, I'll consider only a monatomic ideal gas, like helium or argon. I'll begin with a gas consisting of just one molecule, then work up to the general case of N molecules.

So suppose we have a single gas atom, with kinetic energy U, in a container of volume V. What is the multiplicity of this system? That is, how many microstates could the molecule be in, given the fixed values of U and V?

Well, a container with twice the volume offers twice as many states to a molecule, so the multiplicity should be proportional to V. Also, the more different momentum vectors the molecule can have, the more states are available, so the multiplicity should also be proportional to the "volume" of available **momentum space**. (Momentum space is an imaginary "space" in which the axes are p_x, p_y, and p_z. Each "point" in momentum space corresponds to a momentum vector for the particle.) So let me write schematically

$$\Omega_1 \propto V \cdot V_p, \tag{2.29}$$

where V is the volume of ordinary space (or **position space**), V_p is the volume of momentum space, and the 1 subscript indicates that this is for a gas of just one molecule.

This formula for Ω_1 is still pretty ambiguous. One problem is in determining the available volume of momentum space, V_p. Since the molecule's kinetic energy must equal U, there is a constraint:

$$U = \frac{1}{2}m(v_x^2 + v_y^2 + v_z^2) = \frac{1}{2m}(p_x^2 + p_y^2 + p_z^2). \tag{2.30}$$

This equation can also be written

$$p_x^2 + p_y^2 + p_z^2 = 2mU, \tag{2.31}$$

which defines the surface of a *sphere* in momentum space with radius $\sqrt{2mU}$ (see Figure 2.8). The "volume" of momentum space is really the *surface area* of this sphere (perhaps multiplied by a small thickness if U is allowed to fluctuate somewhat).

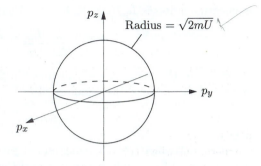

Figure 2.8. A sphere in momentum space with radius $\sqrt{2mU}$. If a molecule has energy U, its momentum vector must lie somewhere on the surface of this sphere.

    The other problem with equation 2.29 is in determining the constant of proportionality. While it seems pretty clear that Ω_1 must be *proportional* to the volumes of position space and momentum space, how can we possibly *count* the various microstates to get a finite number for the multiplicity? It would seem that the number of allowed microstates, even for a gas of just one molecule, is infinite.

    To actually count the number of microstates we must invoke quantum mechanics. (For a systematic overview of quantum mechanics, see Appendix A.) In quantum mechanics, the state of a system is described by a *wavefunction*, which is spread out in both position space and momentum space. The less spread out the wavefunction is in position space, the more spread out it must be in momentum space, and vice versa. This is the famous **Heisenberg uncertainty principle**:

$$(\Delta x)(\Delta p_x) \approx h, \tag{2.32}$$

where Δx is the spread in x, Δp_x is the spread in p_x, and h is Planck's constant. (The product of Δx and Δp_x can also be *more* than h, but we are interested in wavefunctions that specify the position and momentum as precisely as possible.) The same limitation applies to y and p_y, and to z and p_z.

    Even in quantum mechanics, the number of allowed wavefunctions is infinite. But the number of *independent* wavefunctions (in a technical sense that's defined in Appendix A) is finite, if the total available position space and momentum space are limited. I like to picture it as in Figure 2.9. In this one-dimensional example, the number of distinct position states is $L/(\Delta x)$, while the number of distinct

Figure 2.9. A number of "independent" position states and momentum states for a quantum-mechanical particle moving in one dimension. If we make the wavefunctions narrower in position space, they become wider in momentum space, and vice versa.

momentum states is $L_p/(\Delta p_x)$. The total number of distinct states is the product,

$$\frac{L}{\Delta x}\frac{L_p}{\Delta p_x} = \frac{L\,L_p}{h}, \tag{2.33}$$

according to the uncertainty principle. In three dimensions, the lengths become volumes and there are three factors of h:

$$\Omega_1 = \frac{V\,V_p}{h^3}. \tag{2.34}$$

This "derivation" of the constant of proportionality in Ω_1 is admittedly not very rigorous. I certainly haven't proved that there are no further factors of 2 or π in equation 2.34. If you prefer, just think of the result in terms of dimensional analysis: The multiplicity must be a unitless number, and you can easily show that h^3 has just the right units to cancel the units of V and V_p.*

So much for a gas of one molecule. If we add a second molecule, we need a factor of the form of equation 2.34 for each molecule, and we multiply them together because for *each* state of molecule 1, there are Ω_1 states for molecule 2. Well, not quite. The V_p factors are more complicated, since only the *total* energy of the two molecules is constrained. Equation 2.31 now becomes

$$p_{1x}^2 + p_{1y}^2 + p_{1z}^2 + p_{2x}^2 + p_{2y}^2 + p_{2z}^2 = 2mU, \tag{2.35}$$

assuming that both molecules have the same mass. This equation defines the surface of a six-dimensional "hypersphere" in six-dimensional momentum space. I can't visualize it, but one can still compute its "surface area" and call that the total volume of allowed momentum space for the two molecules.

So the multiplicity function for an ideal gas of two molecules should be

$$\Omega_2 = \frac{V^2}{h^6} \times \text{(area of momentum hypersphere)}. \tag{2.36}$$

This formula *is* correct, but only if the two molecules are *distinguishable* from each other. If they're *indistinguishable*, then we've overcounted the microstates by a factor of 2, since interchanging the molecules with each other does not give us a distinct state (see Figure 2.10).[†] Thus the multiplicity for a gas of two *indistinguishable* molecules is

$$\Omega_2 = \frac{1}{2}\frac{V^2}{h^6} \times \text{(area of momentum hypersphere)}. \tag{2.37}$$

---

*Don't worry about the fact that V_p is really a surface area, not a volume. We can always allow the sphere in momentum space to have a tiny thickness, and multiply its area by this thickness to get something with units of momentum cubed. When we get to a gas of N molecules, the multiplicity will be such a huge number that it doesn't matter if we're off a little in the units.

†This argument assumes that the individual states of the two molecules are always different. The two molecules *could* be in a state where they both have the same position and the same momentum, and such a state is not double-counted in equation 2.36. Unless the gas is *very* dense, however, such states hardly ever occur.

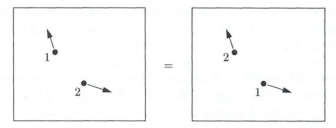

Figure 2.10. In a gas of two identical molecules, interchanging the states of the molecules leaves the system in the same state as before.

For an ideal gas of N indistinguishable molecules, the multiplicity function contains N factors of V, divided by $3N$ factors of h. The factor that compensates for the overcounting is $1/N!$, the number of ways of interchanging the molecules. And the momentum-space factor is the "surface area" of a $3N$-dimensional hypersphere whose radius is (still) $\sqrt{2mU}$:

$$\Omega_N = \frac{1}{N!}\frac{V^N}{h^{3N}} \times \text{(area of momentum hypersphere)}. \qquad (2.38)$$

To make this result more explicit, we need a general formula for the "surface area" of a d-dimensional hypersphere of radius r. For $d = 2$, the "area" is just the circumference of a circle, $2\pi r$. For $d = 3$, the answer is $4\pi r^2$. For general d, the answer should be proportional to r^{d-1}, but the coefficient is not easy to guess. The full formula is

$$\text{"area"} = \frac{2\pi^{d/2}}{(\frac{d}{2} - 1)!}\, r^{d-1}. \qquad (2.39)$$

This formula is derived in Appendix B. For now, you can at least check the coefficient for the case $d = 2$. To check it for $d = 3$, you need to know that $(1/2)! = \sqrt{\pi}/2$.

Plugging equation 2.39 (with $d = 3N$ and $r = \sqrt{2mU}$) into equation 2.38, we obtain

$$\Omega_N = \frac{1}{N!}\frac{V^N}{h^{3N}}\frac{2\pi^{3N/2}}{(\frac{3N}{2} - 1)!}(\sqrt{2mU})^{3N-1} \approx \frac{1}{N!}\frac{V^N}{h^{3N}}\frac{\pi^{3N/2}}{(3N/2)!}(\sqrt{2mU})^{3N}. \qquad (2.40)$$

In the last expression I've thrown away some large factors, which is ok since Ω_N is a very large number.[*]

This formula for the multiplicity of a monatomic ideal gas is a mess, but its dependence on U and V is pretty simple:

$$\Omega(U, V, N) = f(N)V^N U^{3N/2}, \qquad (2.41)$$

where $f(N)$ is a complicated function of N.

---

[*]If you're not happy with my sloppy derivation of equation 2.40, please be patient. In Section 6.7 I'll do a much better job, using a very different method.

Notice that the exponent on U in formula 2.41 is 1/2 times the total number of degrees of freedom ($3N$) in the monatomic gas. The same is true of the multiplicity of an Einstein solid in the high-temperature limit, equation 2.21. These results are special cases of a more general theorem: For any system with only quadratic "degrees of freedom," having so many units of energy that energy quantization is unnoticeable, the multiplicity is proportional to $U^{Nf/2}$, where Nf is the total number of degrees of freedom. A general proof of this theorem is given in Stowe (1984).

Problem 2.26. Consider an ideal monatomic gas that lives in a two-dimensional universe ("flatland"), occupying an area A instead of a volume V. By following the same logic as above, find a formula for the multiplicity of this gas, analogous to equation 2.40.

Interacting Ideal Gases

Suppose now that we have *two* ideal gases, separated by a partition that allows energy to pass through (see Figure 2.11). If each gas has N molecules (of the same species), then the total multiplicity of this system is

$$\Omega_{\text{total}} = [f(N)]^2 (V_A V_B)^N (U_A U_B)^{3N/2}. \tag{2.42}$$

This expression has essentially the same form as the corresponding result for a pair of Einstein solids (equation 2.22): Both energies are raised to a *large* exponent. Following exactly the same reasoning as in Section 2.4, we can conclude that the multiplicity function, plotted as a function of U_A, has a *very* sharp peak:

$$\text{width of peak} = \frac{U_{\text{total}}}{\sqrt{3N/2}}. \tag{2.43}$$

Provided that N is large, only a *tiny* fraction of the macrostates have a reasonable chance of occurring, assuming that the system is in equilibrium.

In addition to exchanging energy, we could allow the gases to exchange *volume*; that is, we could allow the partition to move back and forth, as one gas expands and the other is compressed. In this case we can apply exactly the same argument to volume that we just applied to energy. The multiplicity, plotted as a function

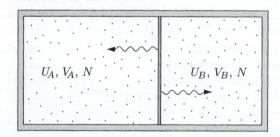

Figure 2.11. Two ideal gases, each confined to a fixed volume, separated by a partition that allows energy to pass through. The total energy of the two gases is fixed.

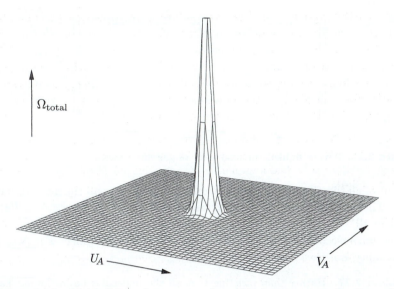

Figure 2.12. Multiplicity of a system of two ideal gases, as a function of the energy and volume of gas A (with the total energy and total volume held fixed). If the number of molecules in each gas is large, the full horizontal scale would stretch far beyond the edge of the page.

of V_A, again has a very sharp peak:

$$\text{width of peak} = \frac{V_{\text{total}}}{\sqrt{N}}. \qquad (2.44)$$

So again, the equilibrium macrostate is essentially determined, to within a tiny fraction of the total volume available (if N is large). In Figure 2.12 I've plotted Ω_{total} as a function of *both* U_A and V_A. Like Figure 2.7, this graph shows only a tiny fraction of the full range of U_A and V_A values. For $N = 10^{20}$, if the full scale were compressed to fit on this page, the spike would be narrower than an atom.

Instead of allowing the partition to move, we could just poke holes in it and let the *molecules* move back and forth between the two sides. Then, to find the equilibrium macrostate, we would want to look at the behavior of Ω_{total} as a function of N_A and U_A. From equation 2.40, you can see that the analysis would be more difficult in this case. But once again, we would find a very sharp peak in the graph, indicating that the equilibrium macrostate is fixed to a very high precision. (As you might expect, the equilibrium macrostate is the one for which the density is the same on both sides of the partition.)

Sometimes you can calculate probabilities of various arrangements of molecules just by looking at the volume dependence of the multiplicity function (2.41). For instance, suppose we want to know the probability of finding the configuration shown in Figure 2.13, where *all* the molecules in a container of gas are somewhere in the left half. This arrangement is just a macrostate with the same energy and number of molecules, but half the original volume. Looking at equation 2.41, we see that replacing V by $V/2$ reduces the multiplicity by a factor of 2^N. In other

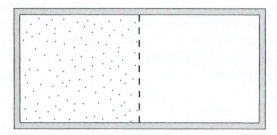

Figure 2.13. A very unlikely arrangement of gas molecules.

words, out of all the allowed microstates, only one in 2^N has all the molecules in the left half. Thus, the probability of this arrangement is 2^{-N}. Even for $N = 100$, this is less than 10^{-30}, so you would have to check a trillion times per second for the age of the universe before finding such an arrangement even once. For $N = 10^{23}$, the probability is a *very* small number.

> **Problem 2.27.** Rather than insisting that all the molecules be in the left half of a container, suppose we only require that they be in the leftmost 99% (leaving the remaining 1% completely empty). What is the probability of finding such an arrangement if there are 100 molecules in the container? What if there are 10,000 molecules? What if there are 10^{23}?

2.6 Entropy

We have now seen that, for a variety of systems, particles and energy tend to rearrange themselves until the multiplicity is at (or very near) its maximum value. In fact, this conclusion seems to be true* for *any* system, provided that it contains enough particles and units of energy for the statistics of very large numbers to apply:

> Any large system in equilibrium will be found in the macrostate with the greatest multiplicity (aside from fluctuations that are normally too small to measure).

This is just a more general statement of the **second law of thermodynamics**. Another way to say it is simply:

> Multiplicity tends to increase.

Even though this law is not "fundamental" (since I essentially derived it by looking at probabilities), I'll treat it as fundamental from now on. If you just remember to look for the macrostate with greatest multiplicity, you can pretty much forget about calculating what the actual probabilities are.

---

*As far as I'm aware, nobody has ever *proved* that it is true for *all* large systems. Perhaps an exception lurks out there somewhere. But the experimental successes of thermodynamics indicate that exceptions must be exceedingly rare.

Since multiplicities tend to be *very* large numbers, which are very cumbersome to work with, we will find it convenient from now on to work with the natural logarithm of the multiplicity instead of the multiplicity itself. For historical reasons, we will also multiply by a factor of Boltzmann's constant. This gives us a quantity called the **entropy**, denoted S:

$$S \equiv k \ln \Omega. \tag{2.45}$$

In words, entropy is just the logarithm of the number of ways of arranging things in the system (times Boltzmann's constant). The logarithm turns a very large number, the multiplicity, into an ordinary large number. If you want to *understand* entropy, my advice is to ignore the factor of k and just think of entropy as a unitless quantity, $\ln \Omega$. When we include the factor of k, however, S has units of energy divided by temperature, or J/K in the SI system. I'll explain the usefulness of these units in Chapter 3.

As a first example, let's go back to the case of a large Einstein solid with N oscillators, q units of energy, and $q \gg N$. Since $\Omega = (eq/N)^N$,

$$S = k \ln(eq/N)^N = Nk[\ln(q/N) + 1]. \tag{2.46}$$

So if $N = 10^{22}$ and $q = 10^{24}$,

$$S = Nk \cdot (5.6) = (5.6 \times 10^{22})k = 0.77 \text{ J/K}. \tag{2.47}$$

Notice also that increasing either q or N increases the entropy of an Einstein solid (though not in direct proportion).

Generally, the more particles there are in a system, and the more energy it contains, the greater its multiplicity and its entropy. Besides adding particles and energy, you can increase the entropy of a system by letting it expand into a larger space, or breaking large molecules apart into small ones, or mixing together substances that were once separate. In each of these cases, the total number of possible arrangements increases.

Some people find it helpful to think of entropy intuitively as being roughly synonymous with "disorder." Whether this idea is accurate, however, depends on exactly what you consider to be disorderly. Most people would agree that a shuffled deck of cards is more disorderly than a sorted deck, and indeed, shuffling increases the entropy because it increases the number of possible arrangements.* However, many people would say that a glass of crushed ice appears more disorderly than a glass of an equal amount of water. In this case, though, the water has much more entropy, since there are so many more ways of arranging the molecules, and so many more ways of arranging the larger amount of energy among them.

---

*This example is actually somewhat controversial: Some physicists would not count these rearrangements into the thermodynamic entropy because cards don't ordinarily rearrange themselves without outside help. Personally, I see no point in being so picky. At worst, my somewhat broad definition of entropy is harmless, because the amount of entropy in dispute is negligible compared to other forms of entropy.

One nice property of entropy is that the total entropy of a composite system is the *sum* of the entropies of its parts. For instance, if there are two parts, A and B, then

$$S_{\text{total}} = k \ln \Omega_{\text{total}} = k \ln(\Omega_A \Omega_B) = k \ln \Omega_A + k \ln \Omega_B = S_A + S_B. \qquad (2.48)$$

I'm assuming here that the macrostates of systems A and B have been specified separately. If these systems can interact, then those macrostates can fluctuate over time, and to compute the entropy over long time scales we should compute Ω_{total} by summing over *all* macrostates for the two systems. Entropy, like multiplicity, is a function of the number of *accessible* microstates, and this number depends on the time scale under consideration. However, in practice, this distinction rarely matters. If we just assume that the composite system is in its *most* likely macrostate, we get essentially the same entropy as if we sum over *all* macrostates (see Problems 2.29 and 2.30).

Since the natural logarithm is a monotonically increasing function of its argument, a macrostate with higher multiplicity also has higher entropy. Therefore we can restate the **second law of thermodynamics** as follows:

Any large system in equilibrium will be found in the macrostate with the greatest entropy (aside from fluctuations that are normally too small to measure).

Or more briefly:

Entropy tends to increase.

Note, however, that a graph of entropy vs. some variable (such as U_A or V_A) that is allowed to fluctuate will generally *not* have a sharp peak. Taking the logarithm smooths out the peak that was present in the multiplicity function. Of course this does not affect our conclusions in the least; it is still true that fluctuations away from the macrostate of greatest entropy will be negligible for any reasonably large system.

Although "spontaneous" processes always occur because of a net increase in entropy, you might wonder whether human intervention could bring about a net decrease in entropy. Common experience seems to suggest that the answer is yes: Anyone can easily turn all the coins in a collection heads-up, or sort a shuffled deck of cards, or clean up a messy room. However, the decreases in entropy in these situations are extremely tiny, while the entropy created by the metabolism of food in our bodies (as we take energy out of chemical bonds and dump most of it into the environment as thermal energy) is always substantial. As far as we can tell, our bodies are just as subject to the laws of thermodynamics as are inanimate objects. So no matter what you do to decrease the entropy in one place, you're bound to create at least as much entropy somewhere else.

Even if *we* can't decrease the total entropy of the universe, isn't it possible that someone (or something) else could? In 1867 James Clerk Maxwell posed this question, wondering whether a "very observant and neat-fingered being"* couldn't

---

*Quoted in Leff and Rex (1990), p. 5.

deflect fast-moving molecules in one direction and slow-moving molecules in an-other, thereby causing heat to flow from a cold object to a hot one. William Thomson later named this mythical creature *Maxwell's Demon*, and physicists and philosophers have been trying to exorcise it ever since. Countless designs for me-chanical "demons" have been drafted, and all have been proven ineffective. Even a hypothetical "intelligent" demon, it turns out, must create entropy as it processes the information needed to sort molecules. Although thinking about demons has taught us much about entropy since Maxwell's time, the verdict seems to be that not even a demon can violate the second law of thermodynamics.

Problem 2.28. How many possible arrangements are there for a deck of 52 playing cards? (For simplicity, consider only the order of the cards, not whether they are turned upside-down, etc.) Suppose you start with a sorted deck and shuffle it repeatedly, so that all arrangements become "accessible." How much entropy do you create in the process? Express your answer both as a pure number (neglecting the factor of k) and in SI units. Is this entropy significant compared to the entropy associated with arranging thermal energy among the molecules in the cards?

Problem 2.29. Consider a system of two Einstein solids, with $N_A = 300$, $N_B = 200$, and $q_{total} = 100$ (as discussed in Section 2.3). Compute the entropy of the most likely macrostate and of the least likely macrostate. Also compute the entropy over long time scales, assuming that *all* microstates are accessible. (Neglect the factor of Boltzmann's constant in the definition of entropy; for systems this small it is best to think of entropy as a pure number.)

Problem 2.30. Consider again the system of two large, identical Einstein solids treated in Problem 2.22.

(a) For the case $N = 10^{23}$, compute the entropy of this system (in terms of Boltzmann's constant), assuming that *all* of the microstates are allowed. (This is the system's entropy over long time scales.)

(b) Compute the entropy again, assuming that the system is in its most likely macrostate. (This is the system's entropy over short time scales, except when there is a large and unlikely fluctuation away from the most likely macrostate.)

(c) Is the issue of time scales really relevant to the entropy of this system?

(d) Suppose that, at a moment when the system is near its most likely macro-state, you suddenly insert a partition between the solids so that they can no longer exchange energy. Now, even over long time scales, the entropy is given by your answer to part (b). Since this number is less than your answer to part (a), you have, in a sense, caused a violation of the second law of thermodynamics. Is this violation significant? Should we lose any sleep over it?

Entropy of an Ideal Gas

The formula for the entropy of a monatomic ideal gas is rather complicated, but extremely useful. If you start with equation 2.40, apply Stirling's approximation, throw away some factors that are merely large, and take the logarithm, you get

$$S = Nk\left[\ln\left(\frac{V}{N}\left(\frac{4\pi mU}{3Nh^2}\right)^{3/2}\right) + \frac{5}{2}\right]. \qquad (2.49)$$

This famous result is known as the **Sackur-Tetrode equation**.

Consider, for instance, a mole of helium at room temperature and atmospheric pressure. The volume is then 0.025 m^3 and the internal energy is $\frac{3}{2}nRT = 3700$ J. Plugging these numbers into the Sackur-Tetrode equation, I find that the argument of the logarithm is 330,000, but the logarithm itself is only 12.7. So the entropy is

$$S = Nk \cdot (15.2) = (9.1 \times 10^{24})k = 126 \text{ J/K}. \tag{2.50}$$

The entropy of an ideal gas depends on its volume, energy, and number of particles. Increasing any of these three variables increases the entropy. The simplest dependence is on the volume; for instance, if the volume changes from V_i to V_f while U and N are held fixed, the entropy changes by

$$\Delta S = Nk \ln \frac{V_f}{V_i} \qquad (U, N \text{ fixed}). \tag{2.51}$$

This formula applies, for instance, to the quasistatic isothermal expansion considered in Section 1.5, where the gas pushes on a piston, doing mechanical work, while we simultaneously supply heat from outside to keep the gas at constant temperature. In this case we can think of the entropy increase as being caused by the heat input. Putting heat into a system *always* increases its entropy; in the following chapter I'll discuss in general the relation between entropy and heat.

A very different way of letting a gas expand is shown in Figure 2.14. Initially, the gas is separated by a partition from an evacuated chamber. We then puncture the partition, letting the gas freely expand to fill the whole available space. This process is called **free expansion**. How much work is done during free expansion? None! The gas isn't pushing *on* anything, so it can't do any work. What about heat? Again, none: No heat whatsoever flowed into or out of the gas. Therefore, by the first law of thermodynamics,

$$\Delta U = Q + W = 0 + 0 = 0. \tag{2.52}$$

The energy content of the gas does not change during free expansion, hence formula 2.51 applies. This time, however, the entropy increase was *not* caused by the input of heat; instead we have manufactured *new* entropy, right here on the spot.

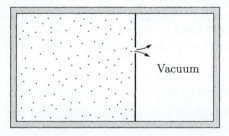

Figure 2.14. Free expansion of a gas into a vacuum. Because the gas neither does work nor absorbs heat, its energy is unchanged. The entropy of the gas increases, however.

Problem 2.31. Fill in the algebraic steps to derive the Sackur-Tetrode equation (2.49).

Problem 2.32. Find an expression for the entropy of the two-dimensional ideal gas considered in Problem 2.26. Express your result in terms of U, A, and N.

Problem 2.33. Use the Sackur-Tetrode equation to calculate the entropy of a mole of argon gas at room temperature and atmospheric pressure. Why is the entropy greater than that of a mole of helium under the same conditions?

Problem 2.34. Show that during the quasistatic isothermal expansion of a monatomic ideal gas, the change in entropy is related to the heat input Q by the simple formula

$$\Delta S = \frac{Q}{T}.$$

In the following chapter I'll prove that this formula is valid for *any* quasistatic process. Show, however, that it is *not* valid for the free expansion process described above.

Problem 2.35. According to the Sackur-Tetrode equation, the entropy of a monatomic ideal gas can become *negative* when its temperature (and hence its energy) is sufficiently low. Of course this is absurd, so the Sackur-Tetrode equation must be invalid at very low temperatures. Suppose you start with a sample of helium at room temperature and atmospheric pressure, then lower the temperature holding the density fixed. Pretend that the helium remains a gas and does not liquefy. Below what temperature would the Sackur-Tetrode equation predict that S is negative? (The behavior of gases at very low temperatures is the main subject of Chapter 7.)

Problem 2.36. For either a monatomic ideal gas or a high-temperature Einstein solid, the entropy is given by Nk times some logarithm. The logarithm is never large, so if all you want is an order-of-magnitude estimate, you can neglect it and just say $S \sim Nk$. That is, the entropy in fundamental units is of the order of the number of particles in the system. This conclusion turns out to be true for most systems (with some important exceptions at low temperatures where the particles are behaving in an orderly way). So just for fun, make a very rough estimate of the entropy of each of the following: this book (a kilogram of carbon compounds); a moose (400 kg of water); the sun (2×10^{30} kg of ionized hydrogen).

Entropy of Mixing

Another way to create entropy is to let two different materials mix with each other. Suppose, for instance, that we start with two different monatomic ideal gases, A and B, each with the same energy, volume, and number of particles. They occupy the two halves of a divided chamber, separated by a partition (see Figure 2.15). If we now remove the partition, the entropy increases. To calculate by how much, we can just treat each gas as a separate system, even after they mix. Since gas A expands to fill twice its initial volume, its entropy increases by

$$\Delta S_A = Nk \ln \frac{V_f}{V_i} = Nk \ln 2, \tag{2.53}$$

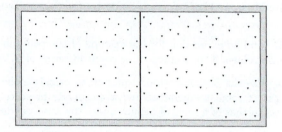

Figure 2.15. Two different gases, separated by a partition. When the partition is removed, each gas expands to fill the whole container, mixing with the other and creating entropy.

while the entropy of gas B increases by the same amount, giving a total increase of

$$\Delta S_{\text{total}} = \Delta S_A + \Delta S_B = 2Nk\ln 2. \tag{2.54}$$

This increase is called the **entropy of mixing**.

It's important to note that this result applies only if the two gases are *different*, like helium and argon. If you start with the *same* gas on both sides, the entropy doesn't increase at all when you remove the partition. (Technically, the total multiplicity does increase, because the distribution of molecules between the two sides can now fluctuate. But the multiplicity increases only by a "large" factor, which has negligible effect on the entropy.)

Let's compare these two situations in a slightly different way. Forget about the partition, and suppose we start with a mole of helium in the chamber. Its total entropy is given by the Sackur-Tetrode equation,

$$S = Nk\left[\ln\left(\frac{V}{N}\left(\frac{4\pi mU}{3Nh^2}\right)^{3/2}\right) + \frac{5}{2}\right]. \tag{2.55}$$

If we now add a mole of argon with the same thermal energy U, the entropy approximately doubles:

$$S_{\text{total}} = S_{\text{helium}} + S_{\text{argon}}. \tag{2.56}$$

(Because the molecular mass enters equation 2.55, the entropy of the argon is actually somewhat greater than the entropy of the helium.) However, if instead we add a second mole of helium, the entropy does *not* double. Look at formula 2.55: If you double the values of both N and U, the ratio U/N inside the logarithm is unchanged, while the N out front becomes $2N$. But there's another N, just inside the logarithm, underneath the V, which also becomes $2N$ and makes the total entropy come out less than you might expect, by a term $2Nk\ln 2$. This "missing" term is precisely the entropy of mixing.

So the difference between adding argon and adding more helium comes from the extra N under the V in the Sackur-Tetrode equation. Where did this N come from? If you look back at the derivation in Section 2.5, you'll see that it came from the $1/N!$ that I slipped into the multiplicity function to account for the fact that molecules in a gas are *indistinguishable* (so interchanging two molecules does not yield a distinct microstate). If I hadn't slipped this factor in, the entropy of a monatomic ideal gas would be

$$S = Nk\left[\ln\left(V\left(\frac{4\pi mU}{3Nh^2}\right)^{3/2}\right) + \frac{3}{2}\right] \qquad \text{(distinguishable molecules).} \qquad (2.57)$$

This formula, if it were correct, would have some rather disturbing consequences. For instance, if you insert a partition into a tank of helium, dividing it in half, this formula predicts that each half would have significantly *less* than half of the original entropy. You could violate the second law of thermodynamics simply by inserting the partition! I don't know an easy way of *proving* that the world isn't like this, but it certainly would be confusing.

This whole issue was first raised by J. Willard Gibbs, and is now known as the **Gibbs paradox**. The best resolution of the paradox is simply to assume that all atoms of a given type are truly indistinguishable. In Chapter 7 we'll see more evidence to support this assumption.

Problem 2.37. Using the same method as in the text, calculate the entropy of mixing for a system of two monatomic ideal gases, A and B, whose relative proportion is arbitrary. Let N be the *total* number of molecules and let x be the fraction of these that are of species B. You should find

$$\Delta S_{\text{mixing}} = -Nk\left[x \ln x + (1-x)\ln(1-x)\right].$$

Check that this expression reduces to the one given in the text when $x = 1/2$.

Problem 2.38. The mixing entropy formula derived in the previous problem actually applies to any ideal gas, and to some dense gases, liquids, and solids as well. For the denser systems, we have to assume that the two types of molecules are the same size and that molecules of different types interact with each other in the same way as molecules of the same type (same forces, etc.). Such a system is called an **ideal mixture**. Explain why, for an ideal mixture, the mixing entropy is given by

$$\Delta S_{\text{mixing}} = k \ln\left(\frac{N}{N_A}\right),$$

where N is the total number of molecules and N_A is the number of molecules of type A. Use Stirling's approximation to show that this expression is the same as the result of the previous problem when both N and N_A are large.

Problem 2.39. Compute the entropy of a mole of helium at room temperature and atmospheric pressure, pretending that all the atoms are distinguishable. Compare to the actual entropy, for indistinguishable atoms, computed in the text.

Reversible and Irreversible Processes

If a physical process increases the total entropy of the universe, that process cannot happen in reverse, since this would violate the second law of thermodynamics. Processes that create new entropy are therefore said to be **irreversible**. By the same token, a process that leaves the total entropy of the universe unchanged would be **reversible**. In practice, no macroscopic process is perfectly reversible, although some processes come close enough for most purposes.

One type of process that creates new entropy is the very sudden expansion of a system, for instance, the free expansion of a gas discussed above. On the other hand, a gradual compression or expansion does not (by itself) change the entropy of a system. In Chapter 3 I'll prove that any reversible volume change must in fact be *quasistatic*, so that $W = -P\Delta V$. (A quasistatic process can still be irreversible, however, if there is also heat flowing in or out or if entropy is being created in some other way.)

It's interesting to think about *why* the slow compression of a gas does not cause its entropy to increase. One way to think about it is to imagine that the molecules in the gas inhabit various quantum-mechanical wavefunctions, each filling the entire box, with discrete (though very closely spaced) energy levels. (See Appendix A for more about the energy levels of particles in a box.) When you compress the gas, each wavefunction gets squeezed, so the energies of all the levels increase, and each molecule's energy increases accordingly. But if the compression is sufficiently slow, molecules will *not* be kicked up into higher energy levels; a molecule that starts in the nth level remains in the nth level (although the energy of that level increases). Thus the number of ways of arranging the molecules among the various energy levels will remain the same, that is, the multiplicity and entropy do not change. On the other hand, if the compression *is* violent enough to kick molecules up into higher levels, then the number of possible arrangements will increase and so will the entropy.

Perhaps the most important type of thermodynamic process is the flow of heat from a hot object to a cold one. We saw in Section 2.3 that this process occurs *because* the total multiplicity of the combined system thereby increases; hence the total entropy increases also, and heat flow is always irreversible. However, we'll see in the next chapter that the increase in entropy becomes negligible in the limit where the temperature difference between the two objects goes to zero. So if you ever hear anyone talking about "reversible heat flow," what they really mean is very slow heat flow, between objects that are at nearly the same temperature. Notice that, in the reversible limit, changing the temperature of one of the objects only infinitesimally can cause the heat to flow in the opposite direction. Similarly, during a quasistatic volume change, an infinitesimal change in the pressure will reverse the direction. In fact, one can *define* a reversible process as one that can be reversed by changing the conditions only infinitesimally.

Most of the processes we observe in life involve large entropy increases and are therefore highly irreversible: sunlight warming the earth, wood burning in the fireplace, metabolism of nutrients in our bodies, mixing ingredients in the kitchen. Because the total entropy of the universe is constantly increasing, and can never decrease, some philosophically inclined physicists have worried that eventually the universe will become a rather boring place: a homogeneous fluid with the maximum possible entropy and no variations in temperature or density anywhere. At the rate we're going, though, this "heat death of the universe" won't occur any time soon; our sun, for instance, should continue to shine brightly for at least another five billion years.*

It may be more fruitful to ask instead about the *beginning* of time. Why did the universe start out in such an improbable, low-entropy state, so that after more than ten billion years it is still so far from equilibrium? Could it have been merely a big coincidence (the biggest of all time)? Or might someone, someday, discover a more satisfying explanation?

Problem 2.40. For each of the following irreversible processes, explain how you can tell that the total entropy of the universe has increased.

(a) Stirring salt into a pot of soup.

(b) Scrambling an egg.

(c) Humpty Dumpty having a great fall.

(d) A wave hitting a sand castle.

(e) Cutting down a tree.

(f) Burning gasoline in an automobile.

Problem 2.41. Describe a few of your favorite, and least favorite, irreversible processes. In each case, explain how you can tell that the entropy of the universe increases.

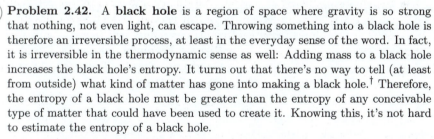

Problem 2.42. A **black hole** is a region of space where gravity is so strong that nothing, not even light, can escape. Throwing something into a black hole is therefore an irreversible process, at least in the everyday sense of the word. In fact, it is irreversible in the thermodynamic sense as well: Adding mass to a black hole increases the black hole's entropy. It turns out that there's no way to tell (at least from outside) what kind of matter has gone into making a black hole.† Therefore, the entropy of a black hole must be greater than the entropy of any conceivable type of matter that could have been used to create it. Knowing this, it's not hard to estimate the entropy of a black hole.

(a) Use dimensional analysis to show that a black hole of mass M should have a radius of order GM/c^2, where G is Newton's gravitational constant and c is the speed of light. Calculate the approximate radius of a one-solar-mass black hole ($M = 2 \times 10^{30}$ kg).

(b) In the spirit of Problem 2.36, explain why the entropy of a black hole, in fundamental units, should be of the order of the maximum number of particles that could have been used to make it.

---

*For a modern analysis of the long-term prospects for our universe, see Steven Frautschi, "Entropy in an Expanding Universe," *Science* **217**, 593–599 (1982).

†This statement is a slight exaggeration. Electric charge and angular momentum are conserved during black hole formation, and these quantities can still be measured from outside a black hole. In this problem I'm assuming for simplicity that both are zero.

(c) To make a black hole out of the maximum possible number of particles, you should use particles with the lowest possible energy: long-wavelength photons (or other massless particles). But the wavelength can't be any longer than the size of the black hole. By setting the total energy of the photons equal to Mc^2, estimate the maximum number of photons that could be used to make a black hole of mass M. Aside from a factor of $8\pi^2$, your result should agree with the exact formula for the entropy of a black hole, obtained* through a much more difficult calculation:

$$S_{\text{b.h.}} = \frac{8\pi^2 G M^2}{hc} k.$$

(d) Calculate the entropy of a one-solar-mass black hole, and comment on the result.

There are 10^{11} stars in the galaxy. That used to be a huge number. But it's only a hundred billion. It's less than the national deficit! We used to call them astronomical numbers. Now we should call them economical numbers.

—Richard Feynman, quoted by David Goodstein, *Physics Today* **42**, 73 (February, 1989).

---

*By Stephen Hawking in 1973. To learn more about black hole thermodynamics, see Stephen Hawking, "The Quantum Mechanics of Black Holes," *Scientific American* **236**, 34–40 (January, 1977); Jacob Beckenstein, "Black Hole Thermodynamics," *Physics Today* **33**, 24–31 (January, 1980); and Leonard Susskind, "Black Holes and the Information Paradox," *Scientific American* **276**, 52–57 (April, 1997).

3 Interactions and Implications

In the previous chapter I argued that whenever two large systems interact, they will evolve toward whatever macrostate has the highest possible entropy. This statement is known as the second law of thermodynamics. The second law is not built into the fundamental laws of nature, though; it arises purely through the laws of probability and the mathematics of very large numbers. But since the probabilities are so overwhelming for any system large enough to see with our eyes, we might as well forget about probabilities and just treat the second law as fundamental. That's what I'll do throughout most of the rest of this book, as we explore the consequences of the second law.

The purpose of the present chapter is twofold. First, we need to figure out how entropy is related to other variables, such as temperature and pressure, that can be measured more directly. I'll derive the needed relations by considering the various ways in which two systems can interact, exchanging energy, volume, and/or particles. In each case, for the second law to apply, entropy must govern the direction of change. Second, we'll use these relations and our various formulas for entropy to predict the thermal properties of a variety of realistic systems, from the heat capacity of a solid to the pressure of a gas to the magnetization of a paramagnetic material.

3.1 Temperature

The second law says that when two objects are in thermal equilibrium, their total entropy has reached its maximum possible value. In Section 1.1, however, I gave another criterion that is met when two objects are in thermal equilibrium: I said that they are then at the same **temperature**. In fact, I *defined* temperature to be the thing that's the same for both objects when they're in thermal equilibrium. So now that we have a more precise understanding of thermal equilibrium in terms of entropy, we are in a position to figure out what temperature is, *really*.

Let's look at a specific example. Consider two Einstein solids, A and B, that are "weakly coupled" so that they can exchange energy (but with the total energy fixed). Suppose (as in Figure 2.5) that the numbers of oscillators in the two solids are $N_A = 300$ and $N_B = 200$, and that they are sharing 100 units of energy: $q_{total} = 100$. Table 3.1 lists the various macrostates and their multiplicities. Now, however, I have also included columns for the entropy of solid A, the entropy of solid B, and the total entropy (which can be obtained either by adding S_A and S_B, or by taking the logarithm of Ω_{total}).

Figure 3.1 shows a graph of S_A, S_B, and S_{total} (in units of Boltzmann's constant), for the same parameters as in the table. The equilibrium point is at $q_A = 60$, where S_{total} reaches its maximum value. At this point, the tangent to the graph of S_{total} is horizontal; that is,

$$\frac{\partial S_{total}}{\partial q_A} = 0 \quad \text{or} \quad \frac{\partial S_{total}}{\partial U_A} = 0 \quad \text{at equilibrium.} \tag{3.1}$$

(Technically it's a *partial* derivative because the number of oscillators in each solid is being held fixed. The energy U_A is just q_A times a constant, the size of each unit of energy.) But the slope of the S_{total} graph is the sum of the slopes of the S_A and S_B graphs. Therefore,

$$\frac{\partial S_A}{\partial U_A} + \frac{\partial S_B}{\partial U_A} = 0 \quad \text{at equilibrium.} \tag{3.2}$$

The second term in this equation is rather awkward, with B in the numerator and A in the denominator. But dU_A is the same thing as $-dU_B$, since adding a bit of energy to solid A is the same as subtracting the same amount from solid B.

q_A	Ω_A	S_A/k	q_B	Ω_B	S_B/k	Ω_{total}	S_{total}/k
0	1	0	100	2.8×10^{81}	187.5	2.8×10^{81}	187.5
1	300	5.7	99	9.3×10^{80}	186.4	2.8×10^{83}	192.1
2	45150	10.7	98	3.1×10^{80}	185.3	1.4×10^{85}	196.0
$\vdots$	$\vdots$	$\vdots$	$\vdots$	$\vdots$	$\vdots$	$\vdots$	$\vdots$
11	5.3×10^{19}	45.4	89	1.1×10^{76}	175.1	5.9×10^{95}	220.5
12	1.4×10^{21}	48.7	88	3.4×10^{75}	173.9	4.7×10^{96}	222.6
13	3.3×10^{22}	51.9	87	1.0×10^{75}	172.7	3.5×10^{97}	224.6
$\vdots$	$\vdots$	$\vdots$	$\vdots$	$\vdots$	$\vdots$	$\vdots$	$\vdots$
59	2.2×10^{68}	157.4	41	3.1×10^{46}	107.0	6.8×10^{114}	264.4
60	1.3×10^{69}	159.1	40	5.3×10^{45}	105.5	6.9×10^{114}	264.4
61	7.7×10^{69}	160.9	39	8.8×10^{44}	103.5	6.8×10^{114}	264.4
$\vdots$	$\vdots$	$\vdots$	$\vdots$	$\vdots$	$\vdots$	$\vdots$	$\vdots$
100	1.7×10^{96}	221.6	0	1	0	1.7×10^{96}	221.6

Table 3.1. Macrostates, multiplicities, and entropies of a system of two Einstein solids, one with 300 oscillators and the other with 200, sharing a total of 100 units of energy.

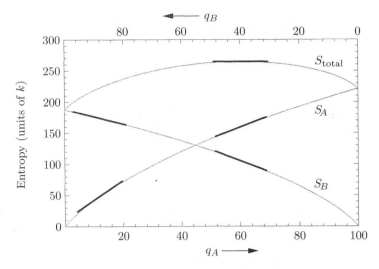

Figure 3.1. A plot of the entropies calculated in Table 3.1. At equilibrium ($q_A = 60$), the total entropy is a maximum so its graph has a horizontal tangent; therefore the slopes of the tangents to the graphs of S_A and S_B are equal in magnitude. Away from equilibrium (for instance, at $q_A = 12$), the solid whose graph has the steeper tangent line tends to gain energy spontaneously; therefore we say that it has the lower temperature.

We can therefore write

$$\frac{\partial S_A}{\partial U_A} = \frac{\partial S_B}{\partial U_B} \qquad \text{at equilibrium.} \tag{3.3}$$

In other words, the thing that's the same for both systems when they're in thermal equilibrium is the *slope* of their entropy vs. energy graphs. This slope must somehow be related to the temperature of a system.

To get a better idea of how temperature is related to the slope of the entropy vs. energy graph, let's look at a point away from equilibrium, for instance, the point $q_A = 12$ in the figure. Here the slope of the S_A graph is considerably steeper than the slope of the S_B graph. This means that if a bit of energy passes from solid B to solid A, the entropy gained by A will be greater than the entropy lost by B. The total entropy will increase, so this process will happen spontaneously, according to the second law. Apparently, the second law tells us that energy will always tend to flow *into* the object with the *steeper* S vs. U graph, and *out of* the object with the *shallower* S vs. U graph. The former really "wants" to gain energy (in order to increase its entropy), while the latter doesn't so much "mind" losing a bit of energy (since its entropy doesn't decrease by much). A steep slope must correspond to a *low* temperature, while a shallow slope must correspond to a *high* temperature.

Now let's look at units. Thanks to the factor of Boltzmann's constant in the definition of entropy, the slope $\partial S/\partial U$ of a system's entropy vs. energy graph has units of $(J/K)/J = 1/K$. If we take the reciprocal of this slope, we get something with units of kelvins, just what we want for temperature. Moreover, we have just seen that when the slope is large the temperature must be small, and vice versa. I therefore propose the following relation:

$$T \equiv \left(\frac{\partial S}{\partial U}\right)^{-1}. \tag{3.4}$$

The **temperature** of a system is the reciprocal of the slope of its entropy vs. energy graph. The partial derivative is to be taken with the system's volume and number of particles held fixed;* more explicitly,

$$\frac{1}{T} \equiv \left(\frac{\partial S}{\partial U}\right)_{N,V}. \tag{3.5}$$

From now on I will take equation 3.5 to be the *definition* of temperature. (To verify that no further factors of 2 or other numbers are needed in equation 3.5, we need to check an example where we already know the answer. I'll do so on page 91.)

You may be wondering why I don't just turn the derivative upside down, and write equation 3.5 as

$$T = \left(\frac{\partial U}{\partial S}\right)_{N,V}. \tag{3.6}$$

The answer is that there's nothing wrong with this, but it's less convenient in practice, because rarely do you ever have a formula for energy in terms of entropy, volume, and number of particles. However, in numerical examples like in Table 3.1, this version of the formula is just fine. For instance, comparing the two lines in the table for $q_A = 11$ and $q_A = 13$ gives for solid A

$$T_A = \frac{13\epsilon - 11\epsilon}{51.9k - 45.4k} = 0.31 \ \epsilon/k, \tag{3.7}$$

where $\epsilon \ (= hf)$ is the size of each energy unit. If $\epsilon = 0.1$ eV, the temperature is about 360 K. This number is the approximate temperature at $q_A = 12$, in the middle of the small interval considered. (Technically, since a difference of one or two energy units is not infinitesimal compared to 12, the derivative is not precisely defined for this small system. For a large system, this ambiguity will never occur.) Similarly, for solid B,

$$T_B = \frac{89\epsilon - 87\epsilon}{175.1k - 172.7k} = 0.83 \ \epsilon/k. \tag{3.8}$$

As expected, solid B is hotter at this point, since it is the one that will tend to *lose* energy.

It's still not obvious that our new definition of temperature (3.5) is in complete agreement with the *operational* definition given in Section 1.1, that is, with the result that we would get by measuring the temperature with a properly calibrated

---

*Volume isn't very relevant for an Einstein solid, although the *size* of the energy units can depend on volume. For some systems there can be other variables, such as magnetic field strength, that must also be held fixed in the partial derivative.

thermometer. If you're skeptical, let me say this: For most practical purposes, the two definitions *are* equivalent. However, any operational definition is of limited scope, since it depends on the physical limitations of the instruments used. In our case, any particular thermometer that you use to "define" temperature will have limitations—it may freeze or melt or something. There are even some systems for which *no* standard thermometer will work; we'll see an example in Section 3.3. So our new definition really is better than the old one, even if it isn't quite the same.

> **Problem 3.1.** Use Table 3.1 to compute the temperatures of solid A and solid B when $q_A = 1$. Then compute both temperatures when $q_A = 60$. Express your answers in terms of ϵ/k, and then in kelvins assuming that $\epsilon = 0.1$ eV.

> **Problem 3.2.** Use the definition of temperature to prove the **zeroth law of thermodynamics**, which says that if system A is in thermal equilibrium with system B, and system B is in thermal equilibrium with system C, then system A is in thermal equilibrium with system C. (If this exercise seems totally pointless to you, you're in good company: Everyone considered this "law" to be completely obvious until 1931, when Ralph Fowler pointed out that it was an unstated assumption of classical thermodynamics.)

A Silly Analogy

To get a better feel for the theoretical definition of temperature (3.5), I like to make a rather silly analogy. Imagine a world, not entirely unlike our own, in which people are constantly exchanging money in their attempts to become happier. They are not merely interested in their *own* happiness, however; each person is actually trying to maximize the *total* happiness of everyone in the community. Now some individuals become much happier when given only a little money. We might call these people "greedy," since they accept money gladly and are reluctant to give any up. Other individuals, meanwhile, become only a little happier when given more money, and only a little sadder upon losing some. These people will be quite generous, giving their money to the more greedy people in order to maximize the total happiness.

The analogy to thermodynamics is as follows. The community corresponds to an isolated system of objects, while the people correspond to the various objects in the system. Money corresponds to *energy*; it is the quantity that is constantly being exchanged, and whose total amount is conserved. Happiness corresponds to *entropy*; the community's overriding goal is to increase its total amount. Generosity corresponds to *temperature*; this is the measure of how willingly someone gives up money (energy). Here is a summary of the analogies:

$$
\begin{array}{ccc}
\text{money} & \leftrightarrow & \text{energy} \\
\text{happiness} & \leftrightarrow & \text{entropy} \\
\text{generosity} & \leftrightarrow & \text{temperature}
\end{array}
$$

One can press this analogy even further. Normally, you would expect that as people acquire more money, they become more generous. In thermodynamics, this would mean that as an object's energy increases, so does its temperature. Indeed, most objects behave in this way. Increasing temperature corresponds to a *decreasing*

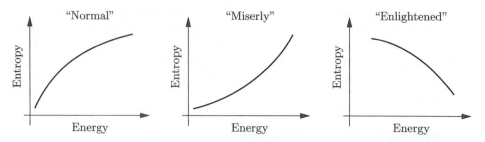

Figure 3.2. Graphs of entropy vs. energy (or happiness vs. money) for a "normal" system that becomes hotter (more generous) as it gains energy; a "miserly" system that becomes colder (less generous) as it gains energy; and an "enlightened" system that doesn't want to gain energy at all.

slope on the entropy vs. energy graph, so the graph for such an object is everywhere concave-down (see Figures 3.1 and 3.2).

However, every community seems to have a few misers who actually become *less* generous as they acquire more money. Similarly, there's no law of physics that prevents an object's temperature from *decreasing* as you add energy. Such an object would have a negative heat capacity; its entropy vs. energy graph would be concave-up. (Collections of particles that are held together by gravity, such as stars and star clusters, behave in exactly this way. Any added energy goes into *potential* energy, as the particles in the system get farther apart and actually slow down. See Problems 1.55, 3.7, and 3.15.)

Even more unusual are those enlightened individuals who become happier as they *lose* money. An analogous thermodynamic system would have an entropy-energy graph with negative slope. This situation is extremely counterintuitive, but does occur in real physical systems, as we'll see in Section 3.3. (The negative-slope portion of the total entropy graph in Figure 3.1 is not an example of "enlightened" behavior; here I'm talking about the equilibrium entropy of a *single* object as a function of its total energy.)

Problem 3.3. Figure 3.3 shows graphs of entropy vs. energy for two objects, A and B. Both graphs are on the same scale. The energies of these two objects initially have the values indicated; the objects are then brought into thermal contact with each other. Explain what happens subsequently and why, *without* using the word "temperature."

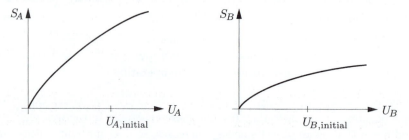

Figure 3.3. Graphs of entropy vs. energy for two objects.

Problem 3.4. Can a "miserly" system, with a concave-up entropy-energy graph, ever be in stable thermal equilibrium with another system? Explain.

Real-World Examples

The theoretical definition of temperature isn't just interesting and intuitive—it is also useful. If you have an explicit formula for the entropy of an object as a function of energy, you can easily calculate its temperature (also as a function of energy).

Perhaps the simplest realistic example is a large Einstein solid, in the limit $q \gg N$ (where N is the number of oscillators). The total energy U is just q times some constant that I'll call ϵ. I computed the entropy in equation 2.46:

$$S = Nk[\ln(q/N) + 1] = Nk \ln U - Nk \ln(\epsilon N) + Nk. \qquad (3.9)$$

Therefore the temperature should be

$$T = \left(\frac{\partial S}{\partial U}\right)^{-1} = \left(\frac{Nk}{U}\right)^{-1}, \qquad (3.10)$$

in other words,

$$U = NkT. \qquad (3.11)$$

But this result is exactly what the equipartition theorem would predict: The total energy should be $\frac{1}{2}kT$ times the number of degrees of freedom, and an Einstein solid has two degrees of freedom for every oscillator. (This result verifies that no factors of 2 or other constants are needed in equation 3.5.)

As another example, let us compute the temperature of a monatomic ideal gas. Recall from equation 2.49 that the entropy is

$$S = Nk \ln V + Nk \ln U^{3/2} + (\text{a function of } N) \qquad (3.12)$$

(where N is the number of molecules). The temperature is therefore

$$T = \left(\frac{\frac{3}{2}Nk}{U}\right)^{-1}. \qquad (3.13)$$

Solving this equation for U gives $U = \frac{3}{2}NkT$, again verifying the equipartition theorem. (At this point we could reverse the logic of Section 1.2 and *derive* the ideal gas law, starting from the formula for U. Instead, however, I'll wait until Section 3.4, and derive the ideal gas law from a much more general formula for pressure.)

Problem 3.5. Starting with the result of Problem 2.17, find a formula for the temperature of an Einstein solid in the limit $q \ll N$. Solve for the energy as a function of temperature to obtain $U = N\epsilon e^{-\epsilon/kT}$ (where ϵ is the size of an energy unit).

Problem 3.6. In Section 2.5 I quoted a theorem on the multiplicity of any system with only quadratic degrees of freedom: In the high-temperature limit where the number of units of energy is much larger than the number of degrees of freedom, the multiplicity of any such system is proportional to $U^{Nf/2}$, where Nf is the total number of degrees of freedom. Find an expression for the energy of such a system in terms of its temperature, and comment on the result. How can you tell that this formula for Ω cannot be valid when the total energy is very small?

Problem 3.7. Use the result of Problem 2.42 to calculate the temperature of a black hole, in terms of its mass M. (The energy is Mc^2.) Evaluate the resulting expression for a one-solar-mass black hole. Also sketch the entropy as a function of energy, and discuss the implications of the shape of the graph.

3.2 Entropy and Heat

Predicting Heat Capacities

In the preceding section we saw how to calculate the temperature as a function of energy (or vice versa) for any system for which we have an explicit formula for the multiplicity. To compare these predictions to experiments, we can differentiate the function $U(T)$ to obtain the heat capacity at constant volume (or simply "energy capacity"):

$$C_V \equiv \left(\frac{\partial U}{\partial T}\right)_{N,V}. \tag{3.14}$$

For an Einstein solid with $q \gg N$ the heat capacity is

$$C_V = \frac{\partial}{\partial T}(NkT) = Nk, \tag{3.15}$$

while for a monatomic ideal gas,

$$C_V = \frac{\partial}{\partial T}(\tfrac{3}{2}NkT) = \tfrac{3}{2}Nk. \tag{3.16}$$

In both of these systems, the heat capacity is independent of the temperature and is simply equal to $k/2$ times the number of degrees of freedom. These results agree with experimental measurements of heat capacities of low-density monatomic gases and of solids at reasonably high temperatures. However, other systems can have much more complicated behavior. One example is the subject of Section 3.3; others are treated in the problems.

Before considering more complicated examples, let me pause and list the steps you have to go through in order to predict the heat capacity of a system using the tools we have developed:

1. Use quantum mechanics and some combinatorics to find an expression for the multiplicity, Ω, in terms of U, V, N, and any other relevant variables.

2. Take the logarithm to find the entropy, S.

3. Differentiate S with respect to U and take the reciprocal to find the temperature, T, as a function of U and other variables.

4. Solve for U as a function of T (and other variables).

5. Differentiate $U(T)$ to obtain a prediction for the heat capacity (with the other variables held fixed).

This procedure is rather intricate, and for most systems, you're likely to get stuck at step 1. In fact, there are very few systems for which I know how to write down an explicit formula for the multiplicity: the two-state paramagnet, the Einstein solid, the monatomic ideal gas, and a few others that are mathematically similar to these. In Chapter 6 I'll show you an alternative route to step 4, yielding a formula for $U(T)$ without the need to know the multiplicity or the entropy. Meanwhile, we can still learn plenty from the simple examples that I've already introduced.

Problem 3.8. Starting with the result of Problem 3.5, calculate the heat capacity of an Einstein solid in the low-temperature limit. Sketch the predicted heat capacity as a function of temperature. (Note: Measurements of heat capacities of actual solids at low temperatures do not confirm the prediction that you will make in this problem. A more accurate model of solids at low temperatures is presented in Section 7.5.)

Measuring Entropies

Even if you can't write down a mathematical *formula* for the entropy of a system, you can still *measure* it, essentially by following steps 3–5 in reverse. According to the theoretical definition (3.5) of temperature, if you add a bit of heat Q to a system while holding its volume constant and doing no other forms of work, its entropy changes by

$$dS = \frac{dU}{T} = \frac{Q}{T} \qquad \text{(constant volume, no work).} \tag{3.17}$$

Since heat and temperature are usually pretty easy to measure, this relation allows us to compute the change in entropy for a wide variety of processes.* In Section 3.4 I'll show that the relation $dS = Q/T$ also applies when the volume *is* changing, provided that the process is quasistatic.

If the temperature of an object remains constant as heat is added to it (as during a phase change), then equation 3.17 can be applied even when Q and dS are not infinitesimal. When T is changing, however, it's usually more convenient to write the relation in terms of the heat capacity at constant volume:

$$dS = \frac{C_V \, dT}{T}. \tag{3.18}$$

Now perhaps you can see what to do if the temperature changes significantly as the heat is added. Imagine the process as a sequence of tiny steps, compute dS for

---

*Equation 3.17 assumes not only fixed volume, but also fixed values of N and any other variables held fixed in equation 3.5. It also assumes that T doesn't vary *within* the system; internal temperature variations would cause internal heat flow and thus further increases in entropy.

each step, and add them up to get the total change in entropy:

$$\Delta S = S_f - S_i = \int_{T_i}^{T_f} \frac{C_V}{T} \, dT. \tag{3.19}$$

Often C_V is fairly constant over the temperature range of interest, and you can take it out of the integral. In other cases, especially at low temperatures, C_V changes quite a bit and must be left inside the integral.

Here's a quick example. Suppose you heat a cup (200 g) of water from 20°C to 100°C. By how much does its entropy increase? Well, the heat capacity of 200 g of water is 200 cal/K or about 840 J/K, and is essentially independent of temperature over this range. Therefore the increase in entropy is

$$\Delta S = (840 \text{ J/K}) \int_{293 \text{ K}}^{373 \text{ K}} \frac{1}{T} \, dT = (840 \text{ J/K}) \ln\left(\frac{373}{293}\right) = 200 \text{ J/K}. \tag{3.20}$$

This may not seem like a huge increase, but in fundamental units (dividing by Boltzmann's constant) it's an increase of 1.5×10^{25}. And this means that the *multiplicity* of the system increases by a *factor* of $e^{1.5 \times 10^{25}}$ (a very large number).

If you're lucky enough to know C_V all the way down to absolute zero, you can calculate a system's *total* entropy simply by taking zero as the lower limit of the integral:

$$S_f - S(0) = \int_0^{T_f} \frac{C_V}{T} \, dT. \tag{3.21}$$

But what is $S(0)$? In principle, zero. At zero temperature a system should settle into its unique lowest-energy state, so $\Omega = 1$ and $S = 0$. This fact is often called the **third law of thermodynamics**.

In practice, however, there can be several reasons why $S(0)$ is effectively nonzero. Most importantly, in some solid crystals it is possible to change the orientations of the molecules with very little change in energy. Water molecules, for example, can orient themselves in several possible ways within an ice crystal. Technically, one particular arrangement will always have a lower energy than any other, but in practice the arrangements are often random or nearly random, and you would have to wait eons for the crystal to rearrange itself into the true ground state. We then say that the solid has a frozen-in **residual entropy**, equal to k times the logarithm of the number of possible molecular arrangements.

Another form of residual entropy comes from the mixing of different nuclear isotopes of an element. Most elements have more than one stable isotope, but in natural systems these isotopes are mixed together randomly, with an associated entropy of mixing. Again, at $T = 0$ there should be a unique lowest-energy state in which the isotopes are unmixed or are distributed in some orderly way, but in practice the atoms are always stuck at their random sites in the crystal lattice.*

---

*An important exception is helium, which remains a liquid at $T = 0$, allowing the two isotopes (^{3}He and ^{4}He) to arrange themselves in an orderly way.

A third type of "residual" entropy comes from the multiplicity of alignments of nuclear spins. At $T = 0$ this entropy *does* disappear as the spins align parallel or antiparallel to their neighbors. But this generally doesn't happen until the temperature is less than a tiny fraction of 1 K, far below the range of routine heat capacity measurements.

Entropies of a wide variety of substances have been computed from measured heat capacities using equation 3.21, and are tabulated in standard reference works. (A few dozen values are included at the back of this book.) By convention, tabulated entropies do include any residual entropy due to molecular orientations, but do not include any entropy of isotopic mixing or of nuclear spin orientations. (The tables are generally compiled by chemists, who don't care much about nuclei.)

You might be worried that the integral in formula 3.21 appears to *diverge* at its lower limit, because of the T in the denominator of the integrand. If it did diverge, either S_f would be infinity or $S(0)$ would be negative infinity. Entropy, however, must always be finite and positive, according to our original definition $S = k \ln \Omega$. The only way out is if C_V also goes to zero at $T = 0$:

$$C_V \to 0 \qquad \text{as} \qquad T \to 0. \tag{3.22}$$

This result is also sometimes called the **third law of thermodynamics**. Apparently, our earlier results (3.15 and 3.16) for the heat capacities of an Einstein solid and an ideal gas *cannot* be correct at very low temperatures. Instead, all degrees of freedom must "freeze out." This is what you should have found in Problem 3.8; we'll see many other examples throughout the rest of this book.

> **Problem 3.9.** In solid carbon monoxide, each CO molecule has two possible orientations: CO or OC. Assuming that these orientations are completely random (not quite true but close), calculate the residual entropy of a mole of carbon monoxide.

The Macroscopic View of Entropy

Historically, the relation $dS = Q/T$ was the original *definition* of entropy. In 1865, Rudolf Clausius defined entropy to be the thing that increases by Q/T whenever heat Q enters a system at temperature T. Although this definition tells us nothing about what entropy actually *is*, it is still sufficient for many purposes, when the microscopic makeup of a system does not concern us.

To illustrate this traditional view of entropy, consider again what happens when a hot object, A, is put in thermal contact with a cold object, B (see Figure 3.4). To be specific, suppose that $T_A = 500$ K and $T_B = 300$ K. From experience we know that heat will flow from A to B. Let's say that during some time interval the amount of heat that flows is 1500 J, and that A and B are large enough objects that their temperatures don't change significantly due to the loss or gain of this amount of energy. Then during this time interval, the entropy of A changes by

$$\Delta S_A = \frac{-1500 \text{ J}}{500 \text{ K}} = -3 \text{ J/K}. \tag{3.23}$$

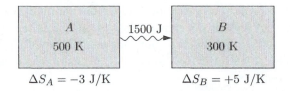

$$\Delta S_A = -3 \text{ J/K} \qquad\qquad \Delta S_B = +5 \text{ J/K}$$

Figure 3.4. When 1500 J of heat leaves a 500 K object, its entropy decreases by 3 J/K. When this same heat enters a 300 K object, its entropy increases by 5 J/K.

Object A *loses* entropy, because heat is flowing *out* of it. Similarly, the entropy of B changes by

$$\Delta S_B = \frac{+1500 \text{ J}}{300 \text{ K}} = +5 \text{ J/K}. \tag{3.24}$$

Object B *gains* entropy, because heat is flowing *into* it. (Notice that the traditional entropy unit of J/K is quite convenient when we compute entropy changes in this way.)

Just as I often visualize energy as a "fluid" that can change forms and move around but never be created or destroyed, I sometimes imagine entropy, as well, to be a fluid. I imagine that, whenever energy enters or leaves a system in the form of heat, it is required (by law) to carry some entropy with it, in the amount Q/T. The weird thing about entropy, though, is that it is only half-conserved: It cannot be destroyed, but it *can* be created, and in fact, new entropy *is* created whenever heat flows between objects at different temperatures. As in the numerical example above, the entropy that is "carried by" the heat is more when it arrives at the cooler object than it was when it left the hotter object (see Figure 3.5). Only in the limit where there is no temperature difference between the two objects will no new entropy be created. In this limit, however, there is no tendency of heat to flow in the first place. It's important to remember that fundamentally, the net increase

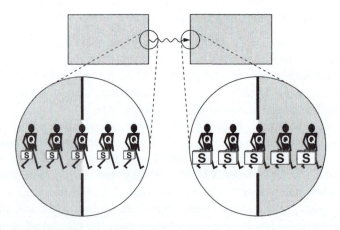

Figure 3.5. Each unit of heat energy (Q) that leaves a hot object is required to carry some entropy (Q/T) with it. When it enters a cooler object, the amount of entropy has increased.

in entropy is the driving force behind the flow of heat. Fundamentally, though, entropy isn't a fluid at all and my model is simply wrong.

Problem 3.10. An ice cube (mass 30 g) at $0°C$ is left sitting on the kitchen table, where it gradually melts. The temperature in the kitchen is $25°C$.

  (a) Calculate the change in the entropy of the ice cube as it melts into water at $0°C$. (Don't worry about the fact that the volume changes somewhat.)

  (b) Calculate the change in the entropy of the water (from the melted ice) as its temperature rises from $0°C$ to $25°C$.

  (c) Calculate the change in the entropy of the kitchen as it gives up heat to the melting ice/water.

  (d) Calculate the net change in the entropy of the universe during this process. Is the net change positive, negative, or zero? Is this what you would expect?

Problem 3.11. In order to take a nice warm bath, you mix 50 liters of hot water at $55°C$ with 25 liters of cold water at $10°C$. How much new entropy have you created by mixing the water?

Problem 3.12. Estimate the change in the entropy of the universe due to heat escaping from your home on a cold winter day.

Problem 3.13. When the sun is high in the sky, it delivers approximately 1000 watts of power to each square meter of earth's surface. The temperature of the surface of the sun is about 6000 K, while that of the earth is about 300 K.

  (a) Estimate the entropy created in one year by the flow of solar heat onto a square meter of the earth.

  (b) Suppose you plant grass on this square meter of earth. Some people might argue that the growth of the grass (or of any other living thing) violates the second law of thermodynamics, because disorderly nutrients are converted into an orderly life form. How would you respond?

Problem 3.14. Experimental measurements of the heat capacity of aluminum at low temperatures (below about 50 K) can be fit to the formula

$$C_V = aT + bT^3,$$

where C_V is the heat capacity of one mole of aluminum, and the constants a and b are approximately $a = 0.00135$ J/K^2 and $b = 2.48 \times 10^{-5}$ J/K^4. From this data, find a formula for the entropy of a mole of aluminum as a function of temperature. Evaluate your formula at $T = 1$ K and at $T = 10$ K, expressing your answers both in conventional units (J/K) and as unitless numbers (dividing by Boltzmann's constant). [Comment: In Chapter 7 I'll explain why the heat capacity of a metal has this form. The linear term comes from energy stored in the conduction electrons, while the cubic term comes from lattice vibrations of the crystal.]

Problem 3.15. In Problem 1.55 you used the virial theorem to estimate the heat capacity of a star. Starting with that result, calculate the entropy of a star, first in terms of its average temperature and then in terms of its total energy. Sketch the entropy as a function of energy, and comment on the shape of the graph.

Problem 3.16. A **bit** of computer memory is some physical object that can be in two different states, often interpreted as 0 and 1. A **byte** is eight bits, a **kilobyte** is 1024 ($= 2^{10}$) bytes, a **megabyte** is 1024 kilobytes, and a **gigabyte** is 1024 megabytes.

(a) Suppose that your computer erases or overwrites one gigabyte of memory, keeping no record of the information that was stored. Explain why this process must create a certain minimum amount of entropy, and calculate how much.

(b) If this entropy is dumped into an environment at room temperature, how much heat must come along with it? Is this amount of heat significant?

3.3 Paramagnetism

At the beginning of the previous section I outlined a five-step procedure for predicting the thermal properties of a material, starting from a combinatoric formula for the multiplicity and applying the definitions of entropy and temperature. I also carried out this procedure for two particular model systems: a monatomic ideal gas, and an Einstein solid in the high-temperature limit ($q \gg N$). Both of these examples, however, were very simple mathematically, and merely verified the equipartition theorem. Next I would like to work out a more complicated example, where the equipartition theorem does not apply at all. This example will be more interesting mathematically, and also rather counterintuitive physically.

The system that I want to discuss is the **two-state paramagnet**, introduced briefly in Section 2.1. I'll start by reviewing the basic microscopic physics.

Notation and Microscopic Physics

The system consists of N spin-1/2 particles, immersed in a constant magnetic field $\vec{B}$ pointing in the $+z$ direction (see Figure 3.6). Each particle behaves like a little compass needle, feeling a torque that tries to align its magnetic dipole moment with the field. Because of this behavior I'll refer to the particles as **dipoles**. For simplicity I'll assume that there are *no* interactions *between* dipoles—each dipole feels only the torque from the external field. In this case we say that the system is an **ideal** paramagnet.

According to quantum mechanics, the component of a particle's dipole moment along a given axis cannot take on just any value—instead it is **quantized**, that is, limited to certain discrete values. For a spin-1/2 particle only *two* values are allowed, which I'll call simply "up" and "down" (along the z axis). The magnetic

Figure 3.6. A two-state paramagnet, consisting of N microscopic magnetic dipoles, each of which is either "up" or "down" at any moment. The dipoles respond only to the influence of the external magnetic field B; they do not interact with their neighbors (except to exchange energy).

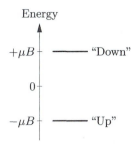

Figure 3.7. The energy levels of a single dipole in an ideal two-state paramagnet are $-\mu B$ (for the "up" state) and $+\mu B$ (for the "down" state).

field, pointing in the $+z$ direction, gives each dipole a preference for the *up* state. To flip a single dipole from up to down we would have to add some energy; the amount of energy required is $2\mu B$, where μ is a constant related to the particle's magnetic moment (essentially the "strength" of the effective compass needle). For the sake of symmetry, I'll say that the energy of a dipole that points up is $-\mu B$, so that the energy of a dipole that points down is $+\mu B$ (see Figure 3.7).

The *total* energy of the system is

$$U = \mu B(N_\downarrow - N_\uparrow) = \mu B(N - 2N_\uparrow), \qquad (3.25)$$

where $N_\uparrow$ and $N_\downarrow$ are the numbers of up and down dipoles, respectively, and $N = N_\uparrow + N_\downarrow$. I'll define the **magnetization**, M, to be the total magnetic moment of the whole system. Each "up" dipole has magnetic moment $+\mu$ and each "down" dipole has magnetic moment $-\mu$, so the magnetization can be written

$$M = \mu(N_\uparrow - N_\downarrow) = -\frac{U}{B}. \qquad (3.26)$$

We would like to know how U and M depend on temperature.

Our first task is to write down a formula for the multiplicity. We will keep N fixed, and consider each different value of $N_\uparrow$ (and hence U and M) to define a different macrostate. Then this system is mathematically equivalent to a collection of N coins with $N_\uparrow$ heads, and the multiplicity is simply

$$\Omega(N_\uparrow) = \binom{N}{N_\uparrow} = \frac{N!}{N_\uparrow!N_\downarrow!}. \qquad (3.27)$$

Numerical Solution

For reasonably small systems, one can just evaluate the multiplicity (3.27) directly, take the logarithm to find the entropy, and so on. Table 3.2 shows part of a computer-generated table of numbers for a paramagnet consisting of 100 elementary dipoles. There is one row in the table for each possible energy value; the rows are written in order of increasing energy, starting with the macrostate with all the dipoles pointing up.

$N_\uparrow$	$U/\mu B$	$M/N\mu$	Ω	S/k	$kT/\mu B$	C/Nk
100	-100	1.00	1	0	0	—
99	-98	.98	100	4.61	.47	.074
98	-96	.96	4950	8.51	.54	.310
97	-94	.94	1.6×10^5	11.99	.60	.365
$\vdots$	$\vdots$	$\vdots$	$\vdots$	$\vdots$	$\vdots$	$\vdots$
52	-4	.04	9.3×10^{28}	66.70	25.2	.001
51	-2	.02	9.9×10^{28}	66.76	50.5	—
50	0	0	1.0×10^{29}	66.78	∞	—
49	2	$-.02$	9.9×10^{28}	66.76	-50.5	—
48	4	$-.04$	9.3×10^{28}	66.70	-25.2	.001
$\vdots$	$\vdots$	$\vdots$	$\vdots$	$\vdots$	$\vdots$	$\vdots$
1	98	$-.98$	100	4.61	$-.47$	.074
0	100	-1.00	1	0	0	—

Table 3.2. Thermodynamic properties of a two-state paramagnet consisting of 100 elementary dipoles. Microscopic physics determines the energy U and total magnetization M in terms of the number of dipoles pointing up, $N_\uparrow$. The multiplicity Ω is calculated from the combinatoric formula 3.27, while the entropy S is $k \ln \Omega$. The last two columns show the temperature and the heat capacity, calculated by taking derivatives as explained in the text.

The behavior of the entropy as a function of energy is particularly interesting, as shown in Figure 3.8. The largest multiplicity and largest entropy occur at $U = 0$, when exactly half of the dipoles point down. As more energy is added to the system, the multiplicity and entropy actually *decrease*, since there are fewer ways to arrange the energy. This behavior is very different from that of a "normal" system such as an Einstein solid (as discussed in Section 3.1).

Let's look at this behavior in more detail. Suppose the system starts out in its minimum-energy state, with all the dipoles pointing up. Here the entropy-energy graph is very steep, so the system has a strong tendency to absorb energy from its environment. As its energy increases (but is still negative), the entropy-energy graph becomes shallower, so the tendency to absorb energy decreases, just as for an Einstein solid or any other "normal" system. However, as the energy of the paramagnet goes to zero, so does the slope of its entropy-energy graph, so its tendency to absorb more energy actually disappears. At this point, exactly half of the dipoles point down, and the system "couldn't care less" whether its energy increases a bit more or not. If we now add a bit more energy to the system, it behaves in a most unusual way. The slope of its entropy-energy graph becomes negative, so it will spontaneously give up energy to any nearby object whose entropy-energy graph has a positive slope. (Remember, any allowed process that increases the *total* entropy will happen spontaneously.)

In the preceding paragraph I have intentionally avoided any mention of "temperature." But now let's think about the temperature of this system as a function of energy. When more than half of the dipoles point up, so the total energy is

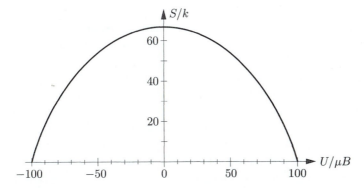

Figure 3.8. Entropy as a function of energy for a two-state paramagnet consisting of 100 elementary dipoles.

negative, this system behaves "normally": Its temperature (the reciprocal of the slope of the entropy-energy graph) increases as energy is added. In the analogy of Section 3.1, the system becomes more "generous" with increasing energy. When $U = 0$, however, the temperature is actually *infinite*, meaning that this system will gladly give up energy to *any* other system whose temperature is finite. The paramagnet is infinitely generous. At still higher energies, we would like to say that its generosity is "higher than infinity," but technically, our definition of temperature says that T is *negative* (since the slope is negative). There's nothing *wrong* with this conclusion, but we have to remember that negative temperatures behave as if they are *higher* than positive temperatures, since a system with negative temperature will give up energy to any system with positive temperature. It would be better, in this example, if we talked about $1/T$ (analogous to "greediness") instead of T. At zero energy, the system has zero greediness, while at higher energies it has negative greediness. A graph of temperature vs. energy is shown in Figure 3.9.

Negative temperatures can occur only for a system whose total energy is limited, so that the multiplicity decreases as the maximum allowed energy is approached. The best examples of such systems are *nuclear* paramagnets, in which the magnetic

Figure 3.9. Temperature as a function of energy for a two-state paramagnet. (This graph was plotted from the analytic formulas derived later in the text; a plot of the data in Table 3.2 would look similar but less smooth.)

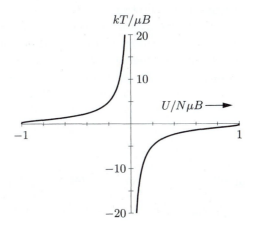

dipoles are the atomic nuclei rather than the electrons. In certain crystals the relaxation time for the nuclear dipoles (exchanging energy with each other) can be much shorter than the relaxation time for the nuclear dipoles to equilibrate with the crystal lattice. Therefore, on short time scales, the dipoles behave as an isolated system with only magnetic energy, no vibrational energy. To give such a system a negative temperature, all you have to do is start at any positive temperature, with most of the dipoles parallel to the magnetic field, then suddenly reverse the field so they're antiparallel. This experiment was first performed by Edward M. Purcell and R. V. Pound in 1951, using the lithium nuclei in a lithium fluoride crystal as the system of dipoles. In their original experiment the nuclear dipoles came to thermal equilibrium among themselves in only 10^{-5} seconds, but required approximately five minutes, after the field reversal, to return to equilibrium with the room-temperature crystal lattice.*

I like the example of the paramagnet, with its negative temperatures and other unusual behavior, because it forces us to think primarily in terms of *entropy* rather than temperature. Entropy is the more fundamental quantity, governed by the second law of thermodynamics. Temperature is less fundamental; it is merely a characterization of a system's "willingness" to give up energy, that is, of the relationship between its energy and entropy.

The sixth column of Table 3.2 lists numerical values of the temperature of this system as a function of energy. I computed each of these using the formula $T = \Delta U/\Delta S$, taking the U and S values from neighboring rows. (To be more precise, I used a "centered-difference" approximation, subtracting the values in the preceding row from those in the following row. So, for instance, the number .47 was computed as $[(-96) - (-100)]/[8.51 - 0]$.) In the last column I've taken another derivative to obtain the heat capacity, $C = \Delta U/\Delta T$. Figure 3.10 shows graphs of the heat capacity and the magnetization vs. temperature. Notice that the heat capacity of this system depends strongly on its temperature, quite unlike the constant values predicted by the equipartition theorem for more familiar systems. At zero temperature the heat capacity goes to zero, as required by the third law of thermodynamics. The heat capacity also goes to zero as T approaches infinity, since at that point only a tiny amount of energy is required to achieve a very large increase in temperature.

The behavior of the magnetization as a function of temperature is also interesting. At zero (positive) temperature the system is "saturated," with all the dipoles pointing up and maximum magnetization. As the temperature increases, random jostling tends to flip more and more dipoles. You might expect that as $T \to \infty$, the energy would be maximized with all the dipoles pointing down, but this is not the

---

*For a more detailed description of this experiment, see the fifth (1968) or sixth (1981) edition of *Heat and Thermodynamics* by Zemansky (with Dittman as coauthor on the sixth edition). The original (very short) letter describing the experiment is published in *Physical Review* **81**, 279 (1951). For an even more dramatic example of negative temperature, see Pertti Hakonen and Olli V. Lounasmaa, *Science* **265**, 1821–1825 (23 September, 1994).

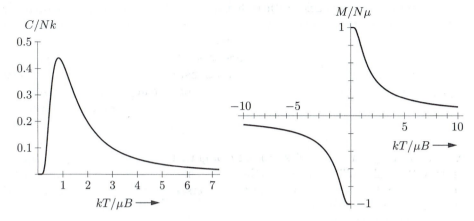

Figure 3.10. Heat capacity and magnetization of a two-state paramagnet (computed from the analytic formulas derived later in the text).

case; instead, $T = \infty$ corresponds to the state of maximum "randomness," with exactly half the dipoles pointing down. The behavior at negative temperature is essentially a mirror image of the positive-T behavior, with the magnetization again saturating, but in the opposite direction, as $T \to 0$ from below.

Problem 3.17. Verify every entry in the third line of Table 3.2 (starting with $N_\uparrow = 98$).

Problem 3.18. Use a computer to reproduce Table 3.2 and the associated graphs of entropy, temperature, heat capacity, and magnetization. (The graphs in this section are actually drawn from the analytic formulas derived below, so your numerical graphs won't be quite as smooth.)

Analytic Solution

Now that we have studied most of the physics of this system through numerical calculations, let us go back and use analytic methods to derive some more general formulas to describe these phenomena.

I will assume that the number of elementary dipoles is *large*, and also that at any given time the numbers of up and down dipoles are separately large. Then we can simplify the multiplicity function (3.27) using Stirling's approximation. Actually, it's easiest to just calculate the entropy:

$$
\begin{aligned}
S/k &= \ln N! - \ln N_\uparrow! - \ln(N - N_\uparrow)! \\
&\approx N \ln N - N - N_\uparrow \ln N_\uparrow + N_\uparrow - (N-N_\uparrow)\ln(N-N_\uparrow) + (N-N_\uparrow) \quad (3.28) \\
&= N \ln N - N_\uparrow \ln N_\uparrow - (N-N_\uparrow)\ln(N-N_\uparrow).
\end{aligned}
$$

From here on the calculations are fairly straightforward but somewhat tedious. I'll outline the logic and the results, but let you fill in some of the algebraic steps (see Problem 3.19).

To find the temperature, we must differentiate S with respect to U. It is simplest to first use the chain rule and equation 3.25 to express the derivative in terms of $N_\uparrow$:

$$\frac{1}{T} = \left(\frac{\partial S}{\partial U}\right)_{N,B} = \frac{\partial N_\uparrow}{\partial U}\frac{\partial S}{\partial N_\uparrow} = -\frac{1}{2\mu B}\frac{\partial S}{\partial N_\uparrow}. \tag{3.29}$$

Now just differentiate the last line of equation 3.28 to obtain

$$\frac{1}{T} = \frac{k}{2\mu B}\ln\left(\frac{N - U/\mu B}{N + U/\mu B}\right). \tag{3.30}$$

Notice from this formula that T and U always have opposite signs.

Equation 3.30 can be solved for U to obtain

$$U = N\mu B\left(\frac{1 - e^{2\mu B/kT}}{1 + e^{2\mu B/kT}}\right) = -N\mu B\tanh\left(\frac{\mu B}{kT}\right), \tag{3.31}$$

where tanh is the hyperbolic tangent function.* The magnetization is therefore

$$M = N\mu\tanh\left(\frac{\mu B}{kT}\right). \tag{3.32}$$

The hyperbolic tangent function is plotted in Figure 3.11; it rises from the origin with a slope of 1, then flattens to an asymptotic value of 1 as its argument goes to infinity. So at very small positive temperatures the system is completely magnetized (as we saw before), while as $T \to \infty$, the magnetization goes to zero. To obtain negative temperature, all we need to do is give the system a negative magnetization, as described above.

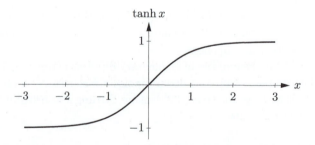

Figure 3.11. The hyperbolic tangent function. In the formulas for the energy and magnetization of a two-state paramagnet, the argument x of the hyperbolic tangent is $\mu B/kT$.

---

*The definitions of the basic hyperbolic functions are $\sinh x = \frac{1}{2}(e^x - e^{-x})$, $\cosh x = \frac{1}{2}(e^x + e^{-x})$, and $\tanh x = (\sinh x)/(\cosh x)$. From these definitions you can easily show that $\frac{d}{dx}\sinh x = \cosh x$ and $\frac{d}{dx}\cosh x = \sinh x$ (with no minus sign).

To calculate the heat capacity of the paramagnet, just differentiate equation 3.31 with respect to T:

$$C_B = \left(\frac{\partial U}{\partial T} \right)_{N,B} = Nk \cdot \frac{(\mu B/kT)^2}{\cosh^2(\mu B/kT)}. \tag{3.33}$$

This function approaches zero at both low and high T, as we also saw in the numerical solution.

In a real-world paramagnet, the individual dipoles can be either electrons or atomic nuclei. Electronic paramagnetism occurs when there are electrons with angular momentum (orbital or spin) that is not compensated by other electrons; the circular currents then give rise to magnetic dipole moments. The number of possible states for each dipole is always some small integer, depending on the total angular momentum of all the electrons in an atom or molecule. The simple case considered here, with just two states, occurs when there is just one electron per atom whose spin is uncompensated. Ordinarily this electron would also have orbital angular momentum, but in some environments the orbital motion is "quenched" by the neighboring atoms, leaving only the spin angular momentum.

For an electronic two-state paramagnet the value of the constant μ is the **Bohr magneton**,

$$\mu_B \equiv \frac{eh}{4\pi m_e} = 9.274 \times 10^{-24} \text{ J/T} = 5.788 \times 10^{-5} \text{ eV/T}. \tag{3.34}$$

(Here e is the electron's charge and m_e is its mass.) If we take $B = 1$ T (a pretty strong magnet), then $\mu B = 5.8 \times 10^{-5}$ eV. But at room temperature, $kT \approx 1/40$ eV. So at ordinary temperatures (more than a few kelvins), we can assume $\mu B/kT \ll 1$. In this limit, $\tanh x \approx x$, so the magnetization becomes

$$M \approx \frac{N\mu^2 B}{kT} \qquad \text{(when } \mu B \ll kT\text{)}. \tag{3.35}$$

The fact that $M \propto 1/T$ was discovered experimentally by Pierre Curie and is known as **Curie's law**; it holds in the high-temperature limit for all paramagnets, even those with more than two angular momentum states. In this limit the heat capacity falls off in proportion to $1/T^2$.

Figure 3.12 shows experimental values of the magnetization of a real two-state paramagnet, an organic free radical known as DPPH.[*] To minimize interactions *between* the elementary dipoles, the DPPH was diluted with benzene to form a 1:1 crystalline complex. Notice that the magnetization follows Curie's law very closely

---

[*]The full name is α, α'-diphenyl-β-picrylhydrazyl, if you really want to know. This rather large molecule is paramagnetic because there is a nitrogen atom in the middle of it with an unpaired electron.

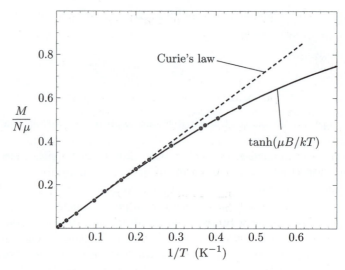

Figure 3.12. Experimental measurements of the magnetization of the organic free radical "DPPH" (in a 1:1 complex with benzene), taken at $B = 2.06$ T and temperatures ranging from 300 K down to 2.2 K. The solid curve is the prediction of equation 3.32 (with $\mu = \mu_B$), while the dashed line is the prediction of Curie's law for the high-temperature limit. (Because the effective number of elementary dipoles in this experiment was uncertain by a few percent, the vertical scale of the theoretical graphs has been adjusted to obtain the best fit.) Adapted from P. Grobet, L. Van Gerven, and A. Van den Bosch, *Journal of Chemical Physics* **68**, 5225 (1978).

down to temperatures of a few kelvins, but then deviates to follow the prediction of equation 3.32 as the total magnetization approaches its maximum possible value.*

For a *nuclear* paramagnet, a typical value of μ can be found by replacing the electron mass with the proton mass in expression 3.34 for the Bohr magneton. Since a proton is nearly 2000 times heavier than an electron, μ is typically smaller for nuclei by a factor of about 2000. This means that to achieve the same degree of magnetization you would need to either make the magnetic field 2000 times stronger, or make the temperature 2000 times lower. Laboratory magnets are

---

*This data is the best I could find for a nearly ideal *two*-state paramagnet. Ideal paramagnets with *more* than two states per dipole turn out to be more common, or at least easier to prepare. The most extensively studied examples are salts in which the paramagnetic ions are either transition metals or rare earths, with unfilled inner electron shells. To minimize interactions between neighboring ions, they are diluted with large numbers of magnetically inert atoms. An example is iron ammonium alum, $Fe_2(SO_4)_3 \cdot (NH_4)_2SO_4 \cdot 24H_2O$, in which there are 23 inert atoms (not counting the very small hydrogens) for each paramagnetic Fe^{3+} ion. The magnetic behavior of this crystal has been shown to be ideal at field strengths up to 5 T and temperatures down to 1.3 K, at which the magnetization is more than 99% complete. See W. E. Henry, *Physical Review* **88**, 561 (1952). The theory of ideal multi-state paramagnets is treated in Problem 6.22.

limited to strengths of a few teslas, so in practice it takes temperatures in the millikelvin range to line up essentially all of the dipoles in a nuclear paramagnet.

Problem 3.19. Fill in the missing algebraic steps to derive equations 3.30, 3.31, and 3.33.

Problem 3.20. Consider an ideal two-state electronic paramagnet such as DPPH, with $\mu = \mu_B$. In the experiment described above, the magnetic field strength was 2.06 T and the minimum temperature was 2.2 K. Calculate the energy, magnetization, and entropy of this system, expressing each quantity as a fraction of its maximum possible value. What would the experimenters have had to do to attain 99% of the maximum possible magnetization?

Problem 3.21. In the experiment of Purcell and Pound, the maximum magnetic field strength was 0.63 T and the initial temperature was 300 K. Pretending that the lithium nuclei have only two possible spin states (in fact they have four), calculate the magnetization per particle, M/N, for this system. Take the constant μ to be 5×10^{-8} eV/T. To detect such a tiny magnetization, the experimenters used resonant absorption and emission of radio waves. Calculate the energy that a radio wave photon should have, in order to flip a single nucleus from one magnetic state to the other. What is the wavelength of such a photon?

Problem 3.22. Sketch (or use a computer to plot) a graph of the entropy of a two-state paramagnet as a function of *temperature*. Describe how this graph would change if you varied the magnetic field strength.

Problem 3.23. Show that the entropy of a two-state paramagnet, expressed as a function of temperature, is $S = Nk[\ln(2 \cosh x) - x \tanh x]$, where $x = \mu B/kT$. Check that this formula has the expected behavior as $T \to 0$ and $T \to \infty$.

*   *   *

The following two problems apply the techniques of this section to a different system, an Einstein solid (or other collection of identical harmonic oscillators) at arbitrary temperature. Both the methods and the results of these problems are extremely important. Be sure to work at least one of them, preferably both.

Problem 3.24. Use a computer to study the entropy, temperature, and heat capacity of an Einstein solid, as follows. Let the solid contain 50 oscillators (initially), and from 0 to 100 units of energy. Make a table, analogous to Table 3.2, in which each row represents a different value for the energy. Use separate columns for the energy, multiplicity, entropy, temperature, and heat capacity. To calculate the temperature, evaluate $\Delta U/\Delta S$ for two nearby rows in the table. (Recall that $U = q\epsilon$ for some constant ϵ.) The heat capacity ($\Delta U/\Delta T$) can be computed in a similar way. The first few rows of the table should look something like this:

q	Ω	S/k	kT/ϵ	C/Nk
0	1	0	0	—
1	50	3.91	.28	.12
2	1275	7.15	.33	.45

(In this table I have computed derivatives using a "centered-difference" approximation. For example, the temperature .28 is computed as $2/(7.15 - 0)$.) Make a graph of entropy vs. energy and a graph of heat capacity vs. temperature. Then change the number of oscillators to 5000 (to "dilute" the system and look at lower temperatures), and again make a graph of heat capacity vs. temperature. Discuss your prediction for the heat capacity, and compare it to the data for lead, aluminum, and diamond shown in Figure 1.14. Estimate the numerical value of ϵ, in electron-volts, for each of those real solids.

Problem 3.25. In Problem 2.18 you showed that the multiplicity of an Einstein solid containing N oscillators and q energy units is approximately

$$\Omega(N, q) \approx \left(\frac{q + N}{q}\right)^q \left(\frac{q + N}{N}\right)^N.$$

(a) Starting with this formula, find an expression for the entropy of an Einstein solid as a function of N and q. Explain why the factors omitted from the formula have no effect on the entropy, when N and q are large.

(b) Use the result of part (a) to calculate the temperature of an Einstein solid as a function of its energy. (The energy is $U = q\epsilon$, where ϵ is a constant.) Be sure to simplify your result as much as possible.

(c) Invert the relation you found in part (b) to find the energy as a function of temperature, then differentiate to find a formula for the heat capacity.

(d) Show that, in the limit $T \to \infty$, the heat capacity is $C = Nk$. (Hint: When x is very small, $e^x \approx 1 + x$.) Is this the result you would expect? Explain.

(e) Make a graph (possibly using a computer) of the result of part (c). To avoid awkward numerical factors, plot C/Nk vs. the dimensionless variable $t = kT/\epsilon$, for t in the range from 0 to about 2. Discuss your prediction for the heat capacity at low temperature, comparing to the data for lead, aluminum, and diamond shown in Figure 1.14. Estimate the value of ϵ, in electron-volts, for each of those real solids.

(f) Derive a more accurate approximation for the heat capacity at high temperatures, by keeping terms through x^3 in the expansions of the exponentials and then carefully expanding the denominator and multiplying everything out. Throw away terms that will be smaller than $(\epsilon/kT)^2$ in the final answer. When the smoke clears, you should find $C = Nk[1 - \frac{1}{12}(\epsilon/kT)^2]$.

Problem 3.26. The results of either of the two preceding problems can also be applied to the vibrational motions of gas molecules. Looking only at the vibrational contribution to the heat capacity graph for H_2 shown in Figure 1.13, estimate the value of ϵ for the vibrational motion of an H_2 molecule.

3.4 Mechanical Equilibrium and Pressure

Next I would like to generalize the ideas of this chapter to include systems whose volumes can change as they interact. Just as the spontaneous exchange of energy between systems is governed by their temperatures, so the exchange of volume between systems is governed by their pressures. Hence, there must be a close relation between pressure and entropy, analogous to the relation $1/T = \partial S/\partial U$.

Consider, then, two systems (perhaps gases) separated by a movable partition (see Figure 3.13). The systems are free to exchange both energy and volume, but the total energy and volume are fixed. The total entropy is a function of two variables,

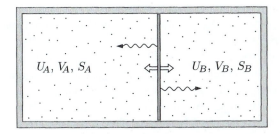

Figure 3.13. Two systems that can exchange both energy and volume with each other. The total energy and total volume are fixed.

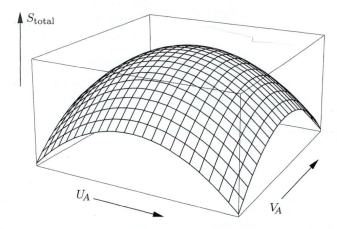

Figure 3.14. A graph of entropy vs. U_A and V_A for the system shown in Figure 3.13. The equilibrium values of U_A and V_A are where the graph reaches its highest point.

U_A and V_A, as shown in Figure 3.14. The equilibrium point is where S_{total} attains its maximum value. At this point, its partial derivatives in both directions vanish:

$$\frac{\partial S_{\text{total}}}{\partial U_A} = 0, \qquad \frac{\partial S_{\text{total}}}{\partial V_A} = 0. \tag{3.36}$$

We studied the first condition already in Section 3.1, where we concluded that this condition is equivalent to saying that the two systems are at the same temperature. Now let us study the second condition in the same way.

The manipulations are exactly analogous to those in Section 3.1:

$$0 = \frac{\partial S_{\text{total}}}{\partial V_A} = \frac{\partial S_A}{\partial V_A} + \frac{\partial S_B}{\partial V_A} = \frac{\partial S_A}{\partial V_A} - \frac{\partial S_B}{\partial V_B}. \tag{3.37}$$

The last step uses the fact that the total volume is fixed, so $dV_A = -dV_B$ (any volume added to A must be subtracted from B). Therefore we can conclude

$$\frac{\partial S_A}{\partial V_A} = \frac{\partial S_B}{\partial V_B} \qquad \text{at equilibrium.} \tag{3.38}$$

The partial derivatives are to be taken with energy (U_A or U_B) held fixed, as well as the number of particles (N_A or N_B). Note, however, that I have assumed that the systems are free to exchange energy, and in fact that they are also in *thermal* equilibrium. (If the partition is allowed to move but does not allow heat to pass through, then the energies of the systems are *not* fixed, and the equilibrium condition is more complicated.)

From experience, though, we know that when two systems are in mechanical equilibrium, their *pressures* must be equal. Therefore pressure must be some function of the derivative $\partial S/\partial V$. To figure out *what* function, let's look at units. Entropy has units of J/K, so $\partial S/\partial V$ has units of (N/m^2)/K, or Pa/K. To get something with units of pressure, we need to multiply by a temperature. But can we? Yes, since we've assumed already that the two systems are in thermal equilibrium, they must be at the same temperature, so the quantity $T(\partial S/\partial V)$ is *also* the same for both systems.

We should also think about whether we want $\partial S/\partial V$ to be large or small when the pressure is large. When $\partial S/\partial V$ is large, the system gains a lot of entropy upon expanding just a little. Since entropy tends to increase, this system really "wants" to expand. Yep, that's exactly what we mean when we say the pressure is large.

I therefore propose the following relation between entropy and pressure:

$$P = T\left(\frac{\partial S}{\partial V}\right)_{U,N}.$$ (3.39)

I won't try to call this the *definition* of pressure, but I hope you agree that this quantity has all the same qualities as pressure, and hence, that it probably is the same thing as force per unit area.

Of course, it's always reassuring to check that the formula works in a case where we already know the answer. So recall the formula for the multiplicity of a monatomic ideal gas,

$$\Omega = f(N)V^N U^{3N/2},$$ (3.40)

where $f(N)$ is a complicated function of N only. Taking the logarithm gives

$$S = Nk\ln V + \tfrac{3}{2}Nk\ln U + k\ln f(N).$$ (3.41)

So according to formula 3.39, the pressure should be

$$P = T\frac{\partial}{\partial V}(Nk\ln V) = \frac{NkT}{V},$$ (3.42)

that is,

$$PV = NkT.$$ (3.43)

Indeed. So if you already believed formula 3.39, then we've just *derived* the ideal gas law. Alternatively, you can think of this calculation as a verification of formula 3.39, and especially of the fact that no additional constant factors are needed in that formula.

The Thermodynamic Identity

There's a nice equation that summarizes both the theoretical definition of temperature and our new formula for pressure. To derive it, let's consider a process in which you change both the energy and the volume of a system by small amounts, ΔU and ΔV. Question: How much does the system's entropy change?

To answer this question, let's mentally divide the process into two steps: In step 1, the energy changes by ΔU but the volume is held fixed. Then, in step 2, the volume changes by ΔV but the energy is held fixed. These two steps are shown graphically in Figure 3.15. The total change in entropy is just the sum of the changes during steps 1 and 2:

$$\Delta S = (\Delta S)_1 + (\Delta S)_2. \tag{3.44}$$

Now multiply and divide the first term by ΔU, and multiply and divide the second term by ΔV:

$$\Delta S = \left(\frac{\Delta S}{\Delta U}\right)_V \Delta U + \left(\frac{\Delta S}{\Delta V}\right)_U \Delta V.$$

The subscripts indicate what quantity is being held fixed, as usual. Now if all of the changes are *small*, the ratios in parentheses become partial derivatives, and the change in entropy can be written

$$
\begin{aligned}
dS &= \left(\frac{\partial S}{\partial U}\right)_V dU + \left(\frac{\partial S}{\partial V}\right)_U dV \\
&= \frac{1}{T} dU + \frac{P}{T} dV,
\end{aligned}
\tag{3.45}
$$

where in the second line I have used the definition of temperature and formula 3.39 for pressure to evaluate the partial derivatives. This result is called the **thermodynamic identity**. It is usually rearranged into the following form:

$$dU = T\,dS - P\,dV. \tag{3.46}$$

This equation is true for any infinitesimal change in any system, provided that T and P are well defined and no other relevant variables are changing. (For instance, I've assumed that the number of particles in the system is fixed.)

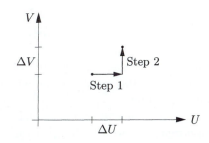

Figure 3.15. To compute the change in entropy when both U and V change, consider the process in two steps: changing U while holding V fixed, then changing V while holding U fixed.

If you memorize only one formula from this chapter, make it the thermodynamic identity, because from it you can recover the formulas for both temperature and pressure as partial derivatives of the entropy. For instance, in a process that takes place at constant volume ($dV = 0$), the thermodynamic identity says $dU = T\,dS$, which can be rearranged to give the definition of temperature (equation 3.5). And for a process in which $dU = 0$, the thermodynamic identity says $T\,dS = P\,dV$, which reproduces equation 3.39 for the pressure.

> **Problem 3.27.** What partial-derivative relation can you derive from the thermodynamic identity by considering a process that takes place at constant entropy? Does the resulting equation agree with what you already knew? Explain.

Entropy and Heat Revisited

The thermodynamic identity looks an awful lot like the first law of thermodynamics,

$$dU = Q + W. \tag{3.47}$$

It is therefore tempting to associate Q with $T\,dS$ and W with $-P\,dV$. However, these associations are not always valid. They *are* valid if any change in volume takes place quasistatically (so the pressure is always uniform throughout the system), if no other forms of work are done, and if no other relevant variables (such as particle numbers) are changing. Then we *know* that $W = -P\,dV$, so equations 3.46 and 3.47 imply

$$Q = T\,dS \qquad \text{(quasistatic)}. \tag{3.48}$$

Thus, under these restricted circumstances, the change in a system's entropy is Q/T, even if work is being done on it during the process. (In the special case of an adiabatic process ($Q = 0$) that is also quasistatic, the entropy is unchanged; such a process is called **isentropic**. In short, adiabatic + quasistatic = isentropic.)

This result (3.48) allows us to go back and repeat much of the discussion in Section 3.2, removing the restriction of constant volume. For example, when a liter of water is boiled at $100°$C and atmospheric pressure, the heat added is 2260 kJ and so the increase in its entropy is

$$\Delta S = \frac{Q}{T} = \frac{2260 \text{ kJ}}{373 \text{ K}} = 6060 \text{ J/K}. \tag{3.49}$$

And for constant-pressure processes in which the temperature changes, we can write $Q = C_P\,dT$, then integrate to obtain

$$(\Delta S)_P = \int_{T_i}^{T_f} \frac{C_P}{T}\,dT. \tag{3.50}$$

Since most tabulated heat capacities are for constant pressure rather than constant volume, this formula is more practical than the analogous equation (3.19) for constant volume.

But even though many familiar processes are approximately quasistatic, it's important to remember that there are exceptions. As an example, suppose you have

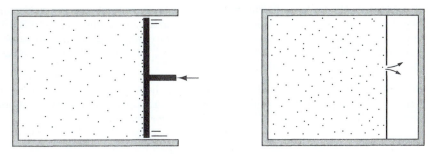

Figure 3.16. Two types of non-quasistatic volume changes: very fast compression that creates internal disequilibrium, and free expansion into a vacuum.

a gas in a cylinder with a piston, and you hit the piston *very* hard, so that it moves inward much faster than the gas molecules themselves are moving (see Figure 3.16). Molecules build up in front of the piston, exerting a very large backward force on it which you must overcome. Let's say that the piston stops after moving only a very small distance, so that after everything settles down, the pressure has increased only infinitesimally. The work you have done on the gas is now *greater* than $-P\,dV$, so any heat that was simultaneously added must be *less* than $T\,dS$. In this example, then,

$$dS > \frac{Q}{T} \qquad \text{(when } W > -P\,dV\text{).} \qquad (3.51)$$

You've created "extra" entropy, because you added extra energy to the gas—more than was needed to accomplish the change in volume.

A related example is the free expansion of a gas into a vacuum, discussed in Section 2.6. Suppose that a membrane partitions a chamber into two parts, one filled with gas and the other containing a vacuum. The membrane is suddenly broken, allowing the gas to expand into the vacuum. Here no work is done on or by the gas, nor does any heat flow into it, so the first law tells us $\Delta U = 0$. Meanwhile, if the increase in the volume of the gas is very small, the thermodynamic identity (3.46) must still apply, so $T\,dS = P\,dV > 0$, that is, there is a positive change in the entropy of the gas. (If it's an ideal gas, you can also see this directly from the Sackur-Tetrode equation for S, as discussed in Section 2.6.)

In both of these examples, there is a mechanical process that creates new entropy, over and above any entropy that might "flow" into the system through heat. It's always possible to create *more* entropy. But the second law says that once we've created it, we can never make it disappear.

Problem 3.28. A liter of air, initially at room temperature and atmospheric pressure, is heated at constant pressure until it doubles in volume. Calculate the increase in its entropy during this process.

Problem 3.29. Sketch a qualitatively accurate graph of the entropy of a substance (perhaps H_2O) as a function of temperature, at fixed pressure. Indicate where the substance is solid, liquid, and gas. Explain each feature of the graph briefly.

Problem 3.30. As shown in Figure 1.14, the heat capacity of diamond near room temperature is approximately linear in T. Extrapolate this function up to 500 K, and estimate the change in entropy of a mole of diamond as its temperature is raised from 298 K to 500 K. Add on the tabulated value at 298 K (from the back of this book) to obtain $S(500 \text{ K})$.

Problem 3.31. Experimental measurements of heat capacities are often represented in reference works as empirical formulas. For graphite, a formula that works well over a fairly wide range of temperatures is (for one mole)

$$C_P = a + bT - \frac{c}{T^2},$$

where $a = 16.86$ J/K, $b = 4.77 \times 10^{-3}$ J/K^2, and $c = 8.54 \times 10^5$ J·K. Suppose, then, that a mole of graphite is heated at constant pressure from 298 K to 500 K. Calculate the increase in its entropy during this process. Add on the tabulated value of $S(298 \text{ K})$ (from the back of this book) to obtain $S(500 \text{ K})$.

Problem 3.32. A cylinder contains one liter of air at room temperature (300 K) and atmospheric pressure (10^5 N/m^2). At one end of the cylinder is a massless piston, whose surface area is 0.01 m^2. Suppose that you push the piston in *very* suddenly, exerting a force of 2000 N. The piston moves only one millimeter, before it is stopped by an immovable barrier of some sort.

(a) How much work have you done on this system?

(b) How much heat has been added to the gas?

(c) Assuming that all the energy added goes into the gas (not the piston or cylinder walls), by how much does the internal energy of the gas increase?

(d) Use the thermodynamic identity to calculate the change in the entropy of the gas (once it has again reached equilibrium).

Problem 3.33. Use the thermodynamic identity to derive the heat capacity formula

$$C_V = T\left(\frac{\partial S}{\partial T}\right)_V,$$

which is occasionally more convenient than the more familiar expression in terms of U. Then derive a similar formula for C_P, by first writing dH in terms of dS and dP.

Problem 3.34. Polymers, like rubber, are made of very long molecules, usually tangled up in a configuration that has lots of entropy. As a very crude model of a rubber band, consider a chain of N links, each of length ℓ (see Figure 3.17). Imagine that each link has only two possible states, pointing either left or right. The total length L of the rubber band is the net displacement from the beginning of the first link to the end of the last link.

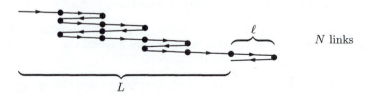

Figure 3.17. A crude model of a rubber band as a chain in which each link can only point left or right.

(a) Find an expression for the entropy of this system in terms of N and N_R, the number of links pointing to the right.

(b) Write down a formula for L in terms of N and N_R.

(c) For a one-dimensional system such as this, the length L is analogous to the volume V of a three-dimensional system. Similarly, the pressure P is replaced by the tension force F. Taking F to be positive when the rubber band is pulling inward, write down and explain the appropriate thermodynamic identity for this system.

(d) Using the thermodynamic identity, you can now express the tension force F in terms of a partial derivative of the entropy. From this expression, compute the tension in terms of L, T, N, and ℓ.

(e) Show that when $L \ll N\ell$, the tension force is directly proportional to L (Hooke's law).

(f) Discuss the dependence of the tension force on temperature. If you increase the temperature of a rubber band, does it tend to expand or contract? Does this behavior make sense?

(g) Suppose that you hold a relaxed rubber band in both hands and suddenly stretch it. Would you expect its temperature to increase or decrease? Explain. Test your prediction with a real rubber band (preferably a fairly heavy one with lots of stretch), using your lips or forehead as a thermometer. (Hint: The entropy you computed in part (a) is not the total entropy of the rubber band. There is additional entropy associated with the vibrational energy of the molecules; this entropy depends on U but is approximately independent of L.)

3.5   Diffusive Equilibrium and Chemical Potential

When two systems are in *thermal* equilibrium, their temperatures are the same. When they're in *mechanical* equilibrium, their pressures are the same. What quantity is the same when they're in *diffusive* equilibrium?

We can find out by applying the same logic as in the previous section. Consider two systems, A and B, that are free to exchange both energy and particles, as shown in Figure 3.18. (The volumes of the systems could also vary, but I'll take these to be fixed for simplicity.) I've drawn a system of two interacting gases, but it could just as well be a gas interacting with a liquid or solid, or even two solids in which atoms gradually migrate around. I'm assuming, though, that both systems are made of the same *species* of particles, for instance, H_2O molecules.

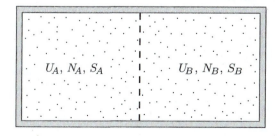

Figure 3.18. Two systems that can exchange both energy and particles.

Assuming that the total energy and total number of particles are fixed, the total entropy of this system is a function of U_A and N_A. At equilibrium, the total entropy is a maximum, so

$$\left(\frac{\partial S_{\text{total}}}{\partial U_A}\right)_{N_A, V_A} = 0 \quad \text{and} \quad \left(\frac{\partial S_{\text{total}}}{\partial N_A}\right)_{U_A, V_A} = 0. \tag{3.52}$$

(If the volumes of the systems are allowed to vary, then $\partial S_{\text{total}}/\partial V_A = 0$ as well.) Again, the first condition says that the two systems must be at the same temperature. The second condition is new, but is entirely analogous to the condition on volume from the previous section. Following the same reasoning as there, we can conclude

$$\frac{\partial S_A}{\partial N_A} = \frac{\partial S_B}{\partial N_B} \qquad \text{at equilibrium}, \tag{3.53}$$

where the partial derivatives are taken at fixed energy and volume. We're free to multiply this equation through by a factor of T, the temperature, since the systems are also in thermal equilibrium. By convention, we also multiply by -1:

$$-T\frac{\partial S_A}{\partial N_A} = -T\frac{\partial S_B}{\partial N_B} \qquad \text{at equilibrium}. \tag{3.54}$$

The quantity $-T(\partial S/\partial N)$ is much less familiar to most of us than temperature or pressure, but it's still extremely important. It is called the **chemical potential**, denoted μ:

$$\mu \equiv -T\left(\frac{\partial S}{\partial N}\right)_{U,V}. \tag{3.55}$$

This is the quantity that's the same for both systems when they're in diffusive equilibrium:

$$\mu_A = \mu_B \qquad \text{at equilibrium}. \tag{3.56}$$

If the two systems are *not* in equilibrium, then the one with the larger value of $\partial S/\partial N$ will tend to gain particles, since it will thereby gain more entropy than the other loses. However, because of the minus sign in definition 3.55, this system has the *smaller* value of μ. Conclusion: Particles tend to flow from the system with higher μ into the system with lower μ (see Figure 3.19).

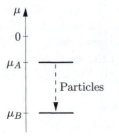

Figure 3.19. Particles tend to flow toward lower values of the chemical potential, even if both values are negative.

It's not hard to generalize the thermodynamic identity to include processes in which N changes. If we imagine changing U by dU, V by dV, and N by dN, then, by the same logic as in the previous section, the total change in the entropy is

$$dS = \left(\frac{\partial S}{\partial U}\right)_{N,V} dU + \left(\frac{\partial S}{\partial V}\right)_{N,U} dV + \left(\frac{\partial S}{\partial N}\right)_{U,V} dN$$
$$= \frac{1}{T} dU + \frac{P}{T} dV - \frac{\mu}{T} dN. \tag{3.57}$$

Solving for dU as before, we obtain

$$dU = T\,dS - P\,dV + \mu\,dN. \tag{3.58}$$

Just as the $-P\,dV$ term is usually associated with mechanical work, the $\mu\,dN$ term is sometimes referred to as "chemical work."

This generalized thermodynamic identity is a great way to remember the various partial-derivative formulas for T, P, and μ, and to generate other similar formulas. Notice that four quantities are changing in this equation: U, S, V, and N. Now just imagine a process in which any two of these are fixed. For instance, in a process with fixed U and V,

$$0 = T\,dS + \mu\,dN, \qquad \text{that is,} \qquad \mu = -T\left(\frac{\partial S}{\partial N}\right)_{U,V}. \tag{3.59}$$

Similarly, in a process with fixed S and V,

$$dU = \mu\,dN, \qquad \text{that is,} \qquad \mu = \left(\frac{\partial U}{\partial N}\right)_{S,V}. \tag{3.60}$$

This last result is another useful formula for the chemical potential. It tells us directly that μ has units of energy; specifically, μ is the amount by which a system's energy changes, when you add one particle and keep the entropy and volume fixed. Normally, to hold the entropy (or multiplicity) fixed, you must *remove* some energy as you add a particle, so μ is negative. However, if you have to give the particle some potential energy (gravitational, if the system lives on a mountain top, or chemical, if the system is a solid crystal) to get it into the system, this energy also contributes to μ. In Chapter 7 we'll see an example where you have to give a particle *kinetic* energy just to get it into a system.

Now let's look at some examples. First consider a very small Einstein solid, with three oscillators and three units of energy. The multiplicity is 10, so the entropy is $k \ln 10$. Now suppose we add one more oscillator (thinking of each oscillator as a "particle"). If we leave all three units of energy in the system, the multiplicity increases to 20 and the entropy increases to $k \ln 20$. To hold the entropy fixed, we need to remove one unit of energy, as shown in Figure 3.20. Thus the chemical potential of this system is

$$\mu = \left(\frac{\Delta U}{\Delta N}\right)_S = \frac{-\epsilon}{1} = -\epsilon, \tag{3.61}$$

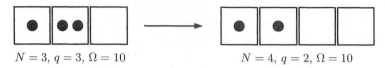

$$N = 3,\ q = 3,\ \Omega = 10 \qquad\qquad N = 4,\ q = 2,\ \Omega = 10$$

Figure 3.20. In order to add an oscillator (represented by a box) to this very small Einstein solid while holding the entropy (or multiplicity) fixed, we must remove one unit of energy (represented by a dot).

if ϵ is the size of a unit of energy. (Because the addition of one particle is not an infinitesimal change for such a small system, this example should be taken with a grain of salt. Strictly speaking, the derivative $\partial U/\partial N$ is not well defined. Besides, in a real solid crystal, adding an atom would entail adding three oscillators, not just one, and we would also have to add some negative potential energy to create the chemical bonds around the added atom.)

As a more realistic example, let's compute μ for a monatomic ideal gas. Here we need the full Sackur-Tetrode equation (2.49) for the entropy,

$$S = Nk\left[\ln\left(V\left(\frac{4\pi mU}{3h^2}\right)^{3/2}\right) - \ln N^{5/2} + \frac{5}{2}\right]. \tag{3.62}$$

Differentiating with respect to N gives

$$
\begin{aligned}
\mu &= -T\left\{k\left[\ln\left(V\left(\frac{4\pi mU}{3h^2}\right)^{3/2}\right) - \ln N^{5/2} + \frac{\cdot 5}{2}\right] - Nk\cdot\frac{5}{2}\frac{1}{N}\right\}\\
&= -kT\ln\left[\frac{V}{N}\left(\frac{4\pi mU}{3Nh^2}\right)^{3/2}\right]\\
&= -kT\ln\left[\frac{V}{N}\left(\frac{2\pi mkT}{h^2}\right)^{3/2}\right].
\end{aligned}
\tag{3.63}
$$

(In the last line I used the relation $U = \frac{3}{2}NkT$.) At room temperature and atmospheric pressure, the volume per molecule, V/N, is 4.2×10^{-26} m^3, while the quantity $(h^2/2\pi mkT)^{3/2}$ is much smaller. For helium, this quantity is 1.3×10^{-31} m^3, so the argument of the logarithm is 3.3×10^5, the logarithm itself is 12.7, and the chemical potential is

$$\mu = -0.32\ \text{eV} \qquad \text{for helium at 300 K, } 10^5 \text{ N/m}^2. \tag{3.64}$$

If the concentration is increased while holding the temperature fixed, μ becomes less negative, indicating that the gas becomes more willing to give up particles to other nearby systems. More generally, increasing the density of particles in a system always increases its chemical potential.

Throughout this section, I've implicitly assumed that each system contains only one type of particle. If a system contains several types of particles (such as air, a mixture of nitrogen and oxygen molecules), then each species has its own chemical potential:

$$\mu_1 \equiv -T\left(\frac{\partial S}{\partial N_1}\right)_{U,V,N_2}, \qquad \mu_2 \equiv -T\left(\frac{\partial S}{\partial N_2}\right)_{U,V,N_1}, \tag{3.65}$$

and so on for each species 1, 2, ... The generalized thermodynamic identity is then

$$dU = T\,dS - P\,dV + \sum_i \mu_i\,dN_i, \tag{3.66}$$

where the sum runs over all species, $i = 1,\ 2,\ \dots$. If two systems are in diffusive equilibrium, the chemical potentials must be separately equal for each species: $\mu_{1A} = \mu_{1B}$, $\mu_{2A} = \mu_{2B}$, and so on, where A and B are the two systems.

The chemical potential is a central concept in the study of equilibrium in chemical reactions and phase transformations. It also plays a central role in "quantum statistics," the study of exotic, dense gases and other related systems. We'll make use of it many times in Chapters 5 and 7.

One more comment: I should mention that chemists usually define the chemical potential in terms of *moles*, not individual particles:

$$\mu_{\text{chemistry}} \equiv -T\left(\frac{\partial S}{\partial n}\right)_{U,V}, \tag{3.67}$$

where $n = N/N_{\text{A}}$ is the number of moles of whatever type of particle is being considered. This means that their chemical potentials are always larger than ours by a factor of Avogadro's number, N_{A}. To translate this section into chemistry conventions, just change every N to an n, except in the examples in equations 3.61 through 3.64, where every formula for μ should be multiplied by N_{A}.

Problem 3.35. In the text I showed that for an Einstein solid with three oscillators and three units of energy, the chemical potential is $\mu = -\epsilon$ (where ϵ is the size of an energy unit and we treat each oscillator as a "particle"). Suppose instead that the solid has three oscillators and *four* units of energy. How does the chemical potential then compare to $-\epsilon$? (Don't try to get an actual *value* for the chemical potential; just explain whether it is more or less than $-\epsilon$.)

Problem 3.36. Consider an Einstein solid for which both N and q are much greater than 1. Think of each oscillator as a separate "particle."

(a) Show that the chemical potential is

$$\mu = -kT\ln\left(\frac{N+q}{N}\right).$$

(b) Discuss this result in the limits $N \gg q$ and $N \ll q$, concentrating on the question of how much S increases when another particle carrying no energy is added to the system. Does the formula make intuitive sense?

Problem 3.37. Consider a monatomic ideal gas that lives at a height z above sea level, so each molecule has potential energy mgz in addition to its kinetic energy.

(a) Show that the chemical potential is the same as if the gas were at sea level, plus an additional term mgz:

$$\mu(z) = -kT\ln\left[\frac{V}{N}\left(\frac{2\pi mkT}{h^2}\right)^{3/2}\right] + mgz.$$

(You can derive this result from either the definition $\mu = -T(\partial S/\partial N)_{U,V}$ or the formula $\mu = (\partial U/\partial N)_{S,V}$.)

(b) Suppose you have two chunks of helium gas, one at sea level and one at height z, each having the same temperature and volume. Assuming that they are in diffusive equilibrium, show that the number of molecules in the higher chunk is

$$N(z) = N(0)e^{-mgz/kT},$$

in agreement with the result of Problem 1.16.

Problem 3.38. Suppose you have a *mixture* of gases (such as air, a mixture of nitrogen and oxygen). The **mole fraction** x_i of any species i is defined as the fraction of all the molecules that belong to that species: $x_i = N_i/N_{\text{total}}$. The **partial pressure** P_i of species i is then defined as the corresponding fraction of the total pressure: $P_i = x_i P$. Assuming that the mixture of gases is ideal, argue that the chemical potential μ_i of species i in this system is the same as if the other gases were not present, at a fixed partial pressure P_i.

3.6 Summary and a Look Ahead

This chapter completes our treatment of the basic principles of thermal physics. The most central principle is the second law: Entropy tends to increase. Because this law governs the tendency of systems to exchange energy, volume, and particles, the derivatives of the entropy with respect to these three variables are of great interest and are relatively easy to measure. Table 3.3 summarizes the three types of interactions and the associated derivatives of the entropy. The three partial-derivative formulas are conveniently summarized in the thermodynamic identity,

$$dU = T\,dS - P\,dV + \mu\,dN. \tag{3.68}$$

These concepts and principles form the foundation of what is called **classical thermodynamics**: the study of systems comprised of large numbers of particles, based on general laws that do not depend on the detailed microscopic behavior of those particles. The formulas that appear here apply to *any* large system whose macrostate is determined by the variables U, V, and N, and these formulas can be generalized with little difficulty to other large systems.

Type of interaction	Exchanged quantity	Governing variable	Formula
thermal	energy	temperature	$\dfrac{1}{T} = \left(\dfrac{\partial S}{\partial U}\right)_{V,N}$
mechanical	volume	pressure	$\dfrac{P}{T} = \left(\dfrac{\partial S}{\partial V}\right)_{U,N}$
diffusive	particles	chemical potential	$\dfrac{\mu}{T} = -\left(\dfrac{\partial S}{\partial N}\right)_{U,V}$

Table 3.3. Summary of the three types of interactions considered in this chapter, and the associated variables and partial-derivative relations.

In addition to these very general concepts, we have also worked with three *specific* model systems: the two-state paramagnet, the Einstein solid, and the monatomic ideal gas. For each of these systems we used the laws of microscopic physics to find explicit formulas for the multiplicity and entropy, and hence computed heat capacities and a variety of other measurable quantities. The business of using microscopic models to derive these kinds of predictions is called **statistical mechanics**.

The remainder of this book explores further applications of thermal physics. Chapters 4 and 5 apply the general laws of classical thermodynamics to a variety of systems of practical interest in engineering, chemistry, and related disciplines. Chapters 6, 7, and 8 then return to statistical mechanics, introducing more sophisticated microscopic models and the mathematical tools needed to derive predictions from them.

Problem 3.39. In Problem 2.32 you computed the entropy of an ideal monatomic gas that lives in a two-dimensional universe. Take partial derivatives with respect to U, A, and N to determine the temperature, pressure, and chemical potential of this gas. (In two dimensions, pressure is defined as force per unit *length*.) Simplify your results as much as possible, and explain whether they make sense.

A good many times I have been present at gatherings of people who, by the standards of the traditional culture, are thought highly educated and who have with considerable gusto been expressing their incredulity at the illiteracy of scientists. Once or twice I have been provoked and have asked the company how many of them could describe the Second Law of Thermodynamics. The response was cold: it was also negative. Yet I was asking something which is about the scientific equivalent of: Have you read a work of Shakespeare's?

—C. P. Snow, *The Two Cultures* (Cambridge University Press, Cambridge, 1959). Reprinted with the permission of Cambridge University Press.

4 Engines and Refrigerators

4.1 Heat Engines

A **heat engine** is any device that absorbs heat and converts part of that energy into work. An important example is the steam turbine, used to generate electricity in most of today's power plants. The familiar internal combustion engine used in automobiles does not actually absorb heat, but we can pretend that the thermal energy comes from outside rather than inside and treat it, also, as a heat engine.

Unfortunately, only *part* of the energy absorbed as heat can be converted to work by a heat engine. The reason is that the heat, as it flows in, brings along entropy, which must somehow be disposed of before the cycle can start over. To get rid of the entropy, every heat engine must dump some waste heat into its environment. The work produced by the engine is the difference between the heat absorbed and the waste heat expelled.

My goal in this section is to make these ideas precise, and to determine exactly how much of the heat absorbed by an engine can be converted into work. Amazingly, we can say a great deal without knowing *anything* about how the engine actually works.

Figure 4.1 shows the flow of energy into and out of a heat engine. The heat absorbed by the engine comes from a place called the **hot reservoir**, while the waste heat is dumped into the **cold reservoir**. The temperatures of these reservoirs, T_h and T_c, are assumed fixed. (In general, a **reservoir** in thermodynamics is anything that's so large that its temperature doesn't change noticeably when heat enters or leaves. For a steam engine, the hot reservoir is the place where the fuel is burned and the cold reservoir is the surrounding environment.) I'll use the symbol Q_h for the heat absorbed from the hot reservoir in some given time period, and Q_c for the heat expelled to the cold reservoir. The net work done by the engine during this time will be W. All three of these symbols will represent positive quantities; in this chapter I'm departing from my earlier sign conventions for heat and work.

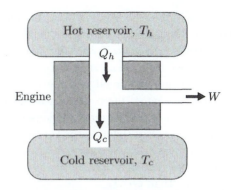

Figure 4.1. Energy-flow diagram for a heat engine. Energy enters as heat from the hot reservoir, and leaves both as work and as waste heat expelled to the cold reservoir.

The benefit of operating a heat engine is the work produced, W. The cost of operation is the heat absorbed, Q_h. Let me therefore define the **efficiency** of an engine, e, as the benefit/cost ratio:

$$e \equiv \frac{\text{benefit}}{\text{cost}} = \frac{W}{Q_h}. \qquad (4.1)$$

The question I would like to ask is this: For given values of T_h and T_c, what is the maximum possible efficiency? To answer this question, all we need are the first and second laws of thermodynamics, plus the assumption that the engine operates in cycles, returning to its original state at the end of each cycle of operation.

The first law of thermodynamics tells us that energy is conserved. Since the state of the engine must be unchanged at the end of a cycle, the energy it absorbs must be precisely equal to the energy it expels. In our notation,

$$Q_h = Q_c + W. \qquad (4.2)$$

If we use this equation to eliminate W in equation 4.1, we have for the efficiency

$$e = \frac{Q_h - Q_c}{Q_h} = 1 - \frac{Q_c}{Q_h}. \qquad (4.3)$$

Thus the efficiency cannot be greater than 1, and can equal 1 only if $Q_c = 0$.

To proceed further we must also invoke the second law, which tells us that the total entropy of the engine plus its surroundings can increase but not decrease. Since the state of the engine must be unchanged at the end of a cycle, the entropy it expels must be at least as much as the entropy it absorbs. (In this context, as in Section 3.2, I like to imagine entropy as a fluid that can be created but never destroyed.) Now the entropy extracted from the hot reservoir is just Q_h/T_h, while the entropy expelled to the cold reservoir is Q_c/T_c. So the second law tells us

$$\frac{Q_c}{T_c} \geq \frac{Q_h}{T_h}, \qquad \text{or} \qquad \frac{Q_c}{Q_h} \geq \frac{T_c}{T_h}. \qquad (4.4)$$

Plugging this result into equation 4.3, we conclude

$$e \leq 1 - \frac{T_c}{T_h}. \qquad (4.5)$$

This is our desired result. So, for instance, if $T_h = 500$ K and $T_c = 300$ K, the maximum possible efficiency is 40%. In general, for the greatest maximum efficiency, you should make the cold reservoir *very* cold, or the hot reservoir *very* hot, or both. The smaller the ratio T_c/T_h, the more efficient your engine can be.

It's easy to make an engine that's *less* efficient than the limit $1 - T_c/T_h$, simply by producing additional entropy during the operation. Then to dispose of this entropy you must dump extra heat into the cold reservoir, leaving less energy to convert to work. The most obvious way of producing new entropy is in the heat transfer processes themselves. For instance, when heat Q_h leaves the hot reservoir, the entropy lost by that reservoir is Q_h/T_h; but if the engine temperature at this time is *less* than T_h, then as the heat enters the engine its associated entropy will be *greater* than Q_h/T_h.

In deriving the limit (4.5) on the efficiency of an engine, we used both the first and second laws of thermodynamics. The first law told us that the efficiency can't be any greater than 1, that is, we can't get more work out than the amount of heat put in. In this context, the first law is often paraphrased, "You can't win." The second law, however, made matters worse. It told us that we can't even achieve $e = 1$ unless $T_c = 0$ or $T_h = \infty$, both of which are impossible in practice. In this context, the second law is often paraphrased, "You can't even break even."

Problem 4.1. Recall Problem 1.34, which concerned an ideal diatomic gas taken around a rectangular cycle on a PV diagram. Suppose now that this system is used as a heat engine, to convert the heat added into mechanical work.

 (a) Evaluate the efficiency of this engine for the case $V_2 = 3V_1$, $P_2 = 2P_1$.

 (b) Calculate the efficiency of an "ideal" engine operating between the same temperature extremes.

Problem 4.2. At a power plant that produces 1 GW (10^9 watts) of electricity, the steam turbines take in steam at a temperature of $500°C$, and the waste heat is expelled into the environment at $20°C$.

 (a) What is the maximum possible efficiency of this plant?

 (b) Suppose you develop a new material for making pipes and turbines, which allows the maximum steam temperature to be raised to $600°C$. Roughly how much money can you make in a year by installing your improved hardware, if you sell the additional electricity for 5 cents per kilowatt-hour? (Assume that the amount of fuel consumed at the plant is unchanged.)

Problem 4.3. A power plant produces 1 GW of electricity, at an efficiency of 40% (typical of today's coal-fired plants).

 (a) At what rate does this plant expel waste heat into its environment?

 (b) Assume first that the cold reservoir for this plant is a river whose flow rate is 100 m^3/s. By how much will the temperature of the river increase?

 (c) To avoid this "thermal pollution" of the river, the plant could instead be cooled by evaporation of river water. (This is more expensive, but in some areas it is environmentally preferable.) At what rate must the water evaporate? What fraction of the river must be evaporated?

Problem 4.4. It has been proposed to use the thermal gradient of the ocean to drive a heat engine. Suppose that at a certain location the water temperature is $22°C$ at the ocean surface and $4°C$ at the ocean floor.

(a) What is the maximum possible efficiency of an engine operating between these two temperatures?

(b) If the engine is to produce 1 GW of electrical power, what minimum volume of water must be processed (to suck out the heat) in every second?

The Carnot Cycle

Let me now explain how to make an engine that *does* achieve the maximum possible efficiency for a given T_h and T_c.

Every engine has a so-called "working substance," which is the material that actually absorbs heat, expels waste heat, and does work. In many heat engines the working substance is a gas. Imagine, then, that we first want the gas to absorb some heat Q_h from the hot reservoir. In the process, the entropy of the reservoir decreases by Q_h/T_h, while the entropy of the gas increases by Q_h/T_{gas}. To avoid making any *new* entropy, we would need to make $T_{gas} = T_h$. This isn't quite possible, because heat won't flow between objects at the same temperature. So let's make T_{gas} just slightly less than T_h, and keep the gas at this temperature (by letting it expand) as it absorbs the heat. This step of the cycle, then, requires that the gas expand isothermally.

Similarly, during the portion of the cycle when the gas is dumping the waste heat into the cold reservoir, we want its temperature to be only infinitesimally greater than T_c, to avoid creating any new entropy. And as the heat leaves the gas, we need to compress it isothermally to keep it at this temperature.

So we have an isothermal expansion at a temperature just less than T_h, and an isothermal compression at a temperature just greater than T_c. The only remaining question is how we *get* the gas from one temperature to the other and back. We don't want any heat to flow in or out when the gas is at intermediate temperatures, so these intermediate steps must be adiabatic. The entire cycle consists of four steps, illustrated in Figures 4.2 and 4.3: isothermal expansion at T_h, adiabatic expansion from T_h to T_c, isothermal compression at T_c, and adiabatic compression from T_c back up to T_h. The theoretical importance of this cycle was first pointed out by Sadi Carnot in 1824, so the cycle is now known as the **Carnot cycle**.

It is possible to prove directly, from the formulas for isothermal and adiabatic processes in Section 1.5, that an ideal gas taken around a Carnot cycle realizes the maximum possible efficiency $1 - T_c/T_h$. But while the proof makes an interesting exercise (see Problem 4.5), it is not really necessary once one understands entropy and the second law. As long as we know that no new entropy was created during the cycle, the strict equality must hold in equation 4.4, and therefore the efficiency must be the maximum allowed by equation 4.5. This conclusion holds even if the gas isn't ideal, and, for that matter, even if the working substance isn't a gas at all.

Although a Carnot cycle is very *efficient*, it's also horribly *impractical*. The heat flows so slowly during the isothermal steps that it takes forever to get a significant amount of work out of the engine. So don't bother installing a Carnot engine in your car; while it would increase your gas mileage, you'd be passed on the highway by pedestrians.

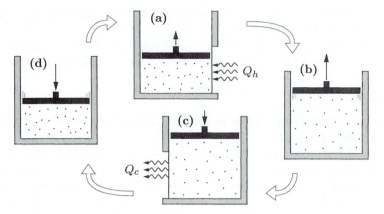

Figure 4.2. The four steps of a Carnot cycle: (a) isothermal expansion at T_h while absorbing heat; (b) adiabatic expansion to T_c; (c) isothermal compression at T_c while expelling heat; and (d) adiabatic compression back to T_h. The system must be put in thermal contact with the hot reservoir during step (a) and with the cold reservoir during step (c).

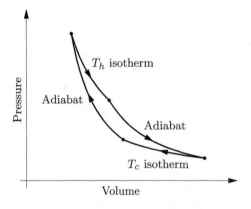

Figure 4.3. *PV* diagram for an ideal monatomic gas undergoing a Carnot cycle.

Problem 4.5. Prove directly (by calculating the heat taken in and the heat expelled) that a Carnot engine using an ideal gas as the working substance has an efficiency of $1 - T_c/T_h$.

Problem 4.6. To get more than an infinitesimal amount of work out of a Carnot engine, we would have to keep the temperature of its working substance below that of the hot reservoir and above that of the cold reservoir by non-infinitesimal amounts. Consider, then, a Carnot cycle in which the working substance is at temperature T_{hw} as it absorbs heat from the hot reservoir, and at temperature T_{cw} as it expels heat to the cold reservoir. Under most circumstances the rates of heat transfer will be directly proportional to the temperature differences:

$$\frac{Q_h}{\Delta t} = K(T_h - T_{hw}) \qquad \text{and} \qquad \frac{Q_c}{\Delta t} = K(T_{cw} - T_c).$$

I've assumed here for simplicity that the constants of proportionality (K) are the same for both of these processes. Let us also assume that both processes take the same amount of time, so the Δt's are the same in both of these equations.[*]

(a) Assuming that no new entropy is created during the cycle except during the two heat transfer processes, derive an equation that relates the four temperatures T_h, T_c, T_{hw}, and T_{cw}.

(b) Assuming that the time required for the two adiabatic steps is negligible, write down an expression for the power (work per unit time) output of this engine. Use the first and second laws to write the power entirely in terms of the four temperatures (and the constant K), then eliminate T_{cw} using the result of part (a).

(c) When the cost of building an engine is much greater than the cost of fuel (as is often the case), it is desirable to optimize the engine for maximum power output, not maximum efficiency. Show that, for fixed T_h and T_c, the expression you found in part (b) has a maximum value at $T_{hw} = \frac{1}{2}(T_h + \sqrt{T_h T_c})$. (Hint: You'll have to solve a quadratic equation.) Find the corresponding expression for T_{cw}.

(d) Show that the efficiency of this engine is $1 - \sqrt{T_c/T_h}$. Evaluate this efficiency numerically for a typical coal-fired steam turbine with $T_h = 600°C$ and $T_c = 25°C$, and compare to the ideal Carnot efficiency for this temperature range. Which value is closer to the actual efficiency, about 40%, of a real coal-burning power plant?

4.2   Refrigerators

A **refrigerator** is a heat engine operated in reverse, more or less. In practice, it may work in a completely different way, but if you only care about what it *does*, not how it works, you can just reverse the arrows in Figure 4.1 to obtain a generalized diagram of a refrigerator, shown in Figure 4.4. Again I'm defining all symbols to stand for positive quantities. The heat sucked out of the cold reservoir (the inside of the fridge) is Q_c, while the electrical energy supplied from the wall outlet is W.

Figure 4.4. Energy-flow diagram for a refrigerator or air conditioner. For a kitchen refrigerator, the space inside it is the cold reservoir and the space outside it is the hot reservoir. An electrically powered compressor supplies the work.

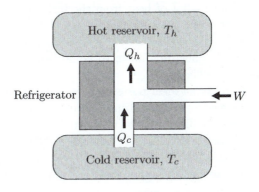

[*]Neither of these assumptions is necessary in order to obtain the final result for the efficiency in part (d). See the article on which this problem is based: F. L. Curzon and B. Ahlborn, "Efficiency of a Carnot engine at maximum power output," *American Journal of Physics* **41**, 22–24 (1975).

There's also some waste heat, Q_h, dumped into your kitchen. By the way, the same diagram could apply to an air conditioner; then the cold reservoir is the inside of your house while the hot reservoir is outside.*

How should we define the "efficiency" of a refrigerator? Again the relevant number is the benefit/cost ratio, but this time the benefit is Q_c while the cost is W. To avoid confusion with equation 4.1, this ratio is called the **coefficient of performance**:

$$\text{COP} = \frac{\text{benefit}}{\text{cost}} = \frac{Q_c}{W}. \tag{4.6}$$

Just as for a heat engine, we can now use the first and second laws to derive a limit on the COP in terms of the temperatures T_h and T_c. The first law tells us $Q_h = Q_c + W$, so

$$\text{COP} = \frac{Q_c}{Q_h - Q_c} = \frac{1}{Q_h/Q_c - 1}. \tag{4.7}$$

Notice that there's no obvious upper limit on this quantity yet; in particular, the first law allows the COP to be greater than 1.

Meanwhile, the second law says that the entropy dumped into the hot reservoir must be at least as much as the entropy absorbed from the cold reservoir:

$$\frac{Q_h}{T_h} \geq \frac{Q_c}{T_c}, \quad \text{or} \quad \frac{Q_h}{Q_c} \geq \frac{T_h}{T_c}. \tag{4.8}$$

(This relation is the reverse of relation 4.4 because the entropy is flowing in the opposite direction.) Plugging this inequality into equation 4.7 gives

$$\text{COP} \leq \frac{1}{T_h/T_c - 1} = \frac{T_c}{T_h - T_c}. \tag{4.9}$$

For a typical kitchen refrigerator (with freezer), T_h might be 298 K while T_c might be 255 K. In this case the coefficient of performance can be as high as 5.9. In other words, for each joule of electrical energy drawn from the wall, the coolant can suck as much as 5.9 J of heat from the inside of the refrigerator/freezer. In this ideal case, the waste heat dumped into the kitchen would be 6.9 J. As you can see from the formula, the COP is largest when T_h and T_c aren't very different. A refrigerator that cools something down to liquid helium temperature (4 K) would have to be *much* less efficient.

To make an ideal refrigerator with the maximum possible COP, one can again use a Carnot cycle, this time operated in reverse. In order to make the heat flow in the opposite direction, the working substance must be slightly *hotter* than T_h while heat is being expelled, and slightly *colder* than T_c while heat is being absorbed. Once again, this is a lousy way to do things in practice, because the heat transfer is much too slow. A more practical refrigerator is described in Section 4.4.

---

*An air conditioner usually also has a fan, which blows air around inside your house to speed up the heat flow on that side. Don't confuse the air (which never leaves the cold reservoir) with the heat (which would flow outward, though more slowly, even without the fan).

Historically, heat engines and refrigerators played a crucial role in the formulation of the second law and the identification of entropy as a quantity of interest. Early versions of the second law, derived from experience, included the statements that all heat engines must produce some waste heat, and that all refrigerators require some work input. Carnot and others invented ingenious arguments to show that these laws could be violated if you could make an engine or a refrigerator whose efficiency exceeded that of a Carnot cycle (see Problems 4.16 and 4.17). Carnot also recognized that for an ideal engine there must be a quantity, associated with heat, that flows in from the hot reservoir and out to the cold reservoir in equal amounts. But Carnot's 1824 memoir did not distinguish carefully enough between this quantity and what we now call simply "heat." At that time the relation between heat and other forms of energy was still controversial, and the simple formula Q/T eluded scientists who had not yet adopted a temperature scale measured from absolute zero. It wasn't until 1865, after these other issues were fully resolved, that Rudolf Clausius brought Carnot's quantity to the full attention of the scientific community and put it on a firm mathematical basis. He coined the term "entropy" for this quantity, after a Greek word meaning "transformation" (and because the word resembles "energy"). Clausius did not explain what entropy actually *is*, however. Ludwig Boltzmann took up that question during the following years, and had it figured out by 1877.

Problem 4.7. Why must you put an air conditioner in the window of a building, rather than in the middle of a room?

Problem 4.8. Can you cool off your kitchen by leaving the refrigerator door open? Explain.

Problem 4.9. Estimate the maximum possible COP of a household air conditioner. Use any reasonable values for the reservoir temperatures.

Problem 4.10. Suppose that heat leaks into your kitchen refrigerator at an average rate of 300 watts. Assuming ideal operation, how much power must it draw from the wall?

Problem 4.11. What is the maximum possible COP for a cyclic refrigerator operating between a high-temperature reservoir at 1 K and a low-temperature reservoir at 0.01 K?

Problem 4.12. Explain why an ideal gas taken around a rectangular PV cycle, as considered in Problems 1.34 and 4.1, cannot be used (in reverse) for refrigeration.

Problem 4.13. Under many conditions, the rate at which heat enters an air conditioned building on a hot summer day is proportional to the difference in temperature between inside and outside, $T_h - T_c$. (If the heat enters entirely by conduction, this statement will certainly be true. Radiation from direct sunlight would be an exception.) Show that, under these conditions, the cost of air conditioning should be roughly proportional to the *square* of the temperature difference. Discuss the implications, giving a numerical example.

Problem 4.14. A **heat pump** is an electrical device that heats a building by pumping heat in from the cold outside. In other words, it's the same as a refrigerator, but its purpose is to warm the hot reservoir rather than to cool the cold reservoir (even though it does both). Let us define the following standard symbols, all taken to be positive by convention:

$$T_h = \text{temperature inside building}$$
$$T_c = \text{temperature outside}$$
$$Q_h = \text{heat pumped into building in 1 day}$$
$$Q_c = \text{heat taken from outdoors in 1 day}$$
$$W = \text{electrical energy used by heat pump in 1 day}$$

(a) Explain why the "coefficient of performance" (COP) for a heat pump should be defined as Q_h/W.

(b) What relation among Q_h, Q_c, and W is implied by energy conservation alone? Will energy conservation permit the COP to be greater than 1?

(c) Use the second law of thermodynamics to derive an upper limit on the COP, in terms of the temperatures T_h and T_c alone.

(d) Explain why a heat pump is better than an electric furnace, which simply converts electrical work directly into heat. (Include some numerical estimates.)

Problem 4.15. In an **absorption refrigerator**, the energy driving the process is supplied not as work, but as heat from a gas flame. (Such refrigerators commonly use propane as fuel, and are used in locations where electricity is unavailable.*) Let us define the following symbols, all taken to be positive by definition:

$$Q_f = \text{heat input from flame}$$
$$Q_c = \text{heat extracted from inside refrigerator}$$
$$Q_r = \text{waste heat expelled to room}$$
$$T_f = \text{temperature of flame}$$
$$T_c = \text{temperature inside refrigerator}$$
$$T_r = \text{room temperature}$$

(a) Explain why the "coefficient of performance" (COP) for an absorption refrigerator should be defined as Q_c/Q_f.

(b) What relation among Q_f, Q_c, and Q_r is implied by energy conservation alone? Will energy conservation permit the COP to be greater than 1?

(c) Use the second law of thermodynamics to derive an upper limit on the COP, in terms of the temperatures T_f, T_c, and T_r alone.

Problem 4.16. Prove that *if* you had a heat engine whose efficiency was better than the ideal value (4.5), you could hook it up to an ordinary Carnot refrigerator to make a refrigerator that requires no work input.

---

*For an explanation of how an absorption refrigerator actually works, see an engineering thermodynamics textbook such as Moran and Shapiro (1995).

Problem 4.17. Prove that *if* you had a refrigerator whose COP was better than the ideal value (4.9), you could hook it up to an ordinary Carnot engine to make an engine that produces no waste heat.

4.3 Real Heat Engines

The previous sections treated heat engines and refrigerators in an idealized way, arriving at theoretical limits on their performance. These theoretical limits are extremely useful, because they tell us generally how the efficiency of an engine or refrigerator tends to depend on its operating temperatures. The limits also serve as benchmarks for judging the efficiency of any real engine or refrigerator. For instance, if you have an engine that operates between $T_c = 300$ K and $T_h = 600$ K, and its efficiency is 45%, you know there's not much point in trying to improve the design further since the highest possible efficiency is only 50%.

You may be wondering, however, how real engines and refrigerators are actually built. This is a vast subject, but in this section and the next I'll describe a few examples of real engines and refrigerators, to alleviate the abstraction of the preceding sections.

Internal Combustion Engines

Let's start with the familiar gasoline engine found in most automobiles. The working substance is a gas, initially a mixture of air and vaporized gasoline. This mixture is first injected into a cylinder and compressed, adiabatically, by a piston. A spark plug then ignites the mixture, raising its temperature and pressure while the volume doesn't change. Next the high-pressure gas pushes the piston outward, expanding adiabatically and producing mechanical work. Finally, the hot exhaust gases are expelled and replaced by a new mixture at lower temperature and pressure. The whole cycle is shown in Figure 4.5, where I've represented the exhaust/replacement step as if it were a simple lowering of pressure due to the extraction of heat. What actually happens is the piston pushes the old mixture out through a valve and pulls a new mixture in through another valve, expelling heat but doing no net work. This cycle is called the **Otto cycle**, after the German inventor Nikolaus August Otto.

Figure 4.5. The idealized Otto cycle, an approximation of what happens in a gasoline engine. In real engines the compression ratio V_1/V_2 is larger than shown here, typically 8 or 10.

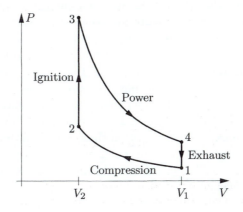

Notice that there is no "hot reservoir" connected to this engine. Instead, thermal energy is produced internally by burning the fuel. The result of this burning, however, is a gas at high temperature and pressure, exactly as if it had absorbed heat from an external source.

The efficiency of a gasoline engine is the net work produced during the cycle divided by the "heat" absorbed during the ignition step. Assuming that the gas is ideal, it's not particularly hard to express these quantities in terms of the various temperatures and volumes (see Problem 4.18). The result is fairly simple:

$$e = 1 - \left(\frac{V_2}{V_1}\right)^{\gamma-1},$$ (4.10)

where V_1/V_2 is the **compression ratio** and γ is the adiabatic exponent introduced in Section 1.5. For air, $\gamma = 7/5$, while a typical compression ratio might be 8, yielding a theoretical efficiency of $1 - (1/8)^{2/5} = 0.56$. This is good, but not as good as a Carnot engine operating between the same extreme temperatures. To compare the two, recall that during an adiabatic process, $TV^{\gamma-1}$ is constant. We can therefore eliminate the volumes in equation 4.10 in favor of the temperatures at the ends of either adiabatic step:

$$e = 1 - \frac{T_1}{T_2} = 1 - \frac{T_4}{T_3}.$$ (4.11)

Either of these temperature ratios is greater than the ratio of the extreme temperatures, T_1/T_3, that appears in the Carnot formula. The Otto engine is therefore less efficient than the Carnot engine. (In practice, a real gasoline engine is still less efficient, because of friction, conductive heat loss, and incomplete combustion of the fuel. Today's automobile engines typically achieve efficiencies of about 20–30%.)

The obvious way to make a gasoline engine more efficient would be to use a higher compression ratio. Unfortunately, if the fuel mixture becomes too hot it will "preignite" spontaneously before the compression step is complete, causing the pressure to jump upward before point 2 in the cycle is reached. Preignition is avoided in the **Diesel engine** by compressing only air, then spraying fuel into the cylinder after the air is hot enough to ignite the fuel. The spraying/ignition is done as the piston begins to move outward, at a rate that is adjusted to maintain approximately constant pressure. An idealized version of the Diesel cycle is shown in Figure 4.6. One can derive a rather complicated formula for the efficiency of the Diesel cycle in terms of the compression ratio V_1/V_2 and the **cutoff ratio**, V_3/V_2. For a given compression ratio the efficiency is actually less than that of the Otto cycle, but Diesel engines generally have higher compression ratios (typically around 20) and hence higher efficiencies (up to about 40% in practice). As far as I know, the only limit on the compression ratio of a Diesel engine comes from the strength and melting point of the material from which it is made. A ceramic engine could in principle withstand higher temperatures and therefore achieve higher efficiency.

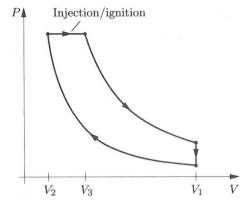

Figure 4.6. *PV* diagram for the Diesel cycle.

Problem 4.18. Derive equation 4.10 for the efficiency of the Otto cycle.

Problem 4.19. The amount of work done by each stroke of an automobile engine is controlled by the amount of fuel injected into the cylinder: the more fuel, the higher the temperature and pressure at points 3 and 4 in the cycle. But according to equation 4.10, the *efficiency* of the cycle depends only on the compression ratio (which is always the same for any particular engine), not on the amount of fuel consumed. Do you think this conclusion still holds when various other effects such as friction are taken into account? Would you expect a real engine to be most efficient when operating at high power or at low power? Explain.

Problem 4.20. Derive a formula for the efficiency of the Diesel cycle, in terms of the compression ratio V_1/V_2 and the cutoff ratio V_3/V_2. Show that for a given compression ratio, the Diesel cycle is less efficient than the Otto cycle. Evaluate the theoretical efficiency of a Diesel engine with a compression ratio of 18 and a cutoff ratio of 2.

Problem 4.21. The ingenious **Stirling engine** is a true heat engine that absorbs heat from an external source. The working substance can be air or any other gas. The engine consists of *two* cylinders with pistons, one in thermal contact with each reservoir (see Figure 4.7). The pistons are connected to a crankshaft in a complicated way that we'll ignore and let the engineers worry about. Between the two cylinders is a passageway where the gas flows past a **regenerator**: a temporary heat reservoir, typically made of wire mesh, whose temperature varies

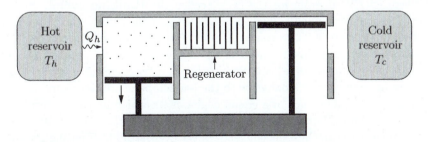

Figure 4.7. A Stirling engine, shown during the power stroke when the hot piston is moving outward and the cold piston is at rest. (For simplicity, the linkages between the two pistons are not shown.)

gradually from the hot side to the cold side. The heat capacity of the regenerator is very large, so its temperature is affected very little by the gas flowing past. The four steps of the engine's (idealized) cycle are as follows:

   i. Power stroke. While in the hot cylinder at temperature T_h, the gas absorbs heat and expands isothermally, pushing the hot piston outward. The piston in the cold cylinder remains at rest, all the way inward as shown in the figure.

  ii. Transfer to the cold cylinder. The hot piston moves in while the cold piston moves out, transferring the gas to the cold cylinder at constant volume. While on its way, the gas flows past the regenerator, giving up heat and cooling to T_c.

 iii. Compression stroke. The cold piston moves in, isothermally compressing the gas back to its original volume as the gas gives up heat to the cold reservoir. The hot piston remains at rest, all the way in.

 iv. Transfer to hot cylinder. The cold piston moves the rest of the way in while the hot piston moves out, transferring the gas back to the hot cylinder at constant volume. While on its way, the gas flows past the regenerator, absorbing heat until it is again at T_h.

(a) Draw a PV diagram for this idealized Stirling cycle.

(b) Forget about the regenerator for the moment. Then, during step 2, the gas will give up heat to the cold reservoir instead of to the regenerator; during step 4, the gas will absorb heat from the hot reservoir. Calculate the efficiency of the engine in this case, assuming that the gas is ideal. Express your answer in terms of the temperature ratio T_c/T_h and the compression ratio (the ratio of the maximum and minimum volumes). Show that the efficiency is less than that of a Carnot engine operating between the same temperatures. Work out a numerical example.

(c) Now put the regenerator back. Argue that, if it works perfectly, the efficiency of a Stirling engine is the same as that of a Carnot engine.

(d) Discuss, in some detail, the various advantages and disadvantages of a Stirling engine, compared to other engines.

The Steam Engine

A very different type of engine is the **steam engine**, ubiquitous in the 19th century and still used today in large power plants. The steam does work by pushing a piston or a turbine, while the heat is provided by burning a fossil fuel or fissioning uranium. A schematic diagram of the cycle is shown in Figure 4.8, along with an idealized PV diagram for the cycle (called the **Rankine cycle**). Starting at point 1, water is pumped to high pressure (2) and then flows into a boiler, where heat is added at constant pressure. At point 3 the steam hits the turbine, where it expands adiabatically, cools, and ends up at the original low pressure (4). Finally the partially condensed fluid is cooled further in a "condenser" (a network of pipes that are in good thermal contact with the low-temperature reservoir).

The working substance in a steam engine is most definitely *not* an ideal gas— it condenses into a liquid during the cycle! Because of this complicated behavior, there's no way to calculate the efficiency of the cycle straight from the PV diagram. However, if you know the pressures everywhere and the temperature at point 3 you can look up the data needed to compute the efficiency in what are called "steam tables."

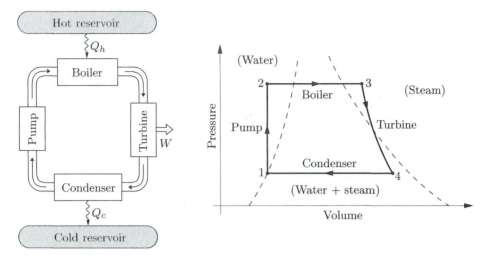

Figure 4.8. Schematic diagram of a steam engine and the associated *PV* cycle (not to scale), called the **Rankine cycle**. The dashed lines show where the fluid is liquid water, where it is steam, and where it is part water and part steam.

Recall from Section 1.6 that the **enthalpy** of a substance is $H = U + PV$, its energy plus the work needed to make room for it in a constant-pressure environment. Therefore the change in enthalpy is equal to the heat absorbed under constant-pressure conditions (assuming that no "other" work is done). In the Rankine cycle, heat is absorbed at constant pressure in the boiler and expelled at constant pressure in the condenser, so we can write the efficiency as

$$e = 1 - \frac{Q_c}{Q_h} = 1 - \frac{H_4 - H_1}{H_3 - H_2} \approx 1 - \frac{H_4 - H_1}{H_3 - H_1}. \tag{4.12}$$

The last approximation, $H_2 \approx H_1$, is pretty good because the pump adds very little energy to the water, while the PV term in H is very small for liquids in comparison to gases.

Two different tables are generally used to look up the needed values of H. The first (see Table 4.1) is for "saturated" water and steam, covering any point between the dashed lines on the PV diagram, where the temperature is determined by the pressure. This table lists the enthalpy and entropy for pure water and for pure steam at the boiling point; for mixtures of water and steam one can interpolate between these two values.

The other table needed (see Table 4.2) is for "superheated" steam, in the right-most region of the PV diagram where the pressure and temperature must be specified separately. Again, the table lists the enthalpy and entropy at each point.

To compute the efficiency of a Rankine cycle, we need the enthalpies at points 1, 3, and 4. The enthalpy at point 1 can be looked up in Table 4.1, while the enthalpy at point 3 can be looked up in Table 4.2. To locate point 4 we use the fact that the expansion of the steam in the turbine is approximately adiabatic ($Q = 0$), so that ideally its entropy does not change during this step. We can look up the entropy

T (°C)	P (bar)	H_{water} (kJ)	H_{steam} (kJ)	S_{water} (kJ/K)	S_{steam} (kJ/K)
0	0.006	0	2501	0	9.156
10	0.012	42	2520	0.151	8.901
20	0.023	84	2538	0.297	8.667
30	0.042	126	2556	0.437	8.453
50	0.123	209	2592	0.704	8.076
100	1.013	419	2676	1.307	7.355

Table 4.1. Properties of saturated water/steam. Pressures are given in bars, where 1 bar $= 10^5$ Pa ≈ 1 atm. All values are for 1 kg of fluid, and are measured relative to liquid water at the triple point (0.01°C and 0.006 bar). Excerpted from Keenan et al. (1978).

P (bar)		Temperature (°C) 200	300	400	500	600
1.0	H (kJ)	2875	3074	3278	3488	3705
	S (kJ/K)	7.834	8.216	8.544	8.834	9.098
3.0	H (kJ)	2866	3069	3275	3486	3703
	S (kJ/K)	7.312	7.702	8.033	8.325	8.589
10	H (kJ)	2828	3051	3264	3479	3698
	S (kJ/K)	6.694	7.123	7.465	7.762	8.029
30	H (kJ)		2994	3231	3457	3682
	S (kJ/K)		6.539	6.921	7.234	7.509
100	H (kJ)			3097	3374	3625
	S (kJ/K)			6.212	6.597	6.903
300	H (kJ)			2151	3081	3444
	S (kJ/K)			4.473	5.791	6.233

Table 4.2. Properties of superheated steam. All values are for 1 kg of fluid, and are measured relative to liquid water at the triple point. Excerpted from Keenan et al. (1978).

at point 3 in Table 4.2, then interpolate in Table 4.1 to find what mixture of liquid and gas has the same entropy at the lower pressure.

For example, suppose that the cycle operates between a minimum pressure of 0.023 bar (where the boiling temperature is 20°C) and a maximum pressure of 300 bars, with a maximum superheated steam temperature of 600°C. Then for each kilogram of water/steam, $H_1 = 84$ kJ and $H_3 = 3444$ kJ. The entropy at point 3 is 6.233 kJ/K, and to obtain this same entropy at point 4 we need a mixture of 29% water and 71% steam. This same mixture has an enthalpy of $H_4 = 1824$ kJ, so the efficiency of the cycle is approximately

$$e \approx 1 - \frac{1824 - 84}{3444 - 84} = 48\%. \tag{4.13}$$

For comparison, an ideal Carnot engine operating over the same temperature range would have an efficiency of 66%. While the temperatures and pressures assumed here are typical of modern fossil-fuel power plants, these plants achieve an actual efficiency of only about 40%, due to a variety of complications that I've neglected. Nuclear power plants operate at lower temperatures for safety reasons, and therefore achieve efficiencies of only about 34%.

Problem 4.22. A small-scale steam engine might operate between the temperatures 20°C and 300°C, with a maximum steam pressure of 10 bars. Calculate the efficiency of a Rankine cycle with these parameters.

Problem 4.23. Use the definition of enthalpy to calculate the change in enthalpy between points 1 and 2 of the Rankine cycle, for the same numerical parameters as used in the text. Recalculate the efficiency using your corrected value of H_2, and comment on the accuracy of the approximation $H_2 \approx H_1$.

Problem 4.24. Calculate the efficiency of a Rankine cycle that is modified from the parameters used in the text in each of the following three ways (one at a time), and comment briefly on the results: (a) reduce the maximum temperature to 500°C; (b) reduce the maximum pressure to 100 bars; (c) reduce the minimum temperature to 10°C.

Problem 4.25. In a real turbine, the entropy of the steam will increase somewhat. How will this affect the percentages of liquid and gas at point 4 in the cycle? How will the efficiency be affected?

Problem 4.26. A coal-fired power plant, with parameters similar to those used in the text above, is to deliver 1 GW (10^9 watts) of power. Estimate the amount of steam (in kilograms) that must pass through the turbine(s) each second.

Problem 4.27. In Table 4.1, why does the entropy of water increase with increasing temperature, while the entropy of steam decreases with increasing temperature?

Problem 4.28. Imagine that your dog has eaten the portion of Table 4.1 that gives entropy data; only the enthalpy data remains. Explain how you could reconstruct the missing portion of the table. Use your method to explicitly check a few of the entries for consistency. How much of Table 4.2 could you reconstruct if it were missing? Explain.

4.4   Real Refrigerators

The operation of an ordinary refrigerator or air conditioner is almost the reverse of the Rankine cycle just discussed. Again the working substance changes back and forth from a gas to a liquid, but here the fluid must have a much lower boiling temperature.

Dozens of fluids have been used as refrigerants, including carbon dioxide (which requires rather high pressures) and ammonia (which is still used in large industrial systems, despite its toxicity). Around 1930, General Motors and du Pont developed and produced the first of the nontoxic chlorofluorocarbon (CFC) refrigerants, giving them the trade name *Freon*. Of these the most familiar is Freon-12 (CCl_2F_2), used in domestic refrigerators and automobile air conditioners. We now know, however, that CFC's that have escaped into the atmosphere are causing the breakdown of the ozone layer. The most damaging CFC's are therefore being replaced with chlorine-free compounds; the usual substitute for Freon-12 is a hydrofluorocarbon, $F_3C_2FH_2$, known by the catchy name *HFC-134a*.

A schematic sketch and *PV* diagram of the standard refrigeration cycle are shown in Figure 4.9. Beginning at point 1, the fluid (here a gas) is first compressed adiabatically, raising its pressure and temperature. It then gives up heat and gradually liquefies in the condenser (a network of pipes in thermal contact with the hot reservoir). Next it passes through a "throttling valve"—a narrow opening or porous plug—emerging on the other side at much lower pressure and temperature. Finally it absorbs heat and turns back into a gas in the evaporator (a network of pipes in thermal contact with the cold reservoir).

It's easy to express the coefficient of performance of a standard refrigerator in terms of the enthalpies of the fluid at various points around the cycle. Since the pressure is constant in the evaporator, the heat absorbed is $Q_c = H_1 - H_4$, the change in enthalpy. Similarly, the heat expelled in the condenser is $Q_h = H_2 - H_3$. So the coefficient of performance is

$$\text{COP} = \frac{Q_c}{Q_h - Q_c} = \frac{H_1 - H_4}{H_2 - H_3 - H_1 + H_4}. \tag{4.14}$$

The enthalpies at points 1, 2, and 3 can be looked up in tables, with point 2 located by assuming that the entropy is constant during the compression stage. To locate point 4 we must analyze the throttling valve in a bit more detail.

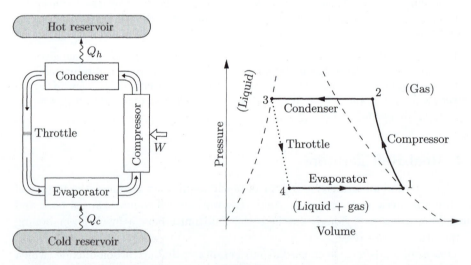

Figure 4.9. A schematic drawing and *PV* diagram (not to scale) of the standard refrigeration cycle. The dashed lines indicate where the refrigerant is liquid, gas, and a combination of the two.

The Throttling Process

The **throttling** process (also known as the **Joule-Thomson process**) is shown in Figure 4.10. I find it helpful to pretend that the fluid is being pushed through the plug by a piston, exerting pressure P_i, while a second piston, exerting pressure P_f, moves backward on the other side to make room. For a particular chunk of fluid, let the initial volume (before going through the plug) be V_i, while the final volume (on the other side) is V_f. Since there is no heat flow during this process, the change in the energy of the fluid is

$$U_f - U_i = Q + W = 0 + W_{\text{left}} + W_{\text{right}}, \tag{4.15}$$

where W_{left} is the (positive) work done by the piston on the left, and W_{right} is the (negative) work done by the piston on the right. (Ultimately, the net work is actually performed by the compressor, way over on the other side of the cycle. Here, however, we're concerned only with what's happening locally.) But the work done from the left in pushing the entire volume V_i through the plug is $P_i V_i$, while the work done from the right is $-P_f V_f$—negative because the piston is moving backwards. Therefore the change in energy is

$$U_f - U_i = P_i V_i - P_f V_f. \tag{4.16}$$

Putting the f's on the left and the i's on the right, this equation becomes

$$U_f + P_f V_f = U_i + P_i V_i, \qquad \text{or} \qquad H_f = H_i. \tag{4.17}$$

The enthalpy is constant during the throttling process.

The *purpose* of the throttling valve is to *cool* the fluid to below the temperature of the cold reservoir, so it can absorb heat as required. If the fluid were an ideal gas, this wouldn't work at all, since

$$H = U + PV = \frac{f}{2}NkT + NkT = \frac{f+2}{2}NkT \qquad \text{(ideal gas)}. \tag{4.18}$$

Constant enthalpy would imply constant temperature! But in a dense gas or liquid, the energy U also contains a potential energy term due to the forces between the molecules:

$$H = U_{\text{potential}} + U_{\text{kinetic}} + PV. \tag{4.19}$$

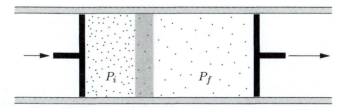

Figure 4.10. The throttling process, in which a fluid is pushed through a porous plug and then expands into a region of lower pressure.

The force between any two molecules is weakly attractive at long distances and strongly repulsive at short distances. Under most (though not all) conditions the attraction dominates; then $U_{\text{potential}}$ is negative, but becomes less negative as the pressure drops and the distance between molecules increases. To compensate for the increase in potential energy, the kinetic energy generally drops, and the fluid cools as desired.

If we use the fact that $H_4 = H_3$ in the refrigeration cycle, the coefficient of performance (4.14) simplifies to

$$\text{COP} = \frac{H_1 - H_3}{H_2 - H_1}. \qquad (4.20)$$

Now one only has to look up three enthalpies. Tables 4.3 and 4.4 give enthalpy and entropy values for the refrigerant HFC-134a.

Problem 4.29. Liquid HFC-134a at its boiling point at 12 bars pressure is throttled to 1 bar pressure. What is the final temperature? What fraction of the liquid vaporizes?

P (bar)	T (°C)	H_{liquid} (kJ)	H_{gas} (kJ)	S_{liquid} (kJ/K)	S_{gas} (kJ/K)
1.0	−26.4	16	231	0.068	0.940
1.4	−18.8	26	236	0.106	0.932
2.0	−10.1	37	241	0.148	0.925
4.0	8.9	62	252	0.240	0.915
6.0	21.6	79	259	0.300	0.910
8.0	31.3	93	264	0.346	0.907
10.0	39.4	105	268	0.384	0.904
12.0	46.3	116	271	0.416	0.902

Table 4.3. Properties of the refrigerant HFC-134a under saturated conditions (at its boiling point for each pressure). All values are for 1 kg of fluid, and are measured relative to an arbitrarily chosen reference state, the saturated liquid at −40°C. Excerpted from Moran and Shapiro (1995).

P (bar)		Temperature (°C) 40	50	60
8.0	H (kJ)	274	284	295
	S (kJ/K)	0.937	0.971	1.003
10.0	H (kJ)	269	280	291
	S (kJ/K)	0.907	0.943	0.977
12.0	H (kJ)		276	287
	S (kJ/K)		0.916	0.953

Table 4.4. Properties of superheated (gaseous) refrigerant HFC-134a. All values are for 1 kg of fluid, and are measured relative to the same reference state as in Table 4.3. Excerpted from Moran and Shapiro (1995).

Problem 4.30. Consider a household refrigerator that uses HFC-134a as the refrigerant, operating between the pressures of 1.0 bar and 10 bars.

(a) The compression stage of the cycle begins with saturated vapor at 1 bar and ends at 10 bars. Assuming that the entropy is constant during compression, find the approximate temperature of the vapor after it is compressed. (You'll have to do an interpolation between the values given in Table 4.4.)

(b) Determine the enthalpy at each of the points 1, 2, 3, and 4, and calculate the coefficient of performance. Compare to the COP of a Carnot refrigerator operating between the same reservoir temperatures. Does this temperature range seem reasonable for a household refrigerator? Explain briefly.

(c) What fraction of the liquid vaporizes during the throttling step?

Problem 4.31. Suppose that the throttling valve in the refrigerator of the previous problem is replaced with a small turbine-generator in which the fluid expands adiabatically, doing work that contributes to powering the compressor. Will this change affect the COP of the refrigerator? If so, by how much? Why do you suppose real refrigerators use a throttle instead of a turbine?

Problem 4.32. Suppose you are told to design a household air conditioner using HFC-134a as its working substance. Over what range of pressures would you have it operate? Explain your reasoning. Calculate the COP for your design, and compare to the COP of an ideal Carnot refrigerator operating between the same reservoir temperatures.

Liquefaction of Gases

If you want to make something *really* cold, you normally don't just stick it into a refrigerator—instead you put it on dry ice (195 K at atmospheric pressure), or immerse it in liquid nitrogen (77 K) or even liquid helium (4.2 K). But how are gases like nitrogen and helium liquefied (or in the case of CO_2, solidified) in the first place? The most common methods all involve the throttling process.

You can liquefy carbon dioxide at room temperature, simply by compressing it isothermally to about 60 atmospheres. Then throttle it back to low pressure and it cools and partially evaporates just as in the refrigeration cycle discussed above. At pressures below 5.1 atm, however, liquid CO_2 cannot exist; instead the condensed phase is a solid, dry ice. So to make dry ice, all you have to do is hook up a tank of liquid CO_2 to a throttling valve and watch the frost form around the nozzle as the gas rushes out.

Liquefying nitrogen (or air) isn't so simple. Compress it all you want at room temperature and it never undergoes a sudden phase transformation to a liquid—it just gets denser and denser in a continuous way. (This behavior is discussed in more detail in Section 5.3.) If you start with nitrogen at, say, 300 K and 100 atm and throttle it down to 1 atm it does cool, but only to about 280 K. To get any liquid, starting from this pressure, the initial temperature must be below about 160 K. At higher initial pressures the initial temperature can be somewhat higher, but must still be well below room temperature. The first liquefaction of oxygen and nitrogen was achieved in 1877 by Louis Cailletet, who used an initial pressure of 300 atm and precooled the gases using other cold liquids. A more convenient method, though, is to use the throttled gas itself to precool the incoming gas. A device that does this

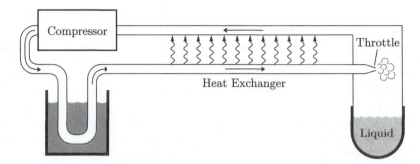

Figure 4.11. Schematic diagram of the Hampson-Linde cycle for gas liquefaction. Compressed gas is first cooled (to room temperature is sufficient if it is nitrogen or oxygen) and then passed through a heat exchanger on its way to a throttling valve. The gas cools upon throttling and returns through the heat exchanger to further cool the incoming gas. Eventually the incoming gas becomes cold enough to partially liquefy upon throttling. From then on, new gas must be added at the compressor to replace what is liquefied.

was invented by William Hampson and Carl von Linde (independently) in 1895; it is shown schematically in Figure 4.11. Instead of being discarded, the throttled gas is sent through a **heat exchanger** where it cools the incoming gas. When that gas passes through the throttle it cools even more, and thus the system gradually gets colder and colder until the throttled gas begins to liquefy.

Starting from room temperature, the Hampson-Linde cycle can be used to liquefy any gas except hydrogen or helium. These gases, when throttled starting at room temperature and any pressure, actually become *hotter*. This happens because the attractive interactions between the molecules are very weak; at high temperatures the molecules are moving too fast to notice much attraction, but they still suffer hard collisions during which there is a large *positive* potential energy. When the gas expands the collisions occur less frequently, so the average potential energy decreases and the average kinetic energy increases.

To liquefy hydrogen or helium, it is therefore necessary to first cool the gas well below room temperature, slowing down the molecules until attraction becomes more important than repulsion. Figure 4.12 shows the range of temperatures and pressures under which hydrogen will cool upon throttling. The temperature below which cooling will occur is called the **inversion temperature**; for hydrogen the maximum inversion temperature is 204 K, while for helium it is only 43 K. Hydrogen was first liquefied in 1898 by James Dewar, using liquid air for precooling. Helium was first liquefied in 1908 by Heike Kamerlingh Onnes, using liquid hydrogen for precooling. Today, however, the precooling of helium is normally accomplished without liquid hydrogen (and sometimes even without liquid nitrogen), by allowing the helium to expand adiabatically as it pushes a piston. This technique is a great advance in safety but is mechanically more challenging. The piston must operate at temperatures as low as 8 K, at which any ordinary lubricant would freeze. The helium itself is therefore used as a lubricant, with extremely small clearances to prevent a significant amount from escaping.

Figure 4.12. Lines of constant enthalpy (approximately horizontal, at intervals of 400 J/mol) and inversion curve (dashed) for hydrogen. In a throttling process the enthalpy is constant, so cooling occurs only to the left of the inversion curve, where the enthalpy lines have positive slopes. The heavy solid line at lower-left is the liquid-gas phase boundary. Data from Vargaftik (1975) and Woolley et al. (1948).

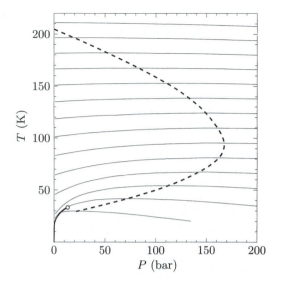

Temperature (K)

	77 (liq.)	77 (gas)	100	200	300	400	500	600
1 bar	−3407	2161	2856	5800	8717	11,635	14,573	17,554
100 bar			−1946	4442	8174	11,392	14,492	17,575

Table 4.5. Molar enthalpy of nitrogen (in joules) at 1 bar and 100 bars. Excerpted from Lide (1994).

Problem 4.33. Table 4.5 gives experimental values of the molar enthalpy of nitrogen at 1 bar and 100 bars. Use this data to answer the following questions about a nitrogen throttling process operating between these two pressures.

(a) If the initial temperature is 300 K, what is the final temperature? (Hint: You'll have to do an interpolation between the tabulated values.)

(b) If the initial temperature is 200 K, what is the final temperature?

(c) If the initial temperature is 100 K, what is the final temperature? What fraction of the nitrogen ends up as a liquid in this case?

(d) What is the highest initial temperature at which some liquefaction takes place?

(e) What would happen if the initial temperature were 600 K? Explain.

Problem 4.34. Consider an ideal Hampson-Linde cycle in which no heat is lost to the environment.

(a) Argue that the combination of the throttling valve and the heat exchanger is a constant-enthalpy device, so that the total enthalpy of the fluid coming out of this combination is the same as the enthalpy of the fluid going in.

(b) Let x be the fraction of the fluid that liquefies on each pass through the cycle. Show that

$$x = \frac{H_{out} - H_{in}}{H_{out} - H_{liq}},$$

where H_{in} is the enthalpy of each mole of compressed gas that goes into the heat exchanger, H_{out} is the enthalpy of each mole of low-pressure gas that comes out of the heat exchanger, and H_{liq} is the enthalpy of each mole of liquid produced.

(c) Use the data in Table 4.5 to calculate the fraction of nitrogen liquefied on each pass through a Hampson-Linde cycle operating between 1 bar and 100 bars, with an input temperature of 300 K. Assume that the heat exchanger works perfectly, so the temperature of the low-pressure gas coming out of it is the same as the temperature of the high-pressure gas going in. Repeat the calculation for an input temperature of 200 K.

Toward Absolute Zero

At atmospheric pressure, liquid helium boils at 4.2 K. The boiling point decreases as the pressure is reduced, so it's not hard to lower the temperature of liquid helium still further, by pumping away the vapor to reduce the pressure; the helium cools through evaporation. Below about 1 K, however, this procedure becomes impractical: Even the smallest heat leak raises the temperature of the helium significantly, and the best vacuum pumps cannot remove the vapor fast enough to compensate. The rare isotope helium-3, whose normal boiling point is only 3.2 K, can be cooled to about 0.3 K by pumping to low pressure.

But isn't 1 K cold enough? Why bother trying to attain still lower temperatures? Perhaps surprisingly, there are a variety of fascinating phenomena that occur only in the millikelvin, microkelvin, and even nanokelvin ranges, including transformations of helium itself, magnetic behavior of atoms and nuclei, and "Bose-Einstein condensation" of dilute gases. To investigate these phenomena, experimenters have developed an equally fascinating array of techniques for reaching extremely low temperatures.[*]

To get from 1 K to a few millikelvins, the method of choice is usually the **helium dilution refrigerator**, shown schematically in Figure 4.13. The cooling occurs by "evaporation" of liquid ^{3}He, but instead of evaporating into a vacuum, it dissolves into a liquid bath of the more common isotope, ^{4}He. At subkelvin temperatures the two isotopes are relatively immiscible, like oil and water. Below about 0.1 K essentially no ^{4}He will dissolve in pure ^{3}He, while a small amount of ^{3}He, about 6%, will dissolve into otherwise pure ^{4}He. This is what happens in the "mixing chamber," where ^{3}He continuously dissolves ("evaporates") into the ^{4}He, absorbing heat in the process. The ^{3}He then diffuses upward through a heat exchanger to a "still" at 0.7 K, where heat is supplied to make it evaporate (in the conventional sense). The ^{4}He is essentially inert during the whole process: It is a "superfluid" in this temperature range, offering negligible resistance to the diffusion of ^{3}He atoms; and it is less volatile than ^{3}He, so it does not evaporate at a significant rate in the still. After evaporating in the still, the gasous ^{3}He is compressed, recooled to a liquid (by a separate bath of ^{4}He), and finally sent back through the heat exchanger to the mixing chamber.

---

[*]For a good overview of methods of reaching subkelvin temperatures, see Olli V. Lounasmaa, "Towards the Absolute Zero," *Physics Today* **32**, 32–41 (December, 1979). For more details on helium dilution refrigerators, see John C. Wheatley, *American Journal of Physics* **36**, 181–210 (1968).

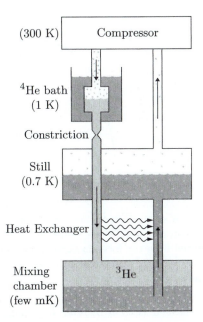

Figure 4.13. Schematic diagram of a helium dilution refrigerator. The working substance is ^{3}He (light gray), which circulates counter-clockwise. The ^{4}He (dark gray) does not circulate.

An alternative method of reaching millikelvin temperatures is **magnetic cooling**, based on the properties of paramagnetic materials. Recall from Section 3.3 that the total magnetization of an ideal two-state paramagnet is a function of the *ratio* of the magnetic field strength B to the temperature:

$$M = \mu(N_\uparrow - N_\downarrow) = N\mu \tanh\left(\frac{\mu B}{kT}\right). \tag{4.21}$$

(For an ideal paramagnet with more than two states per particle, the formula is more complicated but has the same qualitative behavior.) For an *electronic* paramagnet, whose elementary dipoles are electrons, the value of μ is such that a magnetic field of 1 T and a temperature of 1 K yield $M/N\mu = 0.59$: A significant majority of the dipoles are pointing up. Suppose, now, that we start with the system in such a state and then reduce the magnetic field strength without allowing any heat to enter. The populations of the up and down states will not change during this process, so the total magnetization is fixed and therefore the temperature must decrease in proportion to the field strength. If B decreases by a factor of 1000, so does T.

A good way to visualize this process is shown in Figure 4.14, where I've plotted the entropy of the system as a function of temperature for two different values of the magnetic field strength. For any nonzero field strength the entropy goes to zero as $T \to 0$ (as all the dipoles line up) and goes to a nonzero constant value at sufficiently high temperatures (as the alignments of the dipoles become random). The higher the field strength, the more gradually the entropy rises as a function of temperature (due to the greater tendency of the dipoles to remain aligned with the field). In a magnetic cooling process, the sample is first put in good thermal contact with a constant-temperature "reservoir" such as a liquid helium bath. The

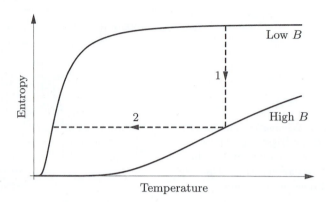

Figure 4.14. Entropy as a function of temperature for an ideal two-state paramagnet, at two different values of the magnetic field strength. (These curves were
plotted from the formula derived in Problem 3.23.) The magnetic cooling process
consists of an isothermal increase in the field strength (step 1), followed by an
adiabatic decrease (step 2).

magnetic field is then increased, causing the entropy of the sample to drop as its
temperature remains fixed. The sample is then insulated from the reservoir and the
field strength reduced, resulting in a drop in temperature at constant entropy. The
process is analogous to the cooling of an ideal gas by adiabatic expansion, following
isothermal compression.

But why should we merely *reduce* the strength of the magnetic field—why not
eliminate it entirely? Then, according to equation 4.21, the temperature of the
paramagnet would have to go to *absolute zero* in order to maintain constant M. As
you might guess, attaining absolute zero isn't so easy. The problem in this case is
that no paramagnet is truly *ideal* at very low temperatures: The elementary dipoles
interact with *each other*, effectively producing a magnetic field that is present even
when the *applied* field is zero. Depending on the details of these interactions, the
dipoles may align parallel or antiparallel to their nearest neighbors. Either way,
their entropy drops almost to zero, as if there were an external magnetic field. To
reach the lowest possible final temperatures, the paramagnetic material should be
one in which the interactions between neighboring dipoles are extremely weak. For
electronic paramagnets, the lowest temperature that can be reached by magnetic
cooling is about 1 mK (see Problem 4.35).

In a *nuclear* paramagnet the dipole-dipole interactions are much weaker, so
much lower temperatures can be attained. The only catch is that you also have
to *start* at a lower temperature, in order to have a significant excess of one spin
alignment over the other. The first nuclear magnetic cooling experiments produced
temperatures of about 1 μK, and it seems that every few years someone improves
the technique to achieve still lower temperatures. In 1993, researchers at Helsinki
University used nuclear magnetic cooling of rhodium to produce temperatures as
low as 280 *pico*kelvins, that is, 2.8×10^{-10} K.*

---

*Pertti Hakonen and Olli V. Lounasmaa, *Science* **265**, 1821–1825 (23 September, 1994).

Meanwhile, other experimenters have reached extremely low temperatures using a completely different technique: **laser cooling**. Here the system is not a liquid or solid but rather a dilute gas—a small cloud of atoms, prevented from condensing into a solid by its very low density.

Imagine that you hit an atom with laser light, tuned to just the right frequency to excite the atom into a higher-energy state. The atom will absorb a photon as it gains energy, then spontaneously emit a photon of the same frequency as it loses the energy a split second later. Photons carry momentum as well as energy, so the atom recoils each time it absorbs or emits a photon. But whereas the absorbed photons all come from the same direction (the laser), the emitted photons exit in all directions (see Figure 4.15). On average, therefore, the atom feels a *force* from the direction of the laser.

Now suppose that we tune the laser to a slightly lower frequency (longer wavelength). An atom at rest will rarely absorb photons of this frequency, so it feels hardly any force. But an atom moving *toward* the laser will see the light Doppler-shifted back to a higher frequency. It therefore absorbs plenty of photons and feels a backward force, opposing its motion. An atom moving away from the laser feels even less force than an atom at rest, but if we aim an identical laser beam at it from the opposite side, then it too feels a backward force. With laser beams coming from all six directions we can exert backward forces that tend to oppose motion in *any* direction. Put thousands or millions of atoms into the region and they'll all slow down, cooling to a very low temperature.

Even at very low speeds, though, the atoms would quickly hit the hot walls of the container (or fall to the bottom) without an additional *trapping* force that pushes them toward the center. Such a force can be created using nonuniform magnetic fields to shift the atomic energy levels and thus vary their tendency to absorb photons depending on where they are. The combination of laser cooling and trapping can readily cool a cloud of atoms to about 1 mK, without any of the hassle of liquid helium or conventional cryogenic equipment. Elaborations of the technique have recently been used to reach temperatures in the microkelvin and even nanokelvin ranges.[*]

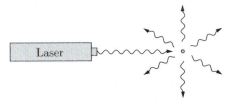

Figure 4.15. An atom that continually absorbs and reemits laser light feels a force from the direction of the laser, because the absorbed photons all come from the same direction while the emitted photons come out in all directions.

---

[*]For an elementary review of laser cooling and its applications, see Steven Chu, "Laser Trapping of Neutral Particles," *Scientific American* **266**, 71–76 (February, 1992). A bibliography of articles on trapping of neutral atoms has been compiled by N. R. Newbury and C. Wieman, *American Journal of Physics* **64**, 18–20 (1996).

Problem 4.35. The magnetic field created *by* a dipole has a strength of approximately $(\mu_0/4\pi)(\mu/r^3)$, where r is the distance from the dipole and μ_0 is the "permeability of free space," equal to exactly $4\pi \times 10^{-7}$ in SI units. (In the formula I'm neglecting the variation of field strength with angle, which is at most a factor of 2.) Consider a paramagnetic salt like iron ammonium alum, in which the magnetic moment μ of each dipole is approximately one Bohr magneton $(9 \times 10^{-24}$ J/T), with the dipoles separated by a distance of 1 nm. Assume that the dipoles interact only via ordinary magnetic forces.

(a) Estimate the strength of the magnetic field at the location of a dipole, due to its neighboring dipoles. This is the effective field strength even when there is no externally applied field.

(b) If a magnetic cooling experiment using this material begins with an external field strength of 1 T, by about what factor will the temperature decrease when the external field is turned off?

(c) Estimate the temperature at which the entropy of this material rises most steeply as a function of temperature, in the absence of an externally applied field.

(d) If the final temperature in a cooling experiment is significantly less than the temperature you found in part (c), the material ends up in a state where $\partial S/\partial T$ is very small and therefore its heat capacity is very small. Explain why it would be impractical to try to reach such a low temperature with this material.

Problem 4.36. An apparent limit on the temperature achievable by laser cooling is reached when an atom's recoil energy from absorbing or emitting a single photon is comparable to its total kinetic energy. Make a rough estimate of this limiting temperature for rubidium atoms that are cooled using laser light with a wavelength of 780 nm.

Problem 4.37. A common (but imprecise) way of stating the third law of thermodynamics is "You can't reach absolute zero." Discuss how the third law, as stated in Section 3.2, puts limits on how low a temperature can be attained by various refrigeration techniques.

According to this principle, the production of heat alone is not sufficient to give birth to the impelling power: it is necessary that there should also be cold; without it, the heat would be useless.

—Sadi Carnot, *Reflections on the Motive Power of Fire*, trans. R. H. Thurston (Macmillan, New York, 1890).

5 Free Energy and Chemical Thermodynamics

The previous chapter applied the laws of thermodynamics to cyclic processes: the operation of engines and refrigerators whose energy and entropy are unchanged over the long term. But many important thermodynamic processes are not cyclic. Chemical reactions, for example, are constrained by the laws of thermodynamics but do not end with the system in the same state where it started.

The purpose of the present chapter is to apply the laws of thermodynamics to chemical reactions and other transformations of matter. One complication that arises immediately is that these transformations most often occur in systems that are not isolated but are interacting with their surroundings, thermally and often mechanically. The energy of the system itself is usually not fixed; rather its *temperature* is held fixed, through interaction with a constant-temperature environment. Similarly, in many cases it is not the volume of the system that is fixed but rather the pressure. Our first task, then, is to develop the conceptual tools needed to understand constant-temperature and constant-pressure processes.

5.1 Free Energy as Available Work

In Section 1.6 I defined the **enthalpy** of a system as its energy plus the work needed to make room for it, in an environment with constant pressure P:

$$H \equiv U + PV. \qquad (5.1)$$

This is the total energy you would need, to create the system out of nothing and put it in such an environment. (Since the initial volume of the system is zero, $\Delta V = V$.) Or, if you could completely annihilate the system, H is the energy you could recover: the system's energy plus the work done by the collapsing atmosphere.

Often, however, we're not interested in the total energy needed or the total energy that can be recovered. If the environment is one of constant temperature,

149

the system can extract heat from this environment for free, so all *we* need to provide, to create the system from nothing, is any additional *work* needed. And if we annihilate the system, we generally can't recover all its energy as work, because we have to dispose of its entropy by dumping some heat into the environment.

So I'd like to introduce two more useful quantities that are related to energy and analogous to H. One is the **Helmholtz free energy**,

$$F \equiv U - TS. \tag{5.2}$$

This is the total energy needed to create the system, minus the heat you can get for free from an environment at temperature T. This heat is given by $T\Delta S = TS$, where S is the system's (final) entropy; the more entropy a system has, the more of its energy can enter as heat. Thus F is the energy that must be provided as work, if you're creating the system out of nothing.* Or if you annihilate the system, the energy that comes out as work is F, since you have to dump some heat, equal to TS, into the environment in order to get rid of the system's entropy. The *available*, or "free," energy is F.

The word "work" in the previous paragraph means *all* work, including any that is done automatically by the system's surroundings. If the system is in an environment with constant pressure P and constant temperature T, then the work *you* need to do to create it, or the work you can recover when you destroy it, is given by the **Gibbs free energy**,

$$G \equiv U - TS + PV. \tag{5.3}$$

This is just the system's energy, minus the heat term that's in F, plus the atmospheric work term that's in H (see Figure 5.1).

Figure 5.1. To create a rabbit out of nothing and place it on the table, the magician need not summon up the entire enthalpy, $H = U + PV$. Some energy, equal to TS, can flow in spontaneously as heat; the magician must provide only the difference, $G = H - TS$, as work.

---

*In the context of creating a system, the term *free* energy is a misnomer. The energy that comes for free is TS, the term we *subtracted* to get F. In this context, F should be called the *costly* energy. The people who named F were instead thinking of the reverse process, where you annihilate the system and recover F as work.

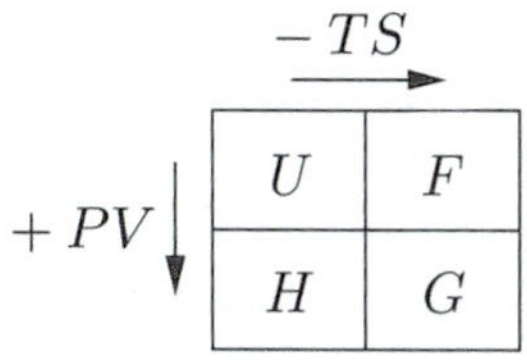

Figure 5.2. To get H from U or G from F, add PV; to get F from U or G from H, subtract TS.

$$- TS$$

U	F
H	G

$+ PV$

The four functions U, H, F, and G are collectively called **thermodynamic potentials**. Figure 5.2 shows a diagram that I use to remember the definitions.

Usually, of course, we deal with processes that are much less dramatic than the creation or annihilation of an entire system. Then instead of F and G themselves, we want to look at the *changes* in these quantities.

For any change in the system that takes place at constant temperature T, the change in F is

$$\Delta F = \Delta U - T \Delta S = Q + W - T \Delta S, \tag{5.4}$$

where Q is the heat added and W is the work done on the system. If no *new* entropy is created during the process, then $Q = T \Delta S$, so the change in F is precisely equal to the work done on the system. If new entropy *is* created, then $T \Delta S$ will be greater than Q, so ΔF will be less than W. In general, therefore,

$$\Delta F \leq W \qquad \text{at constant } T. \tag{5.5}$$

This W includes *all* work done on the system, including any work done automatically by its expanding or collapsing environment.

If the environment is one of constant pressure, and if we're not interested in keeping track of the work that the environment does automatically, then we should think about G instead of F. For any change that takes place at constant T and P, the change in G is

$$\Delta G = \Delta U - T \Delta S + P \Delta V = Q + W - T \Delta S + P \Delta V. \tag{5.6}$$

Again, the difference $Q - T \Delta S$ is always zero or negative. Meanwhile, W includes the work done by the environment, $-P \Delta V$, plus any "other" work (such as electrical work) done on the system:

$$W = -P \Delta V + W_{\text{other}}. \tag{5.7}$$

This $P \Delta V$ cancels the one in equation 5.6, leaving

$$\Delta G \leq W_{\text{other}} \qquad \text{at constant } T, P. \tag{5.8}$$

Because free energy is such a useful quantity, values of ΔG for an enormous variety of chemical reactions and other processes have been measured and tabulated. There are many ways to measure ΔG. The easiest conceptually is to first measure ΔH for the reaction, by measuring the heat absorbed when the reaction takes place at constant pressure and no "other" work is done. Then calculate ΔS from the entropies of the initial and final states of the system, determined separately from heat capacity data as described in Sections 3.2 and 3.4. Finally, compute

$$\Delta G = \Delta H - T\,\Delta S. \tag{5.9}$$

Values of ΔG for the formation of selected compounds and solutions (at $T = 298$ K and $P = 1$ bar) are given in the table at the back of this book. You can compute ΔG values for other reactions by imagining first that each reactant is converted to elemental form and then that these elements are converted into the products.

As with U and H, the actual *value* of F or G is unambiguous only if we include *all* the energy of the system, including the rest energy (mc^2) of every particle. In everyday situations this would be ridiculous, so instead we measure U from some other convenient but arbitrary reference point, and this arbitrary choice also fixes the zero points for H, F, and G. *Changes* in these quantities are unaffected by our choice of reference point, and changes are all we usually talk about anyway, so in practice we can often avoid choosing a reference point.

Problem 5.1. Let the system be one mole of argon gas at room temperature and atmospheric pressure. Compute the total energy (kinetic only, neglecting atomic rest energies), entropy, enthalpy, Helmholtz free energy, and Gibbs free energy. Express all answers in SI units.

Problem 5.2. Consider the production of ammonia from nitrogen and hydrogen,

$$N_2 + 3H_2 \longrightarrow 2NH_3,$$

at 298 K and 1 bar. From the values of ΔH and S tabulated at the back of this book, compute ΔG for this reaction and check that it is consistent with the value given in the table.

Electrolysis, Fuel Cells, and Batteries

As an example of using ΔG, consider the chemical reaction

$$H_2O \longrightarrow H_2 + \tfrac{1}{2}O_2, \tag{5.10}$$

the electrolysis of liquid water into hydrogen and oxygen gas (see Figure 5.3). Assume that we start with one mole of water, so we end with a mole of hydrogen and half a mole of oxygen.

According to standard reference tables, ΔH for this reaction (at room temperature and atmospheric pressure) is 286 kJ. This is the amount of heat you would get out if you burned a mole of hydrogen, running the reaction in reverse. When we *form* hydrogen and oxygen out of water, we need to put 286 kJ of energy into the system in some way or other. Of the 286 kJ, a small amount goes into pushing the atmosphere away to make room for the gases produced; this amount is $P\,\Delta V = 4$ kJ. The other 282 kJ remains in the system itself (see Figure 5.4). But of the 286 kJ needed, must we supply all as work, or can some enter as heat?

To answer this question we must determine the change in the system's entropy. The measured and tabulated entropy values for one mole of each species are

$$S_{H_2O} = 70 \text{ J/K}; \quad S_{H_2} = 131 \text{ J/K}; \quad S_{O_2} = 205 \text{ J/K}. \tag{5.11}$$

Figure 5.3. To separate water into hydrogen and oxygen, just run an electric current through it. In this home experiment the electrodes are mechanical pencil leads (graphite). Bubbles of hydrogen (too small to see) form at the negative electrode (left) while bubbles of oxygen form at the positive electrode (right).

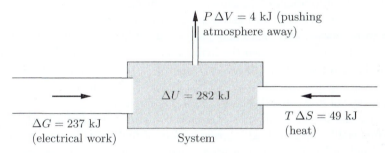

Figure 5.4. Energy-flow diagram for electrolysis of one mole of water. Under ideal conditions, 49 kJ of energy enter as heat ($T\Delta S$), so the electrical work required is only 237 kJ: $\Delta G = \Delta H - T\Delta S$. The difference between ΔH and ΔU is $P\Delta V = 4$ kJ, the work done to make room for the gases produced.

Subtract 70 from $(131 + \frac{1}{2} \cdot 205)$ and you get $+163$ J/K—the system's entropy *increases* by this amount. The maximum amount of heat that can enter the system is therefore $T\Delta S = (298 \text{ K})(163 \text{ J/K}) = 49$ kJ. The amount of energy that must enter as electrical work is the difference between 49 and 286, that is, 237 kJ.

This number, 237 kJ, is the change in the system's Gibbs free energy; it is the minimum "other" work required to make the reaction go. To summarize the computation,

$$\Delta G = \Delta H - T \Delta S,$$
$$237 \text{ kJ} = 286 \text{ kJ} - (298 \text{ K})(163 \text{ J/K}). \tag{5.12}$$

For convenience, standard tables (like the one at the back of this book) generally include ΔG values, saving you from having to do this kind of arithmetic.

We can also apply ΔG to the reverse reaction. If you can combine hydrogen and oxygen gas to produce water in a controlled way, you can, in principle, extract 237 kJ of electrical work for every mole of hydrogen consumed. This is the principle

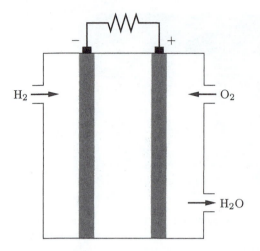

Figure 5.5. In a hydrogen fuel cell, hydrogen and oxygen gas pass through porous electrodes and react to form water, removing electrons from one electrode and depositing electrons on the other.

of the **fuel cell** (see Figure 5.5), a device that might replace the internal combustion engine in future automobiles.* In the process of producing this electrical work, the fuel cell will also expel 49 kJ of waste heat, in order to get rid of the excess entropy that was in the gases. But this waste heat is only 17% of the 286 kJ of heat that would be produced if you burned the hydrogen and tried to run a heat engine from it. So an ideal hydrogen fuel cell has an "efficiency" of 83%, much better than any practical heat engine. (In practice, the waste heat will be more and the efficiency less, but a typical fuel cell still beats almost any engine.)

A similar analysis can tell you the electrical energy output of a **battery**, which is like a fuel cell but has a fixed internal supply of fuel (usually not gaseous). For example, the familiar lead-acid cell used in car batteries runs on the reaction

$$\text{Pb} + \text{PbO}_2 + 4\text{H}^+ + 2\text{SO}_4^{2-} \longrightarrow 2\text{PbSO}_4 + 2\text{H}_2\text{O}. \tag{5.13}$$

According to thermodynamic tables, ΔG for this reaction is -394 kJ/mol, at standard pressure, temperature, and concentration of the solution. So the electrical work produced under these conditions, per mole of metallic lead, is 394 kJ. Meanwhile, ΔH for this reaction is -316 kJ/mol, so the energy that comes out of the chemicals is actually *less* than the work done, by 78 kJ. This extra energy comes from heat, absorbed from the environment. Along with this heat comes some entropy, but that's fine, since the entropy of the products is greater than the entropy of the reactants, by $(78\text{ kJ})/(298\text{ K}) = 260$ J/K (per mole). These energy flows are shown in Figure 5.6. When you *charge* the battery, the reaction runs in reverse, taking the system back to its initial state. Then you have to put the 78 kJ of heat back into the environment, to get rid of the excess entropy.

You can also calculate the *voltage* of a battery or fuel cell, provided that you know how many electrons it pushes around the circuit for each molecule that reacts. To determine this number, it helps to look at the chemistry in more detail. For a

---

*See Sivan Kartha and Patrick Grimes, "Fuel Cells: Energy Conversion for the Next Century," *Physics Today* **47**, 54–61 (November, 1994).

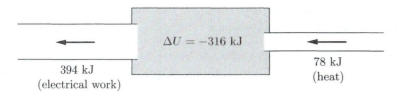

Figure 5.6. Energy-flow diagram for a lead-acid cell operating ideally. For each mole that reacts, the system's energy decreases by 316 kJ and its entropy increases by 260 J/K. Because of the entropy increase, the system can absorb 78 kJ of heat from the environment; the maximum work performed is therefore 394 kJ. (Because no gases are involved in this reaction, volume changes are negligible so $\Delta U \approx \Delta H$ and $\Delta F \approx \Delta G$.)

lead-acid cell, the reaction (5.13) takes place in three steps:

$$\begin{aligned}
\text{in solution:} \quad & 2SO_4^{2-} + 2H^+ \longrightarrow 2HSO_4^-; \\
\text{at } - \text{ electrode:} \quad & Pb + HSO_4^- \longrightarrow PbSO_4 + H^+ + 2e^-; \\
\text{at } + \text{ electrode:} \quad & PbO_2 + HSO_4^- + 3H^+ + 2e^- \longrightarrow PbSO_4 + 2H_2O.
\end{aligned} \qquad (5.14)$$

Thus, two electrons are pushed around the circuit each time the full reaction occurs. The electrical work produced *per electron* is

$$\frac{394 \text{ kJ}}{2 \cdot 6.02 \times 10^{23}} = 3.27 \times 10^{-19} \text{ J} = 2.04 \text{ eV}. \qquad (5.15)$$

But 1 volt is just the voltage needed to give each electron 1 eV of energy, so the cell has a voltage of 2.04 V. In practice the voltage may be slightly different, because the concentrations used are different from the standard concentration (one mole per kilogram of water) assumed in thermodynamic tables. (By the way, a car battery contains six lead-acid cells, giving a total of about 12 V.)

Problem 5.3. Use the data at the back of this book to verify the values of ΔH and ΔG quoted above for the lead-acid reaction 5.13.

Problem 5.4. In a hydrogen fuel cell, the steps of the chemical reaction are

$$\begin{aligned}
\text{at } - \text{ electrode:} \quad & H_2 + 2OH^- \longrightarrow 2H_2O + 2e^-; \\
\text{at } + \text{ electrode:} \quad & \tfrac{1}{2}O_2 + H_2O + 2e^- \longrightarrow 2OH^-.
\end{aligned}$$

Calculate the voltage of the cell. What is the minimum voltage required for electrolysis of water? Explain briefly.

Problem 5.5. Consider a fuel cell that uses methane ("natural gas") as fuel. The reaction is

$$CH_4 + 2O_2 \longrightarrow 2H_2O + CO_2.$$

(a) Use the data at the back of this book to determine the values of ΔH and ΔG for this reaction, for one mole of methane. Assume that the reaction takes place at room temperature and atmospheric pressure.

(b) Assuming ideal performance, how much electrical work can you get out of the cell, for each mole of methane fuel?

(c) How much waste heat is produced, for each mole of methane fuel?

(d) The steps of this reaction are

$$\text{at} - \text{electrode:} \quad CH_4 + 2H_2O \longrightarrow CO_2 + 8H^+ + 8e^-;$$
$$\text{at} + \text{electrode:} \quad 2O_2 + 8H^+ + 8e^- \longrightarrow 4H_2O.$$

What is the voltage of the cell?

Problem 5.6. A muscle can be thought of as a fuel cell, producing work from the metabolism of glucose:

$$C_6H_{12}O_6 + 6O_2 \longrightarrow 6CO_2 + 6H_2O.$$

(a) Use the data at the back of this book to determine the values of ΔH and ΔG for this reaction, for one mole of glucose. Assume that the reaction takes place at room temperature and atmospheric pressure.

(b) What is the maximum amount of work that a muscle can perform, for each mole of glucose consumed, assuming ideal operation?

(c) Still assuming ideal operation, how much heat is absorbed or expelled by the chemicals during the metabolism of a mole of glucose? (Be sure to say which direction the heat flows.)

(d) Use the concept of entropy to explain why the heat flows in the direction it does.

(e) How would your answers to parts (b) and (c) change, if the operation of the muscle is not ideal?

Problem 5.7. The metabolism of a glucose molecule (see previous problem) occurs in many steps, resulting in the synthesis of 38 molecules of ATP (adenosine triphosphate) out of ADP (adenosine diphosphate) and phosphate ions. When the ATP splits back into ADP and phosphate, it liberates energy that is used in a host of important processes including protein synthesis, active transport of molecules across cell membranes, and muscle contraction. In a muscle, the reaction ATP → ADP + phosphate is catalyzed by an enzyme called myosin that is attached to a muscle filament. As the reaction takes place, the myosin molecule pulls on an adjacent filament, causing the muscle to contract. The force it exerts averages about 4 piconewtons and acts over a distance of about 11 nm. From this data and the results of the previous problem, compute the "efficiency" of a muscle, that is, the ratio of the actual work done to the maximum work that the laws of thermodynamics would allow.

Thermodynamic Identities

If you're given the enthalpy or free energy of a substance under one set of conditions, but need to know its value under some other conditions, there are some handy formulas that are often useful. These formulas resemble the thermodynamic identity,

$$dU = T\,dS - P\,dV + \mu\,dN, \tag{5.16}$$

but are written for H or F or G instead of U.

I'll start by deriving the formula for the change in H. If we imagine changing

H, U, P, and V by infinitesimal amounts, then the definition $H = U + PV$ tells us that

$$dH = dU + P\,dV + V\,dP. \tag{5.17}$$

The last two terms give the change in the product PV, according to the product rule for derivatives. Now use the thermodynamic identity 5.16 to eliminate dU, and cancel the $P\,dV$ terms to obtain

$$dH = T\,dS + V\,dP + \mu\,dN. \tag{5.18}$$

This "thermodynamic identity for H" tells you how H changes as you change the entropy, pressure, and/or number of particles.*

Similar logic can be applied to F or G. From the definition of the Helmholtz free energy $(F = U - TS)$, we have

$$dF = dU - T\,dS - S\,dT. \tag{5.19}$$

Plugging in equation 5.16 for dU and canceling the $T\,dS$ terms gives

$$dF = -S\,dT - P\,dV + \mu\,dN. \tag{5.20}$$

I'll call this result the "thermodynamic identity for F." From it one can derive a variety of formulas for partial derivatives. For instance, holding V and N fixed yields the identity

$$S = -\left(\frac{\partial F}{\partial T}\right)_{V,N}. \tag{5.21}$$

Similarly, holding T and either N or V fixed gives

$$P = -\left(\frac{\partial F}{\partial V}\right)_{T,N}, \qquad \mu = \left(\frac{\partial F}{\partial N}\right)_{T,V}. \tag{5.22}$$

Finally, you can derive the thermodynamic identity for G,

$$dG = -S\,dT + V\,dP + \mu\,dN, \tag{5.23}$$

and from it the following partial derivative formulas:

$$S = -\left(\frac{\partial G}{\partial T}\right)_{P,N}, \qquad V = \left(\frac{\partial G}{\partial P}\right)_{T,N}, \qquad \mu = \left(\frac{\partial G}{\partial N}\right)_{T,P}. \tag{5.24}$$

These formulas are especially useful for computing Gibbs free energies at nonstandard temperatures and pressures. For example, since the volume of a mole of

---

*Because of the thermodynamic identity for U, it is most natural to think of U as a function of the variables S, V, and N. Similarly, it is most natural to think of H as a function of S, P, and N. Adding the PV term to U is therefore a kind of change of variables, from V to P. Similarly, subtracting TS changes variables from S to T. The technical name for such a change is **Legendre transformation**.

graphite is 5.3×10^{-6} m^3, its Gibbs free energy increases by 5.3×10^{-6} J for each pascal (N/m^2) of additional pressure.

In all of these formulas I have implicitly assumed that the system contains only one type of particles. If it is a mixture of several types, then you need to replace $\mu \, dN$ with $\sum \mu_i \, dN_i$ in every thermodynamic identity. In the partial-derivative formulas with N fixed, *all* the N's must be held fixed. And each formula with $\partial/\partial N$ becomes several formulas; so for a mixture of two types of particles,

$$\mu_1 = \left(\frac{\partial G}{\partial N_1} \right)_{T,P,N_2} \quad \text{and} \quad \mu_2 = \left(\frac{\partial G}{\partial N_2} \right)_{T,P,N_1}. \quad (5.25)$$

Problem 5.8. Derive the thermodynamic identity for G (equation 5.23), and from it the three partial derivative relations 5.24.

Problem 5.9. Sketch a qualitatively accurate graph of G vs. T for a pure substance as it changes from solid to liquid to gas at fixed pressure. Think carefully about the slope of the graph. Mark the points of the phase transformations and discuss the features of the graph briefly.

Problem 5.10. Suppose you have a mole of water at 25°C and atmospheric pressure. Use the data at the back of this book to determine what happens to its Gibbs free energy if you raise the temperature to 30°C. To compensate for this change, you could increase the pressure on the water. How much pressure would be required?

Problem 5.11. Suppose that a hydrogen fuel cell, as described in the text, is to be operated at 75°C and atmospheric pressure. We wish to estimate the maximum electrical work done by the cell, using only the room-temperature data at the back of this book. It is convenient to first establish a zero-point for each of the three substances, H$_2$, O$_2$, and H$_2$O. Let us take G for both H$_2$ and O$_2$ to be zero at 25°C, so that G for a mole of H$_2$O is -237 kJ at 25°C.

(a) Using these conventions, estimate the Gibbs free energy of a mole of H$_2$ at 75°C. Repeat for O$_2$ and H$_2$O.

(b) Using the results of part (a), calculate the maximum electrical work done by the cell at 75°C, for one mole of hydrogen fuel. Compare to the ideal performance of the cell at 25°C.

Problem 5.12. Functions encountered in physics are generally well enough behaved that their mixed partial derivatives do not depend on which derivative is taken first. Therefore, for instance,

$$\frac{\partial}{\partial V} \left(\frac{\partial U}{\partial S} \right) = \frac{\partial}{\partial S} \left(\frac{\partial U}{\partial V} \right),$$

where each $\partial/\partial V$ is taken with S fixed, each $\partial/\partial S$ is taken with V fixed, and N is always held fixed. From the thermodynamic identity (for U) you can evaluate the partial derivatives in parentheses to obtain

$$\left(\frac{\partial T}{\partial V} \right)_S = -\left(\frac{\partial P}{\partial S} \right)_V,$$

a nontrivial identity called a **Maxwell relation**. Go through the derivation of this relation step by step. Then derive an analogous Maxwell relation from each of the other three thermodynamic identities discussed in the text (for H, F, and G). Hold N fixed in all the partial derivatives; other Maxwell relations *can* be derived by considering partial derivatives with respect to N, but after you've done four of them the novelty begins to wear off. For applications of these Maxwell relations, see the next four problems.

Problem 5.13. Use a Maxwell relation from the previous problem and the third law of thermodynamics to prove that the thermal expansion coefficient β (defined in Problem 1.7) must be zero at $T = 0$.

Problem 5.14. The partial-derivative relations derived in Problems 1.46, 3.33, and 5.12, plus a bit more partial-derivative trickery, can be used to derive a completely general relation between C_P and C_V.

(a) With the heat capacity expressions from Problem 3.33 in mind, first consider S to be a function of T and V. Expand dS in terms of the partial derivatives $(\partial S/\partial T)_V$ and $(\partial S/\partial V)_T$. Note that one of these derivatives is related to C_V.

(b) To bring in C_P, consider V to be a function of T and P and expand dV in terms of partial derivatives in a similar way. Plug this expression for dV into the result of part (a), then set $dP = 0$ and note that you have derived a nontrivial expression for $(\partial S/\partial T)_P$. This derivative is related to C_P, so you now have a formula for the difference $C_P - C_V$.

(c) Write the remaining partial derivatives in terms of measurable quantities using a Maxwell relation and the result of Problem 1.46. Your final result should be
$$C_P = C_V + \frac{TV\beta^2}{\kappa_T}.$$

(d) Check that this formula gives the correct value of $C_P - C_V$ for an ideal gas.

(e) Use this formula to argue that C_P cannot be less than C_V.

(f) Use the data in Problem 1.46 to evaluate $C_P - C_V$ for water and for mercury at room temperature. By what percentage do the two heat capacities differ?

(g) Figure 1.14 shows measured values of C_P for three elemental solids, compared to predicted values of C_V. It turns out that a graph of β vs. T for a solid has same general appearance as a graph of heat capacity. Use this fact to explain why C_P and C_V agree at low temperatures but diverge in the way they do at higher temperatures.

Problem 5.15. The formula for $C_P - C_V$ derived in the previous problem can also be derived starting with the definitions of these quantities in terms of U and H. Do so. Most of the derivation is very similar, but at one point you need to use the relation $P = -(\partial F/\partial V)_T$.

Problem 5.16. A formula analogous to that for $C_P - C_V$ relates the isothermal and isentropic compressibilities of a material:
$$\kappa_T = \kappa_S + \frac{TV\beta^2}{C_P}.$$
(Here $\kappa_S = -(1/V)(\partial V/\partial P)_S$ is the reciprocal of the adiabatic bulk modulus considered in Problem 1.39.) Derive this formula. Also check that it is true for an ideal gas.

Problem 5.17. The enthalpy and Gibbs free energy, as defined in this section, give special treatment to mechanical (compression-expansion) work, $-P\,dV$. Analogous quantities can be defined for other kinds of work, for instance, magnetic work.[*] Consider the situation shown in Figure 5.7, where a long solenoid (N turns, total length L) surrounds a magnetic specimen (perhaps a paramagnetic solid). If the magnetic field inside the specimen is $\vec{B}$ and its total magnetic moment is $\vec{M}$, then we define an auxilliary field $\vec{\mathcal{H}}$ (often called simply the magnetic field) by the relation

$$\vec{\mathcal{H}} \equiv \frac{1}{\mu_0}\vec{B} - \frac{\vec{M}}{V},$$

where μ_0 is the "permeability of free space," $4\pi \times 10^{-7}\,\mathrm{N/A^2}$. Assuming cylindrical symmetry, all vectors must point either left or right, so we can drop the $\vec{}$ symbols and agree that rightward is positive, leftward negative. From Ampere's law, one can also show that when the current in the wire is I, the $\mathcal{H}$ field inside the solenoid is NI/L, whether or not the specimen is present.

(a) Imagine making an infinitesimal change in the current in the wire, resulting in infinitesimal changes in B, M, and $\mathcal{H}$. Use Faraday's law to show that the work required (from the power supply) to accomplish this change is $W_{\text{total}} = V\mathcal{H}\,dB$. (Neglect the resistance of the wire.)

(b) Rewrite the result of part (a) in terms of $\mathcal{H}$ and M, then subtract off the work that would be required even if the specimen were not present. If we define W, the work done on the *system*,[†] to be what's left, show that $W = \mu_0\mathcal{H}\,dM$.

(c) What is the thermodynamic identity for this system? (Include magnetic work but not mechanical work or particle flow.)

(d) How would you define analogues of the enthalpy and Gibbs free energy for a magnetic system? (The Helmholtz free energy is defined in the same way as for a mechanical system.) Derive the thermodynamic identities for each of these quantities, and discuss their interpretations.

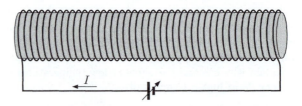

Figure 5.7. A long solenoid, surrounding a magnetic specimen, connected to a power supply that can change the current, performing magnetic work.

---

[*]This problem requires some familiarity with the theory of magnetism in matter. See, for instance, David J. Griffiths, *Introduction to Electrodynamics*, third edition (Prentice-Hall, Englewood Cliffs, NJ, 1999), Chapter 6.

[†]This is not the only possible definition of the "system." Different definitions are suitable for different physical situations, unfortunately leading to much confusion in terminology. For a more complete discussion of the thermodynamics of magnetism see Mandl (1988), Carrington (1994), and/or Pippard (1957).

5.2 Free Energy as a Force toward Equilibrium

For an *isolated* system, the *entropy* tends to increase; the system's entropy is what governs the direction of spontaneous change. But what if a system is *not* isolated? Suppose, instead, that our system is in good thermal contact with its environment (see Figure 5.8). Now energy can pass between the system and the environment, and the thing that tends to increase is not the *system's* entropy but rather the *total* entropy of system plus environment. In this section I'd like to restate this rule in a more useful form.

I'll assume that the environment acts as a "reservoir" of energy, large enough that it can absorb or release unlimited amounts of energy without changing its temperature. The total entropy of the universe can be written as $S + S_R$, where a subscript R indicates a property of the reservoir, while a quantity without a subscript refers to the system alone. The fundamental rule is that the total entropy of the universe tends to increase, so let's consider a small change in the total entropy:

$$dS_{\text{total}} = dS + dS_R. \tag{5.26}$$

I would like to write this quantity entirely in terms of system variables. To do so, I'll apply the thermodynamic identity, in the form

$$dS = \frac{1}{T}\,dU + \frac{P}{T}\,dV - \frac{\mu}{T}\,dN, \tag{5.27}$$

to the reservoir. First I'll assume that V and N for the reservoir are fixed—only energy travels in and out of the system. Then $dS_R = dU_R/T_R$, so equation 5.26 can be written

$$dS_{\text{total}} = dS + \frac{1}{T_R}\,dU_R. \tag{5.28}$$

But the temperature of the reservoir is the same as the temperature of the system, while the change dU_R in the reservoir's energy is minus the change dU in the system's energy. Therefore,

$$dS_{\text{total}} = dS - \frac{1}{T}\,dU = -\frac{1}{T}(dU - T\,dS) = -\frac{1}{T}\,dF. \tag{5.29}$$

Aha! Under these conditions (fixed T, V, and N), an increase in the total entropy of the universe is the same thing as a *decrease* in the Helmholtz free energy of the

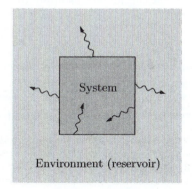

Figure 5.8. For a system that can exchange energy with its environment, the total entropy of both tends to increase.

Environment (reservoir)

system. So we can forget about the reservoir, and just remember that the system will do whatever it can to *minimize* its Helmholtz free energy. By the way, we could have guessed this result from equation 5.5, $\Delta F \leq W$. If *no* work is done on the system, F can only decrease.

If instead we let the volume of the system change but keep it at the same constant *pressure* as the reservoir, then the same line of reasoning gives

$$dS_{\text{total}} = dS - \frac{1}{T}\,dU - \frac{P}{T}\,dV = -\frac{1}{T}(dU - T\,dS + P\,dV) = -\frac{1}{T}\,dG, \qquad (5.30)$$

so it is the *Gibbs* free energy that tends to decrease. Again, we could have guessed this from equation 5.8, $\Delta G \leq W_{\text{other}}$.

Let me summarize these points, just for emphasis:

- At constant energy and volume, S tends to increase.
- At constant temperature and volume, F tends to decrease.
- At constant temperature and pressure, G tends to decrease.

All three statements assume that no particles are allowed to enter or leave the system (but see Problem 5.23).

We can understand these tendencies intuitively by looking again at the definitions of the Helmholtz and Gibbs free energies. Recall that

$$F \equiv U - TS. \qquad (5.31)$$

So in a constant-temperature environment, saying that F tends to decrease is the same as saying that U tends to decrease while S tends to increase. Well, we already know that S tends to increase. But does a system's energy tend to spontaneously decrease? Your intuition probably says yes, and this is correct, but *only* because when the system loses energy, its environment gains that energy, and therefore the *entropy* of the environment increases. At low temperature, this effect tends to be more important, since the entropy transferred to the environment for a given energy transfer is large, proportional to $1/T$. But at high temperature, the environment doesn't gain as much entropy, so the entropy of the system becomes more important in determining the behavior of F.

Similar considerations apply to the Gibbs free energy,

$$G \equiv U + PV - TS. \qquad (5.32)$$

Now, however, the entropy of the environment can increase in two ways: It can acquire energy from the system, or it can acquire volume from the system. So the system's U and V "want" to decrease, while S "wants" to increase, all in the interest of maximizing the total entropy of the universe.

Problem 5.18. Imagine that you drop a brick on the ground and it lands with a thud. Apparently the energy of this system tends to spontaneously decrease. Explain why.

Problem 5.19. In the previous section I derived the formula $(\partial F/\partial V)_T = -P$. Explain why this formula makes intuitive sense, by discussing graphs of F vs. V with different slopes.

Problem 5.20. The first excited energy level of a hydrogen atom has an energy of 10.2 eV, if we take the ground-state energy to be zero. However, the first excited level is really four independent states, all with the same energy. We can therefore assign it an entropy of $S = k \ln 4$, since for this given value of the energy, the multiplicity is 4. Question: For what temperatures is the Helmholtz free energy of a hydrogen atom in the first excited level positive, and for what temperatures is it negative? (Comment: When F for the level is negative, the atom will spontaneously go from the ground state into that level, since $F = 0$ for the ground state and F always tends to decrease. However, for a system this small, the conclusion is only a probabilistic statement; random fluctuations will be very significant.)

Extensive and Intensive Quantities

The number of potentially interesting thermodynamic variables has been growing lately. We now have U, V, N, S, T, P, μ, H, F, and G, among others. One way to organize all these quantities is to pick out the ones that double if you simply double the amount of stuff, adding the new alongside what you had originally (see Figure 5.9). Under this hypothetical operation, you end up with twice the energy and twice the volume, but *not* twice the temperature. Those quantities that *do* double are called **extensive quantities**. Those quantities that are *unchanged* when the amount of stuff doubles are called **intensive quantities**. Here's a list, divided according to this classification:

$$\text{Extensive: } V, \; N, \; S, \; U, \; H, \; F, \; G, \; \text{mass}$$
$$\text{Intensive: } T, \; P, \; \mu, \; \text{density}$$

If you multiply an extensive quantity by an intensive quantity, you end up with an extensive quantity; for example, volume $\times$ density $=$ mass. By the same token, if you divide one extensive quantity by another, you get an intensive quantity. If you multiply two extensive quantities together, you get something that is *neither*; if you're confronted with such a product in one of your calculations, there's a good chance you did something wrong. Adding two quantities of the same type

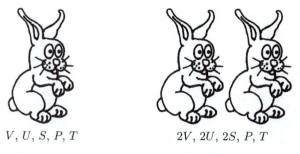

V, U, S, P, T $2V, 2U, 2S, P, T$

Figure 5.9. Two rabbits have twice as much volume, energy, and entropy as one rabbit, but not twice as much pressure or temperature.

yields another quantity of that type; for instance, $H = U + PV$. Adding an extensive quantity to an intensive one isn't allowed at all, so (for instance) you'll never encounter the sum $G + \mu$, even though G and μ have the same units. There's nothing wrong with exponentiating an extensive quantity, however; then you get a quantity that is *multiplicative*, like $\Omega = e^{S/k}$.

It's a good exercise to go back over the various equations involving F and G and show that they make sense in terms of extensiveness and intensiveness. For instance, in the thermodynamic identity for G,

$$dG = -S\,dT + V\,dP + \sum_i \mu_i\,dN_i, \tag{5.33}$$

each term is extensive, because each product involves one extensive and one intensive quantity.

Problem 5.21. Is heat capacity (C) extensive or intensive? What about specific heat (c)? Explain briefly.

Gibbs Free Energy and Chemical Potential

Using the idea of extensive and intensive quantities, we can now derive another useful relation involving the Gibbs free energy. First recall the partial-derivative relation

$$\mu = \left(\frac{\partial G}{\partial N}\right)_{T,P}. \tag{5.34}$$

This equation says that if you add one particle to a system, holding the temperature and pressure fixed, the Gibbs free energy of the system increases by μ (see Figure 5.10). If you keep adding more particles, each one again adds μ to the Gibbs free energy. Now you might think that during this procedure the value of μ could gradually change, so that by the time you've doubled the number of particles, μ has a very different value from when you started. But in fact, if T and P are held fixed, this can't happen: Each additional particle must add exactly the *same* amount to G, because G is an extensive quantity that must simply grow in proportion to the number of particles. The constant of proportionality, according to equation 5.34, is simply μ:

$$G = N\mu. \tag{5.35}$$

This amazingly simple equation gives us a new interpretation of the chemical potential, at least for a pure system with only one type of particle: μ is just the Gibbs free energy per particle.

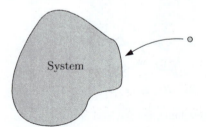

Figure 5.10. When you add a particle to a system, holding the temperature and pressure fixed, the system's Gibbs free energy increases by μ.

The preceding argument is subtle, so please think it through carefully. Perhaps the best way to understand it is to think about why the same logic can't be applied to the Helmholtz free energy, starting with the true relation

$$\mu = \left(\frac{\partial F}{\partial N}\right)_{T,V}. \tag{5.36}$$

The problem here is that to increase F by an amount μ, you have to add a particle while holding the temperature and *volume* fixed. Now, as you add more and more particles, μ *does* gradually change, because the system is becoming more dense. It's true that F is an extensive quantity, but this does *not* imply that F doubles when you double the density of the system, holding its volume fixed. In the previous paragraph it was crucial that the two variables being held fixed in equation 5.34, T and P, were both intensive, so that all extensive quantities could grow in proportion to N.

For a system containing more than one type of particle, equation 5.35 generalizes in a natural way:

$$G = N_1 \mu_1 + N_2 \mu_2 + \cdots = \sum_i N_i \mu_i. \tag{5.37}$$

The proof is the same as before, except that we imagine building up the system in infinitesimal increments keeping the proportions of the various species fixed throughout the process. This result does *not* imply, however, that G for a mixture is simply equal to the sum of the G's for the pure components. The μ's in equation 5.37 are generally *different* from their values for the corresponding pure substances.

As a first application of equation 5.35, let me now derive a very general formula for the chemical potential of an ideal gas. Consider a fixed amount of gas at a fixed temperature, as we vary the pressure. By equations 5.35 and 5.24,

$$\frac{\partial \mu}{\partial P} = \frac{1}{N}\frac{\partial G}{\partial P} = \frac{V}{N}. \tag{5.38}$$

But by the ideal gas law this quantity is just kT/P. Integrating both sides from $P°$ up to P therefore gives

$$\mu(T, P) - \mu(T, P°) = kT \ln(P/P°). \tag{5.39}$$

Here $P°$ can be any convenient reference pressure. Usually we take $P°$ to be atmospheric pressure (1 bar, to be precise). The standard symbol for μ for a gas at atmospheric pressure is $\mu°$, so we can write

$$\mu(T, P) = \mu°(T) + kT \ln(P/P°). \tag{5.40}$$

Values of $\mu°$ (at least at room temperature) can be gotten from tables of Gibbs free energies ($\mu = G/N$). Equation 5.40 then tells you how μ varies as the pressure (or equivalently, the density) changes. And in a *mixture* of ideal gases, equation 5.40 applies to each species separately, if you take P to be the *partial* pressure of that species. This works because ideal gases are mostly empty space: How an ideal gas exchanges particles with its environment isn't going to be affected by the presence of another ideal gas.

Problem 5.22. Show that equation 5.40 is in agreement with the explicit formula for the chemical potential of a monatomic ideal gas derived in Section 3.5. Show how to calculate μ° for a monatomic ideal gas.

Problem 5.23. By subtracting μN from U, H, F, or G, one can obtain four new thermodynamic potentials. Of the four, the most useful is the **grand free energy** (or **grand potential**),

$$\Phi \equiv U - TS - \mu N.$$

(a) Derive the thermodynamic identity for Φ, and the related formulas for the partial derivatives of Φ with respect to T, V, and μ.

(b) Prove that, for a system in thermal and diffusive equilibrium (with a reservoir that can supply both energy and particles), Φ tends to decrease.

(c) Prove that $\Phi = -PV$.

(d) As a simple application, let the system be a single proton, which can be "occupied" either by a single electron (making a hydrogen atom, with energy -13.6 eV) or by none (with energy zero). Neglect the excited states of the atom and the two spin states of the electron, so that both the occupied and unoccupied states of the proton have zero entropy. Suppose that this proton is in the atmosphere of the sun, a reservoir with a temperature of 5800 K and an electron concentration of about 2×10^{19} per cubic meter. Calculate Φ for both the occupied and unoccupied states, to determine which is more stable under these conditions. To compute the chemical potential of the electrons, treat them as an ideal gas. At about what temperature would the occupied and unoccupied states be equally stable, for this value of the electron concentration? (As in Problem 5.20, the prediction for such a small system is only a probabilistic one.)

5.3 Phase Transformations of Pure Substances

A **phase transformation** is a discontinuous change in the properties of a substance, as its environment is changed only infinitesimally. Familiar examples include melting ice and boiling water, either of which can be accomplished with only a very small change in temperature. The different forms of the substance—in this case ice, water, and steam—are called **phases**.

Often there is more than one variable that can affect the phase of a substance. For instance, you can condense steam either by lowering the temperature or by raising the pressure. A graph showing the equilibrium phases as a function of temperature and pressure is called a **phase diagram**.

Figure 5.11 shows a qualitative phase diagram for H_2O, along with some quantitative data on its phase transformations. The diagram is divided into three regions, indicating the conditions under which ice, water, or steam is the most stable phase. It's important to realize, though, that "metastable" phases can still exist; for instance, liquid water can be "supercooled" below the freezing point yet remain a liquid for some time. At high pressures there are actually several different phases of ice, with differing crystal structures and other physical properties.

The lines on a phase diagram represent conditions under which two different phases can coexist in equilibrium; for instance, ice and water can coexist stably at

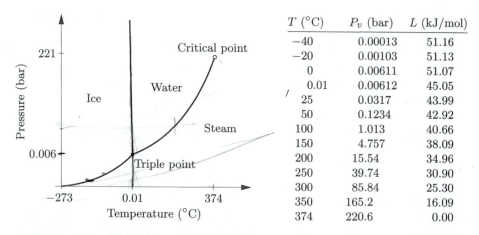

T (°C)	P_v (bar)	L (kJ/mol)
−40	0.00013	51.16
−20	0.00103	51.13
0	0.00611	51.07
0.01	0.00612	45.05
25	0.0317	43.99
50	0.1234	42.92
100	1.013	40.66
150	4.757	38.09
200	15.54	34.96
250	39.74	30.90
300	85.84	25.30
350	165.2	16.09
374	220.6	0.00

Figure 5.11. Phase diagram for H_2O (not to scale). The table gives the vapor pressure and molar latent heat for the solid-gas transformation (first three entries) and the liquid-gas transformation (remaining entries). Data from Keenan et al. (1978) and Lide (1994).

0°C and 1 atm ($\approx$ 1 bar). The pressure at which a gas can coexist with its solid or liquid phase is called the **vapor pressure**; thus the vapor pressure of water at room temperature is approximately 0.03 bar. At $T = 0.01$°C and $P = 0.006$ bar, all three phases can coexist; this point is called the **triple point**. At lower pressures, liquid water cannot exist (in equilibrium): ice "sublimates" directly into vapor.

You have probably observed sublimation of "dry ice," frozen carbon dioxide. Evidently, the triple point of carbon dioxide lies above atmospheric pressure; in fact it is at 5.2 bars. A qualitative phase diagram for carbon dioxide is shown in Figure 5.12. Another difference between CO_2 and H_2O is the slope of the solid-liquid phase boundary. Most substances are like carbon dioxide: Applying more pressure *raises* the melting temperature. Ice, however, is unusual: Applying

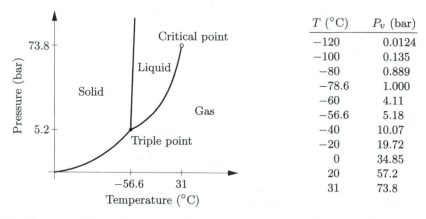

T (°C)	P_v (bar)
−120	0.0124
−100	0.135
−80	0.889
−78.6	1.000
−60	4.11
−56.6	5.18
−40	10.07
−20	19.72
0	34.85
20	57.2
31	73.8

Figure 5.12. Phase diagram for carbon dioxide (not to scale). The table gives the vapor pressure along the solid-gas and liquid-gas equilibrium curves. Data from Lide (1994) and Reynolds (1979).

pressure *lowers* its melting temperature. We will soon see that this is a result of the fact that ice is less dense than water.

The liquid-gas phase boundary always has a positive slope: If you have liquid and gas in equilibrium and you raise the temperature, you must apply *more* pressure to keep the liquid from vaporizing. As the pressure increases, however, the gas becomes more dense, so the difference between liquid and gas grows less. Eventually a point is reached where there is no longer any discontinuous change from liquid to gas. This point is called the **critical point**, and occurs at 374°C and 221 bars for H_2O. The critical point of carbon dioxide is more accessible, at 31°C and 74 bars, while that of nitrogen is at only 126 K and 34 bars. Close to the critical point, it's best to hedge and simply call the substance a "fluid." There's no critical point on the solid-liquid phase boundary, since the distinction between solids and liquids is a qualitative issue (solids having crystal structure and liquids having randomly arranged molecules), not just a matter of degree. Some materials made of long molecules can, however, form a **liquid crystal** phase, in which the molecules move around randomly as in a liquid but still tend to be oriented parallel to each other.

Helium has the most exotic phase behavior of any element. Figure 5.13 shows the phase diagrams of the two isotopes of helium, the common isotope ^4He and the rare isotope ^3He. The boiling point of ^4He at atmospheric pressure is only 4.2 K, and the critical point is only slightly higher, at 5.2 K and 2.3 bars; for ^3He these parameters are somewhat lower still. Helium is the only element that remains a liquid at absolute zero temperature: It *will* form a solid phase, but only at rather high pressures, about 25 bars for ^4He and 30 bars for ^3He. The solid-liquid phase boundary for ^4He is almost horizontal below 1 K, while for ^3He this boundary has a *negative* slope below 0.3 K. Even more interesting, ^4He has two distinct liquid phases: a "normal" phase called helium I, and a **superfluid**

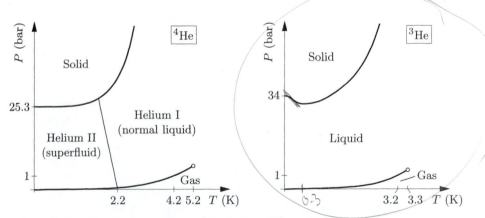

Figure 5.13. Phase diagrams of ^4He (left) and ^3He (right). Neither diagram is to scale, but qualitative relations between the diagrams are shown correctly. Not shown are the three different solid phases (crystal structures) of each isotope, or the superfluid phases of ^3He below 3 mK.

phase, below about 2 K, called helium II. The superfluid phase has a number of remarkable properties including zero viscosity and very high thermal conductivity. Helium-3 actually has two distinct superfluid phases, but only at temperatures below 3 mK.

Besides temperature and pressure, changing other variables such as composition and magnetic field strength can also cause phase transformations. Figure 5.14 shows phase diagrams for two different magnetic systems. At left is the diagram for a typical **type-I superconductor**, such as tin or mercury or lead. The superconducting phase, with zero electrical resistance, exists only when both the temperature and the external magnetic field strength are sufficiently low. At right is the diagram for a **ferromagnet** such as iron, which has *magnetized* phases pointing either up or down, depending on the direction of the applied field. (For simplicity, this diagram assumes that the applied field always points either up or down along a given axis.) When the applied field is zero, phases that are magnetized in both directions can coexist. As the temperature is raised, however, the magnetization of both phases becomes weaker. Eventually, at the **Curie temperature** (1043 K for iron), the magnetization disappears completely, so the phase boundary ends at a critical point.[*]

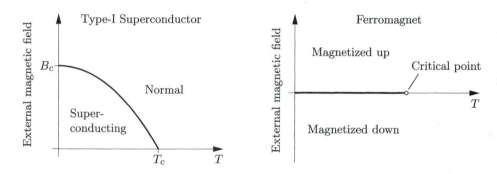

Figure 5.14. Left: Phase diagram for a typical type-I superconductor. For lead, $T_c = 7.2$ K and $B_c = 0.08$ T. Right: Phase diagram for a ferromagnet, assuming that the applied field and magnetization are always along a given axis.

---

[*]For several decades people have tried to classify phase transformations according to the abruptness of the change. Solid-liquid and liquid-gas transformations are classified as "first-order," because S and V, the *first* derivatives of G, are discontinuous at the phase boundary. Less abrupt transitions (such as critical points and the helium I to helium II transition) used to be classified as "second-order" and so on, depending on how many successive derivatives you had to take before getting a discontinuous quantity. Because of various problems with this classification scheme, the current fashion is to simply call all the higher-order transitions "continuous."

Diamonds and Graphite

Elemental carbon has two familiar phases, diamond and graphite (both solids, but with different crystal structures). At ordinary pressures the more stable phase is graphite, so diamonds will spontaneously convert to graphite, although this process is extremely slow at room temperature. (At high temperatures the conversion proceeds more rapidly, so if you own any diamonds, be sure not to throw them into the fireplace.[*])

The fact that graphite is more stable than diamond under standard conditions is reflected in their Gibbs free energies: The Gibbs free energy of a mole of diamond is greater, by 2900 J, than the Gibbs free energy of a mole of graphite. At a given temperature and pressure, the stable phase is always the one with the lower Gibbs free energy, according to the analysis of Section 5.2.

But the difference of 2900 J is for standard conditions, 298 K and atmospheric pressure (1 bar). What happens at higher pressures? The pressure dependence of the Gibbs free energy is determined by the volume of the substance,

$$\left(\frac{\partial G}{\partial P}\right)_{T,N} = V, \tag{5.41}$$

and since a mole of graphite has a greater volume than a mole of diamond, its Gibbs free energy will grow more rapidly as the pressure is raised. Figure 5.15 shows a graph of G vs. P for both substances. If we treat the volumes as constant (neglecting the compressibility of both substances), then each curve is a straight line. The slopes are $V = 5.31 \times 10^{-6}$ m^3 for graphite and $V = 3.42 \times 10^{-6}$ m^3 for diamond. As you can see, the two lines intersect at a pressure of about 15 kilobars. Above this very high pressure, diamond should be more stable than graphite. Apparently,

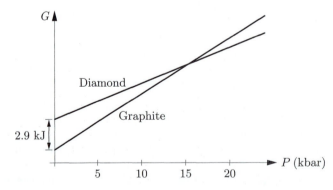

Figure 5.15. Molar Gibbs free energies of diamond and graphite as functions of pressure, at room temperature. These straight-line graphs are extrapolated from low pressures, neglecting the changes in volume as pressure increases.

---

[*]The temperature required to convert diamond to graphite quickly is actually quite high, about 1500°C. But in the presence of oxygen, either diamond or graphite will easily burn to form carbon dioxide.

natural diamonds must form at very great depths. Taking rock to be about three times as dense as water, it's easy to estimate that underground pressures normally increase by 3 bars for every 10 meters of depth. So a pressure of 15 kbar requires a depth of about 50 kilometers.

The temperature dependence of the Gibbs free energies can be determined in a similar way, using the relation

$$\left(\frac{\partial G}{\partial T}\right)_{P,N} = -S. \tag{5.42}$$

As the temperature is raised the Gibbs free energy of either substance decreases, but this decrease is more rapid for graphite since it has more entropy. Thus, raising the temperature tends to make graphite more stable relative to diamond; the higher the temperature, the more pressure is required before diamond becomes the stable phase.

Analyses of this type are extremely useful to geochemists, whose job is to look at rocks and determine the conditions under which they formed. More generally, the Gibbs free energy is the key to attaining a quantitative understanding of phase transformations.

Problem 5.24. Go through the arithmetic to verify that diamond becomes more stable than graphite at approximately 15 kbar.

Problem 5.25. In working high-pressure geochemistry problems it is usually more convenient to express volumes in units of kJ/kbar. Work out the conversion factor between this unit and m^3.

Problem 5.26. How can diamond ever be more stable than graphite, when it has less entropy? Explain how at high pressures the conversion of graphite to diamond can increase the *total* entropy of the carbon plus its environment.

Problem 5.27. Graphite is more compressible than diamond.

(a) Taking compressibilities into account, would you expect the transition from graphite to diamond to occur at higher or lower pressure than that predicted in the text?

(b) The isothermal compressibility of graphite is about 3×10^{-6} bar^{-1}, while that of diamond is more than ten times less and hence negligible in comparison. (Isothermal compressibility is the fractional reduction in volume per unit increase in pressure, as defined in Problem 1.46.) Use this information to make a revised estimate of the pressure at which diamond becomes more stable than graphite (at room temperature).

Problem 5.28. Calcium carbonate, $CaCO_3$, has two common crystalline forms, calcite and aragonite. Thermodynamic data for these phases can be found at the back of this book.

(a) Which is stable at earth's surface, calcite or aragonite?

(b) Calculate the pressure (still at room temperature) at which the other phase should become stable.

Problem 5.29. Aluminum silicate, Al_2SiO_5, has three different crystalline forms: kyanite, andalusite, and sillimanite. Because each is stable under a different set of temperature-pressure conditions, and all are commonly found in metamorphic rocks, these minerals are important indicators of the geologic history of rock bodies.

(a) Referring to the thermodynamic data at the back of this book, argue that at 298 K the stable phase should be kyanite, regardless of pressure.

(b) Now consider what happens at fixed pressure as we vary the temperature. Let ΔG be the difference in Gibbs free energies between any two phases, and similarly for ΔS. Show that the T dependence of ΔG is given by

$$\Delta G(T_2) = \Delta G(T_1) - \int_{T_1}^{T_2} \Delta S(T)\, dT.$$

Although the entropy of any given phase will increase significantly as the temperature increases, above room temperature it is often a good approximation to take ΔS, the *difference* in entropies between two phases, to be independent of T. This is because the temperature dependence of S is a function of the heat capacity (as we saw in Chapter 3), and the heat capacity of a solid at high temperature depends, to a good approximation, only on the number of atoms it contains.

(c) Taking ΔS to be independent of T, find the range of temperatures over which kyanite, andalusite, and sillimanite should be stable (at 1 bar).

(d) Referring to the room-temperature heat capacities of the three forms of Al_2SiO_5, discuss the accuracy the approximation ΔS = constant.

Problem 5.30. Sketch qualitatively accurate graphs of G vs. T for the three phases of H_2O (ice, water, and steam) at atmospheric pressure. Put all three graphs on the same set of axes, and label the temperatures $0°C$ and $100°C$. How would the graphs differ at a pressure of 0.001 bar?

Problem 5.31. Sketch qualitatively accurate graphs of G vs. P for the three phases of H_2O (ice, water, and steam) at $0°C$. Put all three graphs on the same set of axes, and label the point corresponding to atmospheric pressure. How would the graphs differ at slightly higher temperatures?

The Clausius-Clapeyron Relation

Since entropy determines the temperature dependence of the Gibbs free energy, while volume determines its pressure dependence, the shape of any phase boundary line on a PT diagram is related in a very simple way to the entropies and volumes of the two phases. Let me now derive this relation.

For definiteness, I'll discuss the phase boundary between a liquid and a gas, although it could just as well be any other phase boundary. Let's consider some fixed amount of the stuff, say one mole. At the phase boundary, this material is equally stable as a liquid or a gas, so its Gibbs free energy must be the same, whether it is in either phase:

$$G_l = G_g \qquad \text{at phase boundary.} \tag{5.43}$$

(You can also think of this condition in terms of the chemical potentials: If some liquid and some gas are in diffusive equilibrium with each other, then their chemical potentials, i.e., Gibbs free energies per molecule, must be equal.)

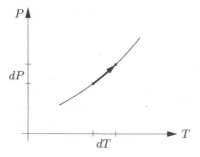

Figure 5.16. Infinitesimal changes in pressure and temperature, related in such a way as to remain on the phase boundary.

Now imagine increasing the temperature by dT and the pressure by dP, in such a way that the two phases remain equally stable (see Figure 5.16). Under this change, the Gibbs free energies must remain equal to each other, so

$$dG_l = dG_g \qquad \text{to remain on phase boundary.} \tag{5.44}$$

Therefore, by the thermodynamic identity for G (equation 5.23),

$$-S_l\,dT + V_l\,dP = -S_g\,dT + V_g\,dP. \tag{5.45}$$

(I've omitted the $\mu\,dN$ terms because I've already assumed that the total amount of stuff is fixed.) Now it's easy to solve for the slope of the phase boundary line, dP/dT:

$$\frac{dP}{dT} = \frac{S_g - S_l}{V_g - V_l}. \tag{5.46}$$

As expected, the slope is determined by the entropies and volumes of the two phases. A large difference in entropy means that a small change in temperature can be very significant in shifting the equilibrium from one phase to the other. This results in a steep phase boundary curve, since a large pressure change is then required to compensate the small temperature change. On the other hand, a large difference in volume means that a small change in pressure can be significant after all, making the phase boundary curve shallower.

It's often more convenient to write the difference in entropies, $S_g - S_l$, as L/T, where L is the (total) latent heat for converting the material (in whatever quantity we're considering) from liquid to gas. Then equation 5.46 takes the form

$$\frac{dP}{dT} = \frac{L}{T\,\Delta V}, \tag{5.47}$$

where $\Delta V = V_g - V_l$. (Notice that, since both L and ΔV are extensive, their ratio is intensive—independent of the amount of material.) This result is known as the **Clausius-Clapeyron relation**. It applies to the slope of any phase boundary line on a PT diagram, not just to the line separating liquid from gas.

As an example, consider again the diamond-graphite system. When a mole of diamond converts to graphite its entropy increases by 3.4 J/K, while its volume increases by 1.9×10^{-6} m^3. (Both of these numbers are for room temperature; at higher temperatures the difference in entropy is somewhat greater.) Therefore the slope of the diamond-graphite phase boundary is

$$\frac{dP}{dT} = \frac{\Delta S}{\Delta V} = \frac{3.4 \text{ J/K}}{1.9 \times 10^{-6} \text{ m}^3} = 1.8 \times 10^6 \text{ Pa/K} = 18 \text{ bar/K}. \tag{5.48}$$

In the previous subsection I showed that at room temperature, diamond is stable at pressures above approximately 15 kbar. Now we see that if the temperature is 100 K higher, we need an additional 1.8 kbar of pressure to make diamond stable. *Rapid* conversion of graphite to diamond requires still higher temperatures, and correspondingly higher pressures, as shown in the phase diagram in Figure 5.17. The first synthesis of diamond from graphite was accomplished at approximately 1800 K and 60 kbar. Natural diamonds are thought to form at similar pressures but somewhat lower temperatures, at depths of 100–200 km below earth's surface.*

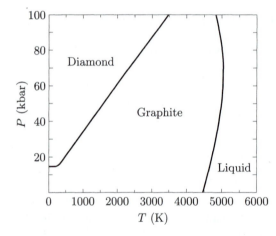

Figure 5.17. The experimental phase diagram of carbon. The stability region of the gas phase is not visible on this scale; the graphite-liquid-gas triple point is at the bottom of the graphite-liquid phase boundary, at 110 bars pressure. From David A. Young, *Phase Diagrams of the Elements* (University of California Press, Berkeley, 1991).

Problem 5.32. The density of ice is 917 kg/m^3.

(a) Use the Clausius-Clapeyron relation to explain why the slope of the phase boundary between water and ice is negative.

(b) How much pressure would you have to put on an ice cube to make it melt at −1°C?

(c) Approximately how deep under a glacier would you have to be before the weight of the ice above gives the pressure you found in part (b)? (Note that the pressure can be greater at some locations, as where the glacier flows over a protruding rock.)

(d) Make a rough estimate of the pressure under the blade of an ice skate, and calculate the melting temperature of ice at this pressure. Some authors have claimed that skaters glide with very little friction because the increased pressure under the blade melts the ice to create a thin layer of water. What do you think of this explanation?

---

*For more on the formation of natural diamonds and the processes that bring them near earth's surface, see Keith G. Cox, "Kimberlite Pipes," *Scientific American* **238**, 120–132 (April, 1978).

Problem 5.33. An inventor proposes to make a heat engine using water/ice as the working substance, taking advantage of the fact that water expands as it freezes. A weight to be lifted is placed on top of a piston over a cylinder of water at $1°C$. The system is then placed in thermal contact with a low-temperature reservoir at $-1°C$ until the water freezes into ice, lifting the weight. The weight is then removed and the ice is melted by putting it in contact with a high-temperature reservoir at $1°C$. The inventor is pleased with this device because it can seemingly perform an unlimited amount of work while absorbing only a finite amount of heat. Explain the flaw in the inventor's reasoning, and use the Clausius-Clapeyron relation to prove that the maximum efficiency of this engine is still given by the Carnot formula, $1 - T_c/T_h$.

Problem 5.34. Below 0.3 K the slope of the ^{3}He solid-liquid phase boundary is negative (see Figure 5.13).

(a) Which phase, solid or liquid, is more dense? Which phase has more entropy (per mole)? Explain your reasoning carefully.

(b) Use the third law of thermodynamics to argue that the slope of the phase boundary must go to zero at $T = 0$. (Note that the ^{4}He solid-liquid phase boundary is essentially horizontal below 1 K.)

(c) Suppose that you compress liquid ^{3}He adiabatically until it becomes a solid. If the temperature just before the phase change is 0.1 K, will the temperature after the phase change be higher or lower? Explain your reasoning carefully.

Problem 5.35. The Clausius-Clapeyron relation 5.47 is a differential equation that can, in principle, be solved to find the shape of the entire phase-boundary curve. To solve it, however, you have to know how both L and ΔV depend on temperature and pressure. Often, over a reasonably small section of the curve, you can take L to be constant. Moreover, if one of the phases is a gas, you can usually neglect the volume of the condensed phase and just take ΔV to be the volume of the gas, expressed in terms of temperature and pressure using the ideal gas law. Making all these assumptions, solve the differential equation explicitly to obtain the following formula for the phase boundary curve:

$$P = (\text{constant}) \times e^{-L/RT}.$$

This result is called the **vapor pressure equation**. Caution: Be sure to use this formula only when all the assumptions just listed are valid.

Problem 5.36. Effect of altitude on boiling water.

(a) Use the result of the previous problem and the data in Figure 5.11 to plot a graph of the vapor pressure of water between $50°C$ and $100°C$. How well can you match the data at the two endpoints?

(b) Reading the graph backwards, estimate the boiling temperature of water at each of the locations for which you determined the pressure in Problem 1.16. Explain why it takes longer to cook noodles when you're camping in the mountains.

(c) Show that the dependence of boiling temperature on altitude is very nearly (though not exactly) a linear function, and calculate the slope in degrees Celsius per thousand feet (or in degrees Celsius per kilometer).

Problem 5.37. Use the data at the back of this book to calculate the slope of the calcite-aragonite phase boundary (at 298 K). You located one point on this phase boundary in Problem 5.28; use this information to sketch the phase diagram of calcium carbonate.

Problem 5.38. In Problems 3.30 and 3.31 you calculated the entropies of diamond and graphite at 500 K. Use these values to predict the slope of the graphite-diamond phase boundary at 500 K, and compare to Figure 5.17. Why is the slope almost constant at still higher temperatures? Why is the slope zero at $T = 0$?

Problem 5.39. Consider again the aluminosilicate system treated in Problem 5.29. Calculate the slopes of all three phase boundaries for this system: kyanite-andalusite, kyanite-sillimanite, and andalusite-sillimanite. Sketch the phase diagram, and calculate the temperature and pressure of the triple point.

Problem 5.40. The methods of this section can also be applied to reactions in which one set of solids converts to another. A geologically important example is the transformation of albite into jadeite + quartz:

$$NaAlSi_3O_8 \longleftrightarrow NaAlSi_2O_6 + SiO_2.$$

Use the data at the back of this book to determine the temperatures and pressures under which a combination of jadeite and quartz is more stable than albite. Sketch the phase diagram of this system. For simplicity, neglect the temperature and pressure dependence of both ΔS and ΔV.

Problem 5.41. Suppose you have a liquid (say, water) in equilibrium with its gas phase, inside some closed container. You then pump in an inert gas (say, air), thus raising the pressure exerted on the liquid. What happens?

(a) For the liquid to remain in diffusive equilibrium with its gas phase, the chemical potentials of each must change by the same amount: $d\mu_l = d\mu_g$. Use this fact and equation 5.40 to derive a differential equation for the equilibrium vapor pressure, P_v, as a function of the total pressure P. (Treat the gases as ideal, and assume that none of the inert gas dissolves in the liquid.)

(b) Solve the differential equation to obtain

$$P_v(P) = P_v(P_v) \cdot e^{(P-P_v)V/NkT},$$

where the ratio V/N in the exponent is that of the *liquid*. (The quantity $P_v(P_v)$ is just the vapor pressure in the absence of the inert gas.) Thus, the presence of the inert gas leads to a slight increase in the vapor pressure: It causes more of the liquid to evaporate.

(c) Calculate the percent increase in vapor pressure when air at atmospheric pressure is added to a system of water and water vapor in equilibrium at 25°C. Argue more generally that the increase in vapor pressure due to the presence of an inert gas will be negligible except under extreme conditions.

Problem 5.42. Ordinarily, the partial pressure of water vapor in the air is less than the equilibrium vapor pressure at the ambient temperature; this is why a cup of water will spontaneously evaporate. The ratio of the partial pressure of water vapor to the equilibrium vapor pressure is called the **relative humidity**. When the relative humidity is 100%, so that water vapor in the atmosphere would be in diffusive equilibrium with a cup of liquid water, we say that the air is **saturated**.[*] The **dew point** is the temperature at which the relative humidity would be 100%, for a given partial pressure of water vapor.

(a) Use the vapor pressure equation (Problem 5.35) and the data in Figure 5.11 to plot a graph of the vapor pressure of water from 0°C to 40°C. Notice that the vapor pressure approximately doubles for every 10° increase in temperature.

(b) The temperature on a certain summer day is 30°C. What is the dew point if the relative humidity is 90%? What if the relative humidity is 40%?

Problem 5.43. Assume that the air you exhale is at 35°C, with a relative humidity of 90%. This air immediately mixes with environmental air at 10°C and unknown relative humidity; during the mixing, a variety of intermediate temperatures and water vapor percentages temporarily occur. If you are able to "see your breath" due to the formation of cloud droplets during this mixing, what can you conclude about the relative humidity of your environment? (Refer to the vapor pressure graph drawn in Problem 5.42.)

Problem 5.44. Suppose that an unsaturated air mass is rising and cooling at the dry adiabatic lapse rate found in Problem 1.40. If the temperature at ground level is 25°C and the relative humidity there is 50%, at what altitude will this air mass become saturated so that condensation begins and a cloud forms (see Figure 5.18)? (Refer to the vapor pressure graph drawn in Problem 5.42.)

Problem 5.45. In Problem 1.40 you calculated the atmospheric temperature gradient required for unsaturated air to spontaneously undergo convection. When a rising air mass becomes saturated, however, the condensing water droplets will give up energy, thus slowing the adiabatic cooling process.

(a) Use the first law of thermodynamics to show that, as condensation forms during adiabatic expansion, the temperature of an air mass changes by

$$dT = \frac{2}{7}\frac{T}{P}\,dP - \frac{2}{7}\frac{L}{nR}\,dn_w,$$

where n_w is the number of moles of water vapor present, L is the latent heat of vaporization per mole, and I've set $\gamma = 7/5$ for air. You may assume that the H_2O makes up only a small fraction of the air mass.

(b) Assuming that the air is always saturated during this process, the ratio n_w/n is a known function of temperature and pressure. Carefully express dn_w/dz in terms of dT/dz, dP/dz, and the vapor pressure $P_v(T)$. Use the Clausius-Clapeyron relation to eliminate dP_v/dT.

(c) Combine the results of parts (a) and (b) to obtain a formula relating the temperature gradient, dT/dz, to the pressure gradient, dP/dz. Eliminate

---

[*]This term is widely used, but is unfortunate and misleading. Air is not a sponge that can hold only a certain amount of liquid; even "saturated" air is mostly empty space. As shown in the previous problem, the density of water vapor that can exist in equilibrium has almost nothing to do with the presence of air.

Figure 5.18. Cumulus clouds form when rising air expands adiabatically and cools to the dew point (Problem 5.44); the onset of condensation slows the cooling, increasing the tendency of the air to rise further (Problem 5.45). These clouds began to form in late morning, in a sky that was clear only an hour before the photo was taken. By mid-afternoon they had developed into thunderstorms.

the latter using the "barometric equation" from Problem 1.16. You should finally obtain

$$\frac{dT}{dz} = -\left(\frac{2}{7}\frac{Mg}{R}\right)\frac{1 + \dfrac{P_v}{P}\dfrac{L}{RT}}{1 + \dfrac{2}{7}\dfrac{P_v}{P}\left(\dfrac{L}{RT}\right)^2},$$

where M is the mass of a mole of air. The prefactor is just the dry adiabatic lapse rate calculated in Problem 1.40, while the rest of the expression gives the correction due to heating from the condensing water vapor. The whole result is called the **wet adiabatic lapse rate**; it is the critical temperature gradient above which saturated air will spontaneously convect.

(d) Calculate the wet adiabatic lapse rate at atmospheric pressure (1 bar) and $25°C$, then at atmospheric pressure and $0°C$. Explain why the results are different, and discuss their implications. What happens at higher altitudes, where the pressure is lower?

Problem 5.46. Everything in this section so far has ignored the *boundary* between two phases, as if each molecule were unequivocally part of one phase or the other. In fact, the boundary is a kind of transition zone where molecules are in an environment that differs from both phases. Since the boundary zone is only a few molecules thick, its contribution to the total free energy of a system is very often negligible. One important exception, however, is the first tiny droplets or bubbles or grains that form as a material begins to undergo a phase transformation. The formation of these initial specks of a new phase is called **nucleation**. In this problem we will consider the nucleation of water droplets in a cloud.

The surface forming the boundary between any two given phases generally has a fixed thickness, regardless of its area. The additional Gibbs free energy of this surface is therefore directly proportional to its area; the constant of proportionality is called the **surface tension**, σ:

$$\sigma \equiv \frac{G_{boundary}}{A}.$$

If you have a blob of liquid in equilibrium with its vapor and you wish to stretch it into a shape that has the same volume but more surface area, then σ is the minimum work that you must perform, per unit of additional area, at fixed temperature and pressure. For water at $20°$C, $\sigma = 0.073$ J/m^2.

(a) Consider a spherical droplet of water containing N_l molecules, surrounded by $N - N_l$ molecules of water vapor. Neglecting surface tension for the moment, write down a formula for the total Gibbs free energy of this system in terms of N, N_l, and the chemical potentials of the liquid and vapor. Rewrite N_l in terms of v_l, the volume per molecule in the liquid, and r, the radius of the droplet.

(b) Now add to your expression for G a term to represent the surface tension, written in terms of r and σ.

(c) Sketch a qualitative graph of G vs. r for both signs of $\mu_g - \mu_l$, and discuss the implications. For which sign of $\mu_g - \mu_l$ does there exist a nonzero equilibrium radius? Is this equilibrium stable?

(d) Let r_c represent the critical equilibrium radius that you discussed qualitatively in part (c). Find an expression for r_c in terms of $\mu_g - \mu_l$. Then rewrite the difference of chemical potentials in terms of the relative humidity (see Problem 5.42), assuming that the vapor behaves as an ideal gas. (The relative humidity is defined in terms of equilibrium of a vapor with a flat surface, or with an infinitely large droplet.) Sketch a graph of the critical radius as a function of the relative humidity, including numbers. Discuss the implications. In particular, explain why it is unlikely that the clouds in our atmosphere would form by spontaneous aggregation of water molecules into droplets. (In fact, cloud droplets form around nuclei of dust particles and other foreign material, when the relative humidity is close to 100%.)

Problem 5.47. For a *magnetic* system held at constant T and $\mathcal{H}$ (see Problem 5.17), the quantity that is minimized is the magnetic analogue of the Gibbs free energy, which obeys the thermodynamic identity

$$dG_m = -S \, dT - \mu_0 M \, d\mathcal{H}.$$

Phase diagrams for two magnetic systems are shown in Figure 5.14; the vertical axis on each of these figures is $\mu_0 \mathcal{H}$.

(a) Derive an analogue of the Clausius-Clapeyron relation for the slope of a phase boundary in the $\mathcal{H}$-T plane. Write your equation in terms of the difference in entropy between the two phases.

(b) Discuss the application of your equation to the ferromagnet phase diagram in Figure 5.14.

(c) In a type-I superconductor, surface currents flow in such a way as to completely cancel the magnetic field (B, not $\mathcal{H}$) inside. Assuming that M is negligible when the material is in its normal (non-superconducting) state, discuss the application of your equation to the superconductor phase diagram in Figure 5.14. Which phase has the greater entropy? What happens to the difference in entropy between the phases at each end of the phase boundary?

The van der Waals Model

To understand phase transformations more deeply, a good approach is to introduce a specific mathematical model. For liquid-gas systems, the most famous model is the **van der Waals equation**,

$$\left(P + \frac{aN^2}{V^2}\right)(V - Nb) = NkT, \tag{5.49}$$

proposed by Johannes van der Waals in 1873. This is a modification of the ideal gas law that takes molecular interactions into account in an approximate way. (Any proposed relation among P, V, and T, like the ideal gas law or the van der Waals equation, is called an **equation of state**.)

The van der Waals equation makes two modifications to the ideal gas law: adding aN^2/V^2 to P and subtracting Nb from V. The second modification is easier to understand: A fluid can't be compressed all the way down to zero volume, so we've limited the volume to a minimum value of Nb, at which the pressure goes to infinity. The constant b then represents the minimum volume occupied by a molecule, when it's "touching" all its neighbors. The first modification, adding aN^2/V^2 to P, accounts for the short-range attractive forces between molecules when they're not touching (see Figure 5.19). Imagine freezing all the molecules in place, so that the only type of energy present is the negative potential energy due to molecular attraction. If we were to double the density of the system, each molecule would then have twice as many neighbors as before, so the potential energy due to all its interactions with neighbors would double. In other words, the potential energy associated with a single molecule's interactions with all its neighbors is proportional to the density of particles, or to N/V. The *total* potential energy associated with all molecules' interactions must then be proportional to N^2/V, since there are N molecules:

$$\text{total potential energy} = -\frac{aN^2}{V}, \tag{5.50}$$

where a is some positive constant of proportionality that depends on the type of molecules. To calculate the pressure, imagine varying the volume slightly while holding the entropy fixed (which isn't a problem if we've frozen all thermal motion); then by the thermodynamic identity, $dU = -P\,dV$ or $P = -(\partial U/\partial V)_S$. The contribution to the pressure from just the potential energy is therefore

$$P_{\text{due to p.e.}} = -\frac{d}{dV}\left(-\frac{aN^2}{V}\right) = -\frac{aN^2}{V^2}. \tag{5.51}$$

If we add this negative pressure to the pressure that the fluid would have in the

Figure 5.19. When two molecules come very close together they repel each other strongly. When they are a short distance apart they attract each other.

absence of attractive forces (namely, $NkT/(V - Nb)$), we obtain the van der Waals equation,

$$P = \frac{NkT}{V - Nb} - \frac{aN^2}{V^2}. \tag{5.52}$$

While the van der Waals equation has the right properties to account for the qualitative behavior of real fluids, I need to emphasize that it is nowhere near exact. In "deriving" it I've neglected a number of effects, most notably the fact that as a gas becomes more dense it can become inhomogeneous on the microscopic scale: Clusters of molecules can begin to form, violating my assertion that the number of neighbors a molecule has will be directly proportional to N/V. So throughout this section, please keep in mind that we won't be making any accurate quantitative predictions. What we're after is qualitative understanding, which can provide a starting point if you later decide to study liquid-gas phase transformations in more depth.

The constants a and b will have different values for different substances, and (since the model isn't exact) will even vary somewhat for the same substance under different conditions. For small molecules like N_2 and H_2O, a good value of b is about 6×10^{-29} m^3 $\approx$ (4 Å)^3, roughly the cube of the average width of the molecule. The constant a is much more variable, because some types of molecules attract each other much more strongly than others. For N_2, a good value of a is about 4×10^{-49} J·m^3, or 2.5 eV·Å^3. If we think of a as being roughly the product of the average interaction energy times the volume over which the interaction can act, then this value is fairly sensible: a small fraction of an electron-volt times a few tens of cubic ångstroms. The value of a for H_2O is about four times as large, because of the molecule's permanent electric polarization. Helium is at the other extreme, with interactions so weak that its value of a is 40 times less than that of nitrogen.

Now let us investigate the consequences of the van der Waals model. A good way to start is by plotting the predicted pressure as a function of volume for a variety of different temperatures (see Figure 5.20). At volumes much greater than Nb the isotherms are concave-up, like those of an ideal gas. At sufficiently high

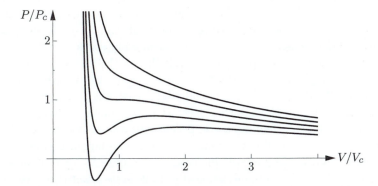

Figure 5.20. Isotherms (lines of constant temperature) for a van der Waals fluid. From bottom to top, the lines are for 0.8, 0.9, 1.0, 1.1, and 1.2 times T_c, the temperature at the critical point. The axes are labeled in units of the pressure and volume at the critical point; in these units the minimum volume (Nb) is 1/3.

temperatures, reducing the volume causes the pressure to rise smoothly, eventually approaching infinity as the volume goes to Nb. At lower temperatures, however, the behavior is much more complicated: As V decreases the isotherm rises, falls, and then rises again, seeming to imply that for some states, compressing the fluid can cause its pressure to decrease. Real fluids don't behave like this. But a more careful analysis shows that the van der Waals model doesn't predict this, either.

At a given temperature and pressure, the true equilibrium state of a system is determined by its Gibbs free energy. To calculate G for a van der Waals fluid, let's start with the thermodynamic identity for G:

$$dG = -S\,dT + V\,dP + \mu\,dN. \tag{5.53}$$

For a fixed amount of material at a given, fixed temperature, this equation reduces to $dG = V\,dP$. Dividing both sides by dV then gives

$$\left(\frac{\partial G}{\partial V}\right)_{N,T} = V\left(\frac{\partial P}{\partial V}\right)_{N,T}. \tag{5.54}$$

The right-hand side can be computed directly from the van der Waals equation (5.52), yielding

$$\left(\frac{\partial G}{\partial V}\right)_{N,T} = -\frac{NkTV}{(V-Nb)^2} + \frac{2aN^2}{V^2}. \tag{5.55}$$

To integrate the right-hand side, write the V in the numerator of the first term as $(V-Nb)+(Nb)$, then integrate each of these two pieces separately. The result is

$$G = -NkT\ln(V - Nb) + \frac{(NkT)(Nb)}{V - Nb} - \frac{2aN^2}{V} + c(T), \tag{5.56}$$

where the integration constant, $c(T)$, can be different for different temperatures but is unimportant for our purposes. This equation allows us to plot the Gibbs free energy for any fixed T.

Instead of plotting G as a function of volume, it's more useful to plot G vertically and P horizontally, calculating each as a function of the parameter V. Figure 5.21 shows an example, for the temperature whose isotherm is shown alongside. Although the van der Waals equation associates some pressures with more than one volume, the thermodynamically stable state is that with the lowest Gibbs free energy; thus the triangular loop in the graph of G (points 2-3-4-5-6) corresponds to unstable states. As the pressure is gradually increased, the system will go straight from point 2 to point 6, with an abrupt decrease in volume: a phase transformation. At point 2 we should call the fluid a gas, because its volume decreases rapidly with increasing pressure. At point 6 we should call the fluid a liquid, because its volume decreases only slightly under a large increase in pressure. At intermediate volumes between these points, the thermodynamically stable state is actually a combination of part gas and part liquid, still at the transition pressure, as indicated by the straight horizontal line on the PV diagram. The curved portion of the isotherm that is cut off by this straight line correctly indicates what the allowed states *would* be if the fluid were homogeneous; but these homogeneous states are unstable, since

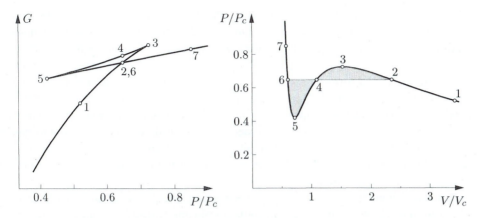

Figure 5.21. Gibbs free energy as a function of pressure for a van der Waals fluid at $T = 0.9T_c$. The corresponding isotherm is shown at right. States in the range 2-3-4-5-6 are unstable.

there is always another state (gas or liquid) at the same pressure with a lower Gibbs free energy.

The pressure at the phase transformation is easy enough to determine from the graph of G, but there is a clever method of reading it straight off the PV diagram, without plotting G at all. To derive this method, note that the net change in G as we go around the triangular loop (2-3-4-5-6) is zero:

$$0 = \int_{\text{loop}} dG = \int_{\text{loop}} \left(\frac{\partial G}{\partial P}\right)_T dP = \int_{\text{loop}} V\, dP. \qquad (5.57)$$

Written in this last form, the integral can be computed from the PV diagram, though it's easier to turn the diagram sideways (see Figure 5.22). The integral from point 2 to point 3 gives the entire area under this segment, but the integral from point 3 to point 4 cancels out all but the shaded region A. The integral from 4 to 5 gives minus the area under that segment, but then the integral from 5 to 6 adds back all but the shaded region B. Thus the entire integral equals the area

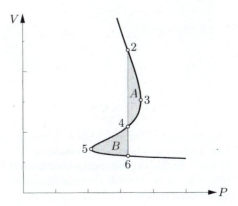

Figure 5.22. The same isotherm as in Figure 5.21, plotted sideways. Regions A and B have equal areas.

of A minus the area of B, and if this is to equal zero, we conclude that the two shaded regions must have equal areas. Drawing the straight line so as to enclose equal areas in this way is called the **Maxwell construction**, after James Clerk Maxwell.

Repeating the Maxwell construction for a variety of temperatures yields the results shown in Figure 5.23. For each temperature there is a well-defined pressure, called the **vapor pressure**, at which the liquid-gas transformation takes place; plotting this pressure vs. temperature gives us a prediction for the entire liquid-gas phase boundary. Meanwhile, the straight segments of the isotherms on the PV diagram fill a region in which the stable state is a combination of gas and liquid, indicated by the shaded area.

But what about the *high*-temperature isotherms, which rise monotonically as V decreases? For these temperatures there is no abrupt transition from low-density states to high-density states: no phase transformation. The phase boundary therefore disappears above a certain temperature, called the **critical temperature**, T_c. The vapor pressure just at T_c is called the **critical pressure**, P_c, while the corresponding volume is called the **critical volume**, V_c. These values define the **critical point**, where the properties of the liquid and gas become identical.

I find it remarkable that a model as simple as the van der Waals equation predicts *all* of the important qualitative properties of real fluids: the liquid-gas phase transformation, the general shape of the phase boundary curve, and even the critical point. Unfortunately, the model fails when it comes to numbers. For example, the experimental phase boundary for H_2O falls more steeply from the critical point than does the predicted boundary shown above; at $T/T_c = 0.8$, the measured vapor pressure is only about $0.2P_c$, instead of $0.4P_c$ as predicted. More

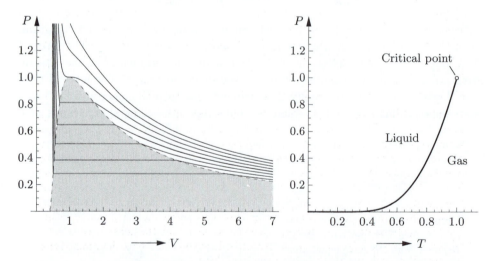

Figure 5.23. Complete phase diagrams predicted by the van der Waals model. The isotherms shown at left are for T/T_c ranging from 0.75 to 1.1 in increments of 0.05. In the shaded region the stable state is a combination of gas and liquid. The full vapor pressure curve is shown at right. All axes are labeled in units of the critical values.

accurate models of the behavior of dense fluids are beyond the scope of this book,* but at least we've taken a first step toward understanding the liquid-gas phase transformation.

Problem 5.48. As you can see from Figure 5.20, the critical point is the unique point on the original van der Walls isotherms (before the Maxwell construction) where both the first and second derivatives of P with respect to V (at fixed T) are zero. Use this fact to show that

$$V_c = 3Nb, \qquad P_c = \frac{1}{27}\frac{a}{b^2}, \qquad \text{and} \qquad kT_c = \frac{8}{27}\frac{a}{b}.$$

Problem 5.49. Use the result of the previous problem and the approximate values of a and b given in the text to estimate T_c, P_c, and V_c/N for N_2, H_2O, and He. (Tabulated values of a and b are often determined by working backward from the measured critical temperature and pressure.)

Problem 5.50. The **compression factor** of a fluid is defined as the ratio PV/NkT; the deviation of this quantity from 1 is a measure of how much the fluid differs from an ideal gas. Calculate the compression factor of a van der Waals fluid at the critical point, and note that the value is independent of a and b. (Experimental values of compression factors at the critical point are generally lower than the van der Waals prediction, for instance, 0.227 for H_2O, 0.274 for CO_2, 0.305 for He.)

Problem 5.51. When plotting graphs and performing numerical calculations, it is convenient to work in terms of **reduced variables**,

$$t \equiv T/T_c, \qquad p \equiv P/P_c, \qquad v \equiv V/V_c.$$

Rewrite the van der Waals equation in terms of these variables, and notice that the constants a and b disappear.

Problem 5.52. Plot the van der Waals isotherm for $T/T_c = 0.95$, working in terms of reduced variables. Perform the Maxwell construction (either graphically or numerically) to obtain the vapor pressure. Then plot the Gibbs free energy (in units of NkT_c) as a function of pressure for this same temperature and check that this graph predicts the same value for the vapor pressure.

Problem 5.53. Repeat the preceding problem for $T/T_c = 0.8$.

Problem 5.54. Calculate the *Helmholtz* free energy of a van der Waals fluid, up to an undetermined function of temperature as in equation 5.56. Using reduced variables, carefully plot the Helmholtz free energy (in units of NkT_c) as a function of volume for $T/T_c = 0.8$. Identify the two points on the graph corresponding to the liquid and gas at the vapor pressure. (If you haven't worked the preceding problem, just read the appropriate values off Figure 5.23.) Then prove that the Helmholtz free energy of a *combination* of these two states (part liquid, part gas) can be represented by a straight line connecting these two points on the graph. Explain why the combination is more stable, at a given volume, than the homogeneous state represented by the original curve, and describe how you could have determined the two transition volumes directly from the graph of F.

---

*Chapter 8 introduces an accurate approximation for treating *weakly* interacting gases, as well as the more general technique of Monte Carlo simulation, which can be applied to dense fluids.

Problem 5.55. In this problem you will investigate the behavior of a van der Waals fluid near the critical point. It is easiest to work in terms of reduced variables throughout.

(a) Expand the van der Waals equation in a Taylor series in $(V - V_c)$, keeping terms through order $(V - V_c)^3$. Argue that, for T sufficiently close to T_c, the term quadratic in $(V - V_c)$ becomes negligible compared to the others and may be dropped.

(b) The resulting expression for $P(V)$ is antisymmetric about the point $V = V_c$. Use this fact to find an approximate formula for the vapor pressure as a function of temperature. (You may find it helpful to plot the isotherm.) Evaluate the slope of the phase boundary, dP/dT, at the critical point.

(c) Still working in the same limit, find an expression for the difference in volume between the gas and liquid phases at the vapor pressure. You should find $(V_g - V_l) \propto (T_c - T)^\beta$, where β is known as a **critical exponent**. Experiments show that β has a universal value of about $1/3$, but the van der Waals model predicts a larger value.

(d) Use the previous result to calculate the predicted latent heat of the transformation as a function of temperature, and sketch this function.

(e) The shape of the $T = T_c$ isotherm defines another critical exponent, called δ: $(P - P_c) \propto (V - V_c)^\delta$. Calculate δ in the van der Waals model. (Experimental values of δ are typically around 4 or 5.)

(f) A third critical exponent describes the temperature dependence of the isothermal compressibility,

$$\kappa \equiv -\frac{1}{V}\left(\frac{\partial V}{\partial P}\right)_T.$$

This quantity diverges at the critical point, in proportion to a power of $(T - T_c)$ that in principle could differ depending on whether one approaches the critical point from above or below. Therefore the critical exponents γ and γ' are defined by the relations

$$\kappa \propto \begin{cases} (T - T_c)^{-\gamma} & \text{as } T \to T_c \text{ from above,} \\ (T_c - T)^{-\gamma'} & \text{as } T \to T_c \text{ from below.} \end{cases}$$

Calculate κ on both sides of the critical point in the van der Waals model, and show that $\gamma = \gamma'$ in this model.

5.4 Phase Transformations of Mixtures

Phase transformations become a lot more complicated when a system contains two or more types of particles. Consider air, for example, a mixture of approximately 79% nitrogen and 21% oxygen (neglecting various minor components for simplicity). What happens when you lower the temperature of this mixture, at atmospheric pressure? You might expect that all the oxygen would liquefy at 90.2 K (the boiling point of pure oxygen), leaving a gas of pure nitrogen which would then liquefy at 77.4 K (the boiling point of pure nitrogen). In fact, however, no liquid at all forms until the temperature drops to 81.6 K, when a liquid consisting of 48% oxygen begins to condense. Similar behavior occurs in liquid-solid transitions, such as the crystallization of alloys and igneous rocks. How can we understand this behavior?

Free Energy of a Mixture

As usual, the key is to look at the (Gibbs) free energy,

$$G = U + PV - TS. \tag{5.58}$$

Let's consider a system of two types of molecules, A and B, and suppose that they are initially separated, sitting side by side at the same temperature and pressure (see Figure 5.24). Imagine varying the proportions of A and B while holding the total number of molecules fixed, say at one mole. Let G_A° be the free energy of a mole of pure A, and G_B° the free energy of a mole of pure B. For an unmixed combination of part A and part B, the total free energy is just the sum of the separate free energies of the two subsystems:

$$G = (1-x)G_A^\circ + xG_B^\circ \qquad \text{(unmixed)}; \tag{5.59}$$

where x is the fraction of B molecules, so that $x = 0$ for pure A and $x = 1$ for pure B. A graph of G vs. x for this unmixed system is a straight line, as shown in Figure 5.25.

Now suppose that we remove the partition between the two sides and stir the A and B molecules together to form a homogeneous mixture. (I'll use the term **mixture** only when the substances are mixed at the molecular level. A "mixture" of salt and pepper does not qualify.) What happens to the free energy? From the definition $G = U + PV - TS$, we see that G can change because of changes in U, V, and/or S. The energy, U, might increase or decrease, depending on how the forces between dissimilar molecules compare to the forces between identical molecules. The volume, as well, may increase or decrease depending on these forces and on the shapes of the molecules. The entropy, however, will most certainly *increase*, because there are now many more possible ways to arrange the molecules.

As a first approximation, therefore, let us neglect any changes in U and V and assume that the entire change in G comes from the entropy of mixing. As a further simplification, let's also assume that the entropy of mixing can be calculated as in Problem 2.38, so that for one mole,

$$\Delta S_{\text{mixing}} = -R\big[x \ln x + (1-x)\ln(1-x)\big]. \tag{5.60}$$

A graph of this expression is shown in Figure 5.25. This expression is correct for ideal gases, and also for liquids and solids when the two types of molecules are the same size and have no "preference" for having like or unlike neighbors. When this

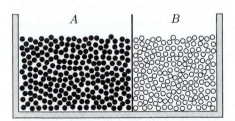
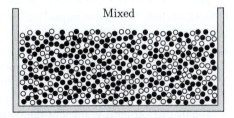

Figure 5.24. A collection of two types of molecules, before and after mixing.

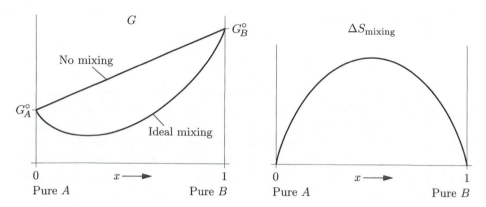

Figure 5.25. Before mixing, the free energy of a collection of A and B molecules is a linear function of $x = N_B/(N_A + N_B)$. After mixing it is a more complicated function; shown here is the case of an "ideal" mixture, whose entropy of mixing is shown at right. Although it isn't obvious on this scale, the graphs of both ΔS_{mixing} and G (after mixing) have vertical slopes at the endpoints.

expression for the mixing entropy holds and when U and V do not change upon mixing, the free energy of the mixture is

$$G = (1-x)G_A^\circ + xG_B^\circ + RT\left[x \ln x + (1-x)\ln(1-x)\right] \quad \text{(ideal mixture).} \quad (5.61)$$

This function is plotted in Figure 5.25. A mixture having this simple free energy function is called an **ideal mixture**. Liquid and solid mixtures rarely come close to being ideal, but the ideal case still makes a good starting point for arriving at some qualitative understanding.

One important property of expression 5.60 for the entropy of mixing is that its derivative with respect to x goes to infinity at $x = 0$ and to minus infinity at $x = 1$. The graph of this expression therefore has a vertical slope at each endpoint. Similarly, expression 5.61 for the Gibbs free energy has an infinite derivative at each endpoint: Adding a tiny amount of impurity to either pure substance lowers the free energy significantly, except when $T = 0$.* Although the precise formulas written above hold only for ideal solutions, the infinite slope at the endpoints is a general property of the free energy of any mixture. Because a system will spontaneously seek out the state of lowest free energy, this property tells us that equilibrium phases almost always contain impurities.

*Non*ideal mixtures often have the same qualitative properties as ideal mixtures, but not always. The most important exception is when mixing the two substances increases the total energy. This happens in liquids when unlike molecules are less attracted to each other than are like molecules, as with oil and water. The energy change upon mixing is then a concave-down function of x, as shown in Figure 5.26. At $T = 0$ the free energy ($G = U + PV - TS$) is also a concave-down function

---

*Hiding one needle in a stack of pure hay increases the entropy a lot more than does adding a needle to a haystack already containing thousands of needles.

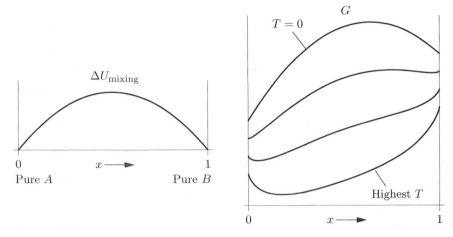

Figure 5.26. Mixing A and B can often increase the energy of the system; shown at left is the simple case where the mixing energy is a quadratic function (see Problem 5.58). Shown at right is the free energy in this case, at four different temperatures.

(if we neglect any change in V upon mixing). At nonzero T, however, there is a competition in G between the concave-down contribution from the mixing energy and the concave-up contribution from $-T$ times the mixing entropy. At sufficiently high T the entropy contribution always wins and G is everywhere concave-up. But even at very low nonzero T, the entropy contribution *still* dominates the shape of G near the endpoints $x = 0$ and $x = 1$. This is because the entropy of mixing has an infinite derivative at the endpoints, while the energy of mixing has only a finite derivative at the endpoints: When there is very little impurity, the mixing energy is simply proportional to the number of impurity molecules. Thus, at small nonzero T, the free energy function is concave-up near the endpoints and concave-down near the middle, as shown in Figure 5.26.

But a concave-down free energy function indicates an unstable mixture. Pick any two points on the graph of G and connect them with a straight line. This line denotes the free energy of an *unmixed* combination of the two phases represented by the endpoints (just as the straight line in Figure 5.25 denotes the free energy of the unmixed pure phases). Whenever the graph of G is concave-*down*, you can draw a straight connecting line that lies *below* the curve, and therefore the unmixed combination has a lower free energy than the homogeneous mixture. The lowest possible connecting line intersects the curve as a tangent at each end (see Figure 5.27). The tangent points indicate the compositions of the two separated phases, denoted x_a and x_b in the figure. Thus, if the composition of the system lies between x_a and x_b, it will spontaneously separate into an A-rich phase of composition x_a and a B-rich phase of composition x_b. We say that the system has a **solubility gap**, or that the two phases are **immiscible**. Decreasing the temperature of this system widens the solubility gap (see Figure 5.26), while increasing the temperature narrows the gap until it disappears when G is everywhere concave-up.

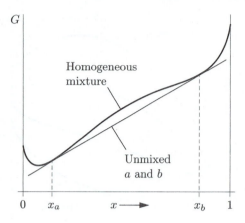

Figure 5.27. To construct the equilibrium free energy curve, draw the lowest possible straight line across the concave-down section, tangent to the curve at both ends. At compositions between the tangent points the mixture will spontaneously separate into phases whose compositions are x_a and x_b, in order to lower its free energy.

If we plot the compositions x_a and x_b at each temperature, we obtain a T vs. x phase diagram like those shown in Figure 5.28. Above the curve the equilibrium state is a single homogeneous mixture, while below the curve the system separates into two phases whose compositions lie on the curve. For the familiar case of oil and water at atmospheric pressure, the critical temperature where complete mixing would occur is far above the boiling point of water. The figure shows a less familiar mixture, water and phenol (C_6H_5OH), whose critical mixing temperature is 67°C.

Solubility gaps occur in solid as well as liquid mixtures. For solids, however, there is often the further complication that the pure-A and pure-B solids have qualitatively different crystal structures. Let us call these structures α and β, respectively. Adding a few B molecules to the α phase, or a few A molecules

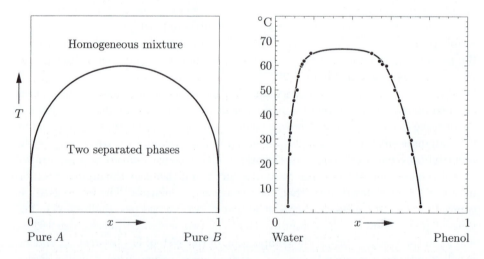

Figure 5.28. Left: Phase diagram for the simple model system whose mixing energy is plotted in Figure 5.26. Right: Experimental data for a real system, water + phenol, that shows qualitatively similar behavior. Adapted with permission from Alan N. Campbell and A. Jean R. Campbell, *Journal of the American Chemical Society* **59**, 2481 (1937). Copyright 1937 American Chemical Society.

to the β phase, is always possible, again because of the infinite slope of the entropy of mixing function. But a large amount of impurity will usually stress the crystal lattice significantly, greatly raising is energy. A free energy diagram for such a system therefore looks like Figure 5.29. Again we can draw a straight line tangent to the two concave-up sections, so there is a solubility gap: The stable configuration at intermediate compositions is an unmixed combination of the two phases indicated by the tangent points. For some solids the situation is even more complicated because other crystal structures are stable at intermediate compositions. For example, brass, an alloy of copper and zinc, has five possible crystal structures, each stable within a certain composition range.

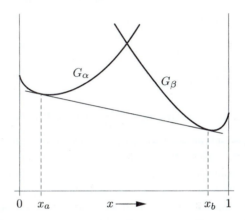

Figure 5.29. Free energy graphs for a mixture of two solids with different crystal structures, α and β. Again, the lowest possible straight connecting line indicates the range of compositions where an unmixed combination of a and b phases is more stable than a homogeneous mixture.

Problem 5.56. Prove that the entropy of mixing of an ideal mixture has an infinite slope, when plotted vs. x, at $x = 0$ and $x = 1$.

Problem 5.57. Consider an ideal mixture of just 100 molecules, varying in composition from pure A to pure B. Use a computer to calculate the mixing entropy as a function of N_A, and plot this function (in units of k). Suppose you start with all A and then convert one molecule to type B; by how much does the entropy increase? By how much does the entropy increase when you convert a second molecule, and then a third, from A to B? Discuss.

Problem 5.58. In this problem you will model the mixing energy of a mixture in a relatively simple way, in order to relate the existence of a solubility gap to molecular behavior. Consider a mixture of A and B molecules that is ideal in every way but one: The potential energy due to the interaction of neighboring molecules depends upon whether the molecules are like or unlike. Let n be the average number of nearest neighbors of any given molecule (perhaps 6 or 8 or 10). Let u_0 be the average potential energy associated with the interaction between neighboring molecules that are the same (A-A or B-B), and let u_{AB} be the potential energy associated with the interaction of a neighboring unlike pair (A-B). There are no interactions beyond the range of the nearest neighbors; the values of u_0 and u_{AB} are independent of the amounts of A and B; and the entropy of mixing is the same as for an ideal solution.

(a) Show that when the system is unmixed, the total potential energy due to all neighbor-neighbor interactions is $\frac{1}{2}Nnu_0$. (Hint: Be sure to count each neighboring pair only once.)

(b) Find a formula for the total potential energy when the system is mixed, in terms of x, the fraction of B. (Assume that the mixing is totally random.)

(c) Subtract the results of parts (a) and (b) to obtain the change in energy upon mixing. Simplify the result as much as possible; you should obtain an expression proportional to $x(1-x)$. Sketch this function vs. x, for both possible signs of $u_{AB} - u_0$.

(d) Show that the slope of the mixing energy function is finite at both end-points, unlike the slope of the mixing *entropy* function.

(e) For the case $u_{AB} > u_0$, plot a graph of the Gibbs free energy of this system vs. x at several temperatures. Discuss the implications.

(f) Find an expression for the maximum temperature at which this system has a solubility gap.

(g) Make a very rough estimate of $u_{AB} - u_0$ for a liquid mixture that has a solubility gap below $100°C$.

(h) Plot the phase diagram (T vs. x) for this system.

Phase Changes of a Miscible Mixture

Now let us return to the process described at the beginning of this section, the liquefaction of a mixture of nitrogen and oxygen. Liquid nitrogen and oxygen are completely miscible, so the free energy function of the liquid mixture is everywhere concave-up. The free energy of the gaseous mixture is also everywhere concave-up. By considering the relation between these two functions at various temperatures, we can understand the behavior of this system and sketch its phase diagram.

Figure 5.30 shows the free energy functions of a model system that behaves as an ideal mixture in both the gaseous and liquid phases. Think of the components A and B as nitrogen and oxygen, whose behavior should be qualitatively similar. The boiling points of pure A and pure B are denoted T_A and T_B, respectively. At temperatures greater than T_B the stable phase is a gas regardless of composition, so the free energy curve of the gas lies entirely below that of the liquid. As the temperature drops, both free energy functions increase ($\partial G/\partial T = -S$), but that of the gas increases more because the gas has more entropy. At $T = T_B$ the curves intersect at $x = 1$, where the liquid and gas phases of pure B are in equilibrium. As T decreases further the intersection point moves to the left, until at $T = T_A$ the curves intersect at $x = 0$. At still lower temperatures the free energy of the liquid is less than that of the gas at all compositions.

At intermediate temperatures, between T_A and T_B, either the liquid or the gas phase may be more stable, depending on composition. But notice, from the shape of the curves, that you can draw a straight line, tangent to both curves, that lies below both curves. Between the two tangent points, therefore, the stable configuration is an *unmixed* combination of a gas whose composition is indicated by the left tangent point and a liquid whose composition is indicated by the right tangent point. The straight line denotes the free energy of this unmixed combination. By drawing such a straight line for every temperature between T_A and T_B, we can construct the T vs. x phase diagram for this system. The mixture is entirely gas in the upper region

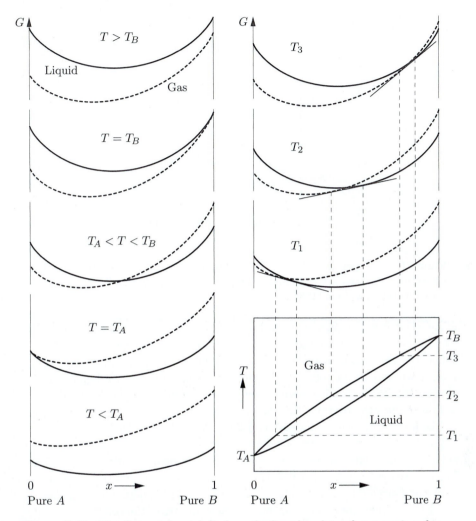

Figure 5.30. The five graphs at left show the liquid and gas free energies of an ideal mixture at temperatures above, below, at, and between the boiling points T_A and T_B. Three graphs at intermediate temperatures are shown at right, along with the construction of the phase diagram.

of the diagram, entirely liquid in the lower region, and an unmixed combination in the region between the two curves.

Figure 5.31 shows the *experimental* phase diagram for the nitrogen-oxygen system. Although this diagram isn't exactly the same as that of the ideal *A*-*B* model system, it has all the same qualitative features. From the diagram you can see that if you start with an air-like mixture of 79% nitrogen and 21% oxygen and lower the temperature, it remains a gas until the temperature reaches 81.6 K. At this point a liquid begins to condense. A horizontal line at this temperature intersects the lower curve at $x = 0.48$, so the liquid is initially 48% oxygen. Because oxygen condenses more easily than nitrogen, the liquid is enriched in oxygen compared to

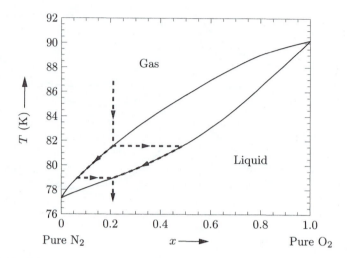

Figure 5.31. Experimental phase diagram for nitrogen and oxygen at atmospheric pressure. Data from *International Critical Tables* (volume 3), with endpoints adjusted to values in Lide (1994).

the gas. But it is not pure oxygen, because the entropy of mixing gives too much of a thermodynamic advantage to impure phases. As the temperature decreases further, the gas becomes depleted of oxygen and its composition follows the upper curve, down and to the left. Meanwhile the composition of the liquid follows the lower curve, also increasing its nitrogen/oxygen ratio. At 79.0 K the liquid composition reaches the overall composition of 21% oxygen, so there can't be any gas left; the last bit of gas to condense contains about 7% oxygen.

The liquid-gas transitions of many other mixtures behave similarly. Furthermore, for some mixtures the solid-liquid transition behaves in this way. Examples of such mixtures include copper-nickel, silicon-germanium, and the common minerals olivene (varying from Fe_2SiO_4 to Mg_2SiO_4) and plagioclase feldspar (see Problem 5.64). In all of these systems, the crystal structure of the solid is essentially the same throughout the entire range of composition, so the two pure solids can form approximately ideal mixtures in all proportions. Such a mixture is called a **solid solution**.

Problem 5.59. Suppose you cool a mixture of 50% nitrogen and 50% oxygen until it liquefies. Describe the cooling sequence in detail, including the temperatures and compositions at which liquefaction begins and ends.

Problem 5.60. Suppose you start with a liquid mixture of 60% nitrogen and 40% oxygen. Describe what happens as the temperature of this mixture increases. Be sure to give the temperatures and compositions at which boiling begins and ends.

Problem 5.61. Suppose you need a tank of oxygen that is 95% pure. Describe a process by which you could obtain such a gas, starting with air.

Problem 5.62. Consider a completely miscible two-component system whose overall composition is x, at a temperature where liquid and gas phases coexist. The composition of the gas phase at this temperature is x_a and the composition of the liquid phase is x_b. Prove the **lever rule**, which says that the proportion of liquid to gas is $(x - x_a)/(x_b - x)$. Interpret this rule graphically on a phase diagram.

Problem 5.63. Everything in this section assumes that the total pressure of the system is fixed. How would you expect the nitrogen-oxygen phase diagram to change if you increase or decrease the pressure? Justify your answer.

Problem 5.64. Figure 5.32 shows the phase diagram of plagioclase feldspar, which can be considered a mixture of albite ($NaAlSi_3O_8$) and anorthite ($CaAl_2Si_2O_8$).

(a) Suppose you discover a rock in which each plagioclase crystal varies in composition from center to edge, with the centers of the largest crystals composed of 70% anorthite and the outermost parts of all crystals made of essentially pure albite. Explain in some detail how this variation might arise. What was the composition of the liquid magma from which the rock formed?

(b) Suppose you discover another rock body in which the crystals near the top are albite-rich while the crystals near the bottom are anorthite-rich. Explain how this variation might arise.

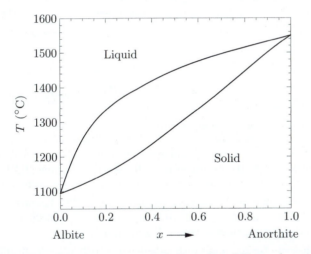

Figure 5.32. The phase diagram of plagioclase feldspar (at atmospheric pressure). From N. L. Bowen, "The Melting Phenomena of the Plagioclase Feldspars," *American Journal of Science* **35**, 577–599 (1913).

Problem 5.65. In constructing the phase diagram from the free energy graphs in Figure 5.30, I assumed that both the liquid and the gas are ideal mixtures. Suppose instead that the liquid has a substantial positive mixing energy, so that its free energy curve, while still concave-up, is much flatter. In this case a portion of the curve may still lie *above* the gas's free energy curve at T_A. Draw a qualitatively accurate phase diagram for such a system, showing how you obtained the phase diagram from the free energy graphs. Show that there is a particular composition at which this gas mixture will condense with no change in composition. This special composition is called an **azeotrope**.

Problem 5.66. Repeat the previous problem for the opposite case where the liquid has a substantial *negative* mixing energy, so that its free energy curve dips below the gas's free energy curve at a temperature higher than T_B. Construct the phase diagram and show that this system also has an azeotrope.

Problem 5.67. In this problem you will derive approximate formulas for the shapes of the phase boundary curves in diagrams such as Figures 5.31 and 5.32, assuming that both phases behave as ideal mixtures. For definiteness, suppose that the phases are liquid and gas.

(a) Show that in an ideal mixture of A and B, the chemical potential of species A can be written

$$\mu_A = \mu_A^\circ + kT \ln(1 - x),$$

where μ_A° is the chemical potential of *pure* A (at the same temperature and pressure) and $x = N_B/(N_A + N_B)$. Derive a similar formula for the chemical potential of species B. Note that both formulas can be written for either the liquid phase or the gas phase.

(b) At any given temperature T, let x_l and x_g be the compositions of the liquid and gas phases that are in equilibrium with each other. By setting the appropriate chemical potentials equal to each other, show that x_l and x_g obey the equations

$$\frac{1 - x_l}{1 - x_g} = e^{\Delta G_A^\circ / RT} \qquad \text{and} \qquad \frac{x_l}{x_g} = e^{\Delta G_B^\circ / RT},$$

where ΔG° represents the change in G for the pure substance undergoing the phase change at temperature T.

(c) Over a limited range of temperatures, we can often assume that the main temperature dependence of $\Delta G^\circ = \Delta H^\circ - T\Delta S^\circ$ comes from the explicit T; both ΔH° and ΔS° are approximately constant. With this simplification, rewrite the results of part (b) entirely in terms of ΔH_A°, ΔH_B°, T_A, and T_B (eliminating ΔG and ΔS). Solve for x_l and x_g as functions of T.

(d) Plot your results for the nitrogen-oxygen system. The latent heats of the pure substances are $\Delta H_{N_2}^\circ = 5570$ J/mol and $\Delta H_{O_2}^\circ = 6820$ J/mol. Compare to the experimental diagram, Figure 5.31.

(e) Show that you can account for the shape of Figure 5.32 with suitably chosen ΔH° values. What are those values?

Phase Changes of a Eutectic System

Most two-component solid mixtures do not maintain the same crystal structure over the entire range of composition. The situation shown in Figure 5.29 is more common: two different crystal structures, α and β, at compositions close to pure A and pure B, with an unmixed combination of α and β being stable at intermediate compositions. Let us now consider the solid-liquid transitions of such a system, assuming that A and B are completely miscible in the liquid phase. Again the idea is to look at the free energy functions at various temperatures (see Figure 5.33). For definiteness, suppose that T_B, the melting temperature of pure B, is higher than T_A, the melting temperature of pure A.

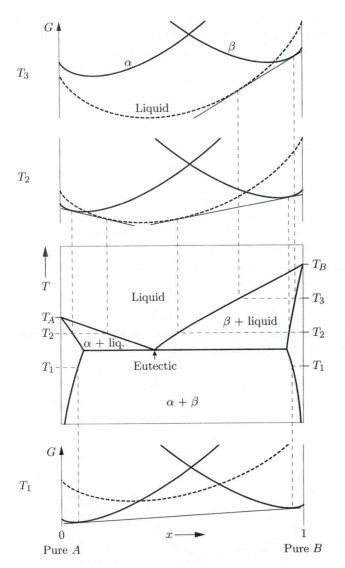

Figure 5.33. Construction of the phase diagram of a eutectic system from free energy graphs.

At high temperatures the free energy of the liquid will be below that of either solid phase. Then, as the temperature decreases, all three free-energy functions will increase ($\partial G/\partial T = -S$), but the free energy of the liquid will increase fastest because it has the most entropy. Below T_B the liquid's free energy curve intersects that of the β phase, so there is a range of compositions for which the stable config-uration is an unmixed combination of liquid and β. As the temperature decreases this range widens and reaches further toward the A side of the diagram. Even-tually the liquid curve intersects the α curve as well and there is an A-rich range of compositions for which the stable phase is an unmixed combination of liquid and α. As T decreases further this range reaches toward the B side of the diagram until finally it intersects the liquid $+ \beta$ range at the **eutectic point**. At still lower temperatures the stable configuration is an unmixed combination of the α and β solids; the free energy of the liquid is higher than that of this combination.

The eutectic point defines a special composition at which the melting temperature is as low as possible, lower than that of either pure substance. A liquid near the eutectic composition remains stable at low temperatures because it has more mixing entropy than the unmixed combination of solids would have. (A solid mixture would have about as much mixing entropy, but is forbidden by the large positive mixing *energy* that results from stressing the crystal lattice.)

A good example of a eutectic mixture is the tin-lead solder used in electrical circuits. Figure 5.34 shows the phase diagram of the tin-lead system. Common electrical solder is very close to the eutectic composition of 38% lead by weight (or 26% by number of atoms). Using this composition has several advantages: the melting temperature is as low as possible (183°C); the solder freezes suddenly rather than gradually; and the cooled metal is relatively strong, with small crystals of the two phases uniformly alternating at the microscopic scale.

Many other mixtures behave in a similar way. Most pure liquid crystals freeze at inconveniently high temperatures, so eutectic mixtures are often used to obtain liquid crystals for use at room temperature. A less exotic mixture is water + table salt (NaCl), which can remain a liquid at temperatures as low as -21°C, at the eutectic composition of 23% NaCl by weight.* Another familiar example is the coolant used in automobile engines, a mixture of water and ethylene glycol (HOCH$_2$CH$_2$OH). Pure water freezes at 0°C, and pure ethylene glycol at -13°C,

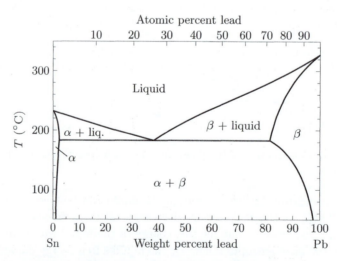

Figure 5.34. Phase diagram for mixtures of tin and lead. From Thaddeus B. Massalski, ed., *Binary Alloy Phase Diagrams*, second edition (ASM International, Materials Park, OH, 1990).

---

*The water + NaCl phase diagram is shown in Zemansky and Dittman (1997).

so neither would remain a liquid on winter nights in a cold climate. Fortunately, a 50-50 mixture (by volume) of these two liquids does not begin to freeze until the temperature reaches $-31°$C. The eutectic point is lower still, at $-49°$C and a composition of 56% ethylene glycol by volume.*

Although the phase diagram of a eutectic system may seem complicated enough, many two-component systems are further complicated by the existence of other crystal structures of intermediate compositions; Problems 5.71 and 5.72 explore some of the possibilities. Then there are *three*-component systems, for which the composition axis of the phase diagram is actually a plane (usually represented by a triangle). You can find hundreds of intricate phase diagrams in books on metallurgy, ceramics, and petrology. All can be understood qualitatively in terms of free energy graphs as we have done here. Because this is an introductory text, though, let us move on and explore the properties of some simple mixtures more quantitatively.

Problem 5.68. Plumber's solder is composed of 67% lead and 33% tin by weight. Describe what happens to this mixture as it cools, and explain why this composition might be more suitable than the eutectic composition for joining pipes.

Problem 5.69. What happens when you spread salt crystals over an icy sidewalk? Why is this procedure rarely used in *very* cold climates?

Problem 5.70. What happens when you add salt to the ice bath in an ice cream maker? How is it possible for the temperature to *spontaneously* drop below $0°$C? Explain in as much detail as you can.

Problem 5.71. Figure 5.35 (left) shows the free energy curves at one particular temperature for a two-component system that has three possible solid phases (crystal structures), one of essentially pure A, one of essentially pure B, and one of intermediate composition. Draw tangent lines to determine which phases are present at which values of x. To determine qualitatively what happens at other temperatures, you can simply shift the liquid free energy curve up or down (since the entropy of the liquid is larger than that of any solid). Do so, and construct

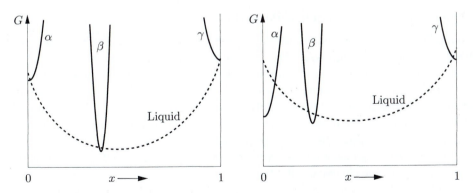

Figure 5.35. Free energy diagrams for Problems 5.71 and 5.72.

---

*For the full phase diagram see J. Bevan Ott, J. Rex Goates, and John D. Lamb, *Journal of Chemical Thermodynamics* **4**, 123–126 (1972).

a qualitative phase diagram for this system. You should find two eutectic points. Examples of systems with this behavior include water + ethylene glycol and tin + magnesium.

Problem 5.72. Repeat the previous problem for the diagram in Figure 5.35 (right), which has an important qualitative difference. In this phase diagram, you should find that β and liquid are in equilibrium only at temperatures *below* the point where the liquid is in equilibrium with infinitesimal amounts of α and β. This point is called a **peritectic point**. Examples of systems with this behavior include water + NaCl and leucite + quartz.

5.5   Dilute Solutions

A **solution** is the same thing as a mixture, except that we think of one component (the **solvent**) as being primary and the other components (the **solutes**) as being secondary. A solution is called **dilute** if the solute molecules are much less abundant than the solvent molecules (see Figure 5.36), so that each solute molecule is "always" surrounded by solvent molecules and "never" interacts directly with other solute molecules. In many ways the solute in a dilute solution behaves like an ideal gas. We can therefore predict many of the properties of a dilute solution (including its boiling and freezing points) quantitatively.

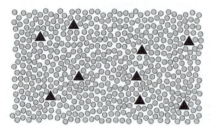

Figure 5.36. A dilute solution, in which the solute is much less abundant than the solvent.

Solvent and Solute Chemical Potentials

To predict the properties of a dilute solution interacting with its environment, we'll need to know something about the chemical potentials of the solvent and solutes. The chemical potential, μ_A, of species A is related to the Gibbs free energy by $\mu_A = \partial G/\partial N_A$, so what we need is a formula for the Gibbs free energy of a dilute solution in terms of the numbers of solvent and solute molecules. Coming up with the correct formula for G is a bit tricky but very worthwhile: Once we have this formula, a host of applications become possible.

Suppose we start with a *pure* solvent of A molecules. Then the Gibbs free energy is just N_A times the chemical potential:

$$G = N_A \mu_0(T, P) \quad \text{(pure solvent)}, \tag{5.62}$$

where μ_0 is the chemical potential of the pure solvent, a function of temperature and pressure.

Now imagine that we add a single B molecule, holding the temperature and pressure fixed. Under this operation the (Gibbs) free energy changes by

$$dG = dU + P \, dV - T \, dS. \tag{5.63}$$

The important thing about dU and $P \, dV$ is that neither depends on N_A: Both depend only on how the B molecule interacts with its immediate neighbors, regardless of how many other A molecules are present. For the $T \, dS$ term, however, the situation is more complicated. Part of dS is independent of N_A, but another part comes from our freedom in choosing *where* to put the B molecule. The number of choices is proportional to N_A, so this operation increases the multiplicity by a factor proportional to N_A, and therefore dS, the increase in entropy, includes a term $k \ln N_A$:

$$dS = k \ln N_A + (\text{terms independent of } N_A). \tag{5.64}$$

The change in the free energy can therefore be written

$$dG = f(T, P) - kT \ln N_A \qquad (\text{adding one } B \text{ molecule}), \tag{5.65}$$

where $f(T, P)$ is a function of temperature and pressure but not of N_A.

Next imagine that we add *two* B molecules to the pure solvent. For this operation we can almost apply the preceding argument twice and conclude that

$$dG = 2f(T, P) - 2kT \ln N_A \qquad (\text{wrong}). \tag{5.66}$$

The problem is that there is a further change in entropy resulting from the fact that the two B molecules are identical. Interchanging these two molecules does not result in a distinct state, so we need to divide the multiplicity by 2, or subtract $k \ln 2$ from the entropy. With this modification,

$$dG = 2f(T, P) - 2kT \ln N_A + kT \ln 2 \qquad (\text{adding two } B \text{ molecules}). \tag{5.67}$$

The generalization to many B molecules is now straightforward. In the free energy we get N_B times $f(T, P)$ and N_B times $-kT \ln N_A$. To account for the interchangeability of the B molecules, we also get a term $kT \ln N_B! \approx kT N_B (\ln N_B - 1)$. Adding all these terms to the free energy of the pure solvent, we finally obtain

$$G = N_A \mu_0(T, P) + N_B f(T, P) - N_B kT \ln N_A + N_B kT \ln N_B - N_B kT. \tag{5.68}$$

This expression is valid in the limit $N_B \ll N_A$, that is, when the solution is dilute. For a nondilute solution the situation would be much more complicated because the B molecules would also interact with each other. If a solution contains more than one solute, all the terms in equation 5.68 except the first get repeated, with N_B replaced by N_C, N_D, and so on.

The solvent and solute chemical potentials follow immediately from equation 5.68:

$$\mu_A = \left(\frac{\partial G}{\partial N_A} \right)_{T, P, N_B} = \mu_0(T, P) - \frac{N_B kT}{N_A}; \tag{5.69}$$

$$\mu_B = \left(\frac{\partial G}{\partial N_B} \right)_{T, P, N_A} = f(T, P) + kT \ln(N_B / N_A). \tag{5.70}$$

As we would expect, adding more solute reduces the chemical potential of A and increases the chemical potential of B. Note also that both of these quantities, being intensive, depend only on the ratio N_B/N_A, not on the absolute number of solvent or solute molecules.

It is conventional to rewrite equation 5.70 in terms of the **molality*** of the solution, which is defined as the number of moles of solute per kilogram of solvent:

$$\text{molality} = m = \frac{\text{moles of solute}}{\text{kilograms of solvent}}. \tag{5.71}$$

The molality is a constant times the ratio N_B/N_A, and the constant can be absorbed into the function $f(T, P)$ to give a new function called $\mu^\circ(T, P)$. The solute chemical potential can then be written

$$\mu_B = \mu^\circ(T, P) + kT \ln m_B, \tag{5.72}$$

where m_B is the molality of the solute (in moles per kilogram) and μ° is the chemical potential under the "standard" condition $m_B = 1$. Values of μ° can be obtained from tables of Gibbs free energies, so equation 5.72 relates the tabulated value to the value at any other molality (so long as the solution is dilute).

> **Problem 5.73.** If expression 5.68 is correct, it must be extensive: Increasing both N_A and N_B by a common factor while holding all intensive variables fixed should increase G by the same factor. Show that expression 5.68 has this property. Show that it would *not* have this property had we not added the term proportional to $\ln N_B!$.

> **Problem 5.74.** Check that equations 5.69 and 5.70 satisfy the identity $G = N_A\mu_A + N_B\mu_B$ (equation 5.37).

> **Problem 5.75.** Compare expression 5.68 for the Gibbs free energy of a dilute solution to expression 5.61 for the Gibbs free energy of an ideal mixture. Under what circumstances should these two expressions agree? Show that they do agree under these circumstances, and identify the function $f(T, P)$ in this case.

Osmotic Pressure

As a first application of equation 5.69, consider a solution that is separated from some pure solvent by a membrane that allows only solvent molecules, not solute molecules, to pass through (see Figure 5.37). One example of such a semipermeable membrane is the membrane surrounding any plant or animal cell, which is permeable to water and other very small molecules but not to larger molecules or charged ions. Other semipermeable membranes are used in industry, for instance, in the desalination of seawater.

According to equation 5.69, the chemical potential of the solvent in the solution is less than that of the pure solvent, at a given temperature and pressure. Particles

---

*Molality is not the same as **molarity**, the number of moles of solute per liter of solution. For dilute solutions in water, however, the two are almost identical.

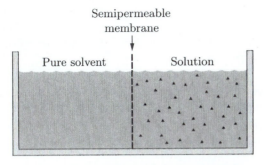

Figure 5.37. When a solution is separated by a semipermeable membrane from pure solvent at the same temperature and pressure, solvent will spontaneously flow into the solution.

tend to flow toward lower chemical potential, so in this situation the solvent molecules will spontaneously flow from the pure solvent into the solution. This flow of molecules is called **osmosis**. That osmosis should happen is hardly surprising: Solvent molecules are constantly bombarding the membrane on both sides, but more frequently on the side where the solvent is more concentrated, so naturally they hit the holes and pass through from that side more often.

If you want to *prevent* osmosis from happening, you can do so by applying some additional pressure to the solution (see Figure 5.38). How much pressure is required? Well, when the pressure is just right to stop the osmotic flow, the chemical potential of the solvent must be the same on both sides of the membrane. Using equation 5.69, this condition is

$$\mu_0(T, P_1) = \mu_0(T, P_2) - \frac{N_B kT}{N_A}, \tag{5.73}$$

where P_1 is the pressure on the side with pure solvent and P_2 is the pressure on the side of the solution. Assuming that these two pressures are not *too* different, we can approximate

$$\mu_0(T, P_2) \approx \mu_0(T, P_1) + (P_2 - P_1)\frac{\partial \mu_0}{\partial P}, \tag{5.74}$$

and plug this expression into equation 5.73 to obtain

$$(P_2 - P_1)\frac{\partial \mu_0}{\partial P} = \frac{N_B kT}{N_A}. \tag{5.75}$$

To evaluate the derivative $\partial \mu_0 / \partial P$, recall that the chemical potential of a pure substance is just the Gibbs free energy per particle, G/N. Since $\partial G / \partial P = V$ (at

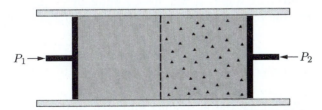

Figure 5.38. To prevent osmosis, P_2 must exceed P_1 by an amount called the **osmotic pressure**.

fixed T and N), our derivative is

$$\frac{\partial \mu_0}{\partial P} = \frac{V}{N},$$ (5.76)

the volume per molecule of the pure solvent. But since the solution is dilute, *its* volume per solvent molecule is essentially the same. Let's therefore take V in equation 5.76 to be the volume of the solution, and N to be the number of solvent molecules in the solution, that is, N_A. Then equation 5.75 becomes

$$(P_2 - P_1)\frac{V}{N_A} = \frac{N_B kT}{N_A},$$ (5.77)

or simply

$$(P_2 - P_1) = \frac{N_B kT}{V} = \frac{n_B RT}{V}$$ (5.78)

(where n_B/V is the number of *moles* of solute per unit volume). This pressure difference is called the **osmotic pressure**. It is the excess pressure required on the side of the solution to prevent osmosis.

Equation 5.78 for the osmotic pressure of a dilute solution is called **van't Hoff's formula**, after Jacobus Hendricus van't Hoff. It says that the osmotic pressure is exactly the same as the pressure of an ideal gas of the same concentration as the solute. In fact, it's tempting to think of the osmotic pressure as being exerted entirely *by* the solute, once we have balanced the pressure of the solvent on both sides. This interpretation is bad physics, but I still use it as a mnemonic aid to remember the formula.

As an example, consider the solution of ions, sugars, amino acids, and other molecules inside a biological cell. In a typical cell there are about 200 water molecules for each molecule of something else, so this solution is reasonably dilute. Since a mole of water has a mass of 18 g and a volume of 18 cm^3, the number of moles of solute per unit volume is

$$\frac{n_B}{V} = \left(\frac{1}{200}\right)\left(\frac{1 \text{ mol}}{18 \text{ cm}^3}\right)\left(\frac{100 \text{ cm}}{1 \text{ m}}\right)^3 = 278 \text{ mol/m}^3.$$ (5.79)

If you put a cell into pure water, it will absorb water by osmosis until the pressure inside exceeds the pressure outside by the osmotic pressure, which at room temperature is

$$(278 \text{ mol/m}^3)(8.3 \text{ J/mol·K})(300 \text{ K}) = 6.9 \times 10^5 \text{ N/m}^2,$$ (5.80)

or about 7 atm. An animal cell membrane subjected to this much pressure will burst, but plant cells have rigid walls that can withstand such a pressure.

Problem 5.76. Seawater has a salinity of 3.5%, meaning that if you boil away a kilogram of seawater, when you're finished you'll have 35 g of solids (mostly NaCl) left in the pot. When dissolved, sodium chloride dissociates into separate Na^+ and Cl^- ions.

   (a) Calculate the osmotic pressure difference between seawater and fresh water. Assume for simplicity that all the dissolved salts in seawater are NaCl.

   (b) If you apply a pressure difference *greater* than the osmotic pressure to a solution separated from pure solvent by a semipermeable membrane, you get **reverse osmosis**: a flow of solvent *out* of the solution. This process can be used to desalinate seawater. Calculate the minimum work required to desalinate one liter of seawater. Discuss some reasons why the actual work required would be greater than the minimum.

Problem 5.77. Osmotic pressure measurements can be used to determine the molecular weights of large molecules such as proteins. For a solution of large molecules to qualify as "dilute," its molar concentration must be very low and hence the osmotic pressure can be too small to measure accurately. For this reason, the usual procedure is to measure the osmotic pressure at a variety of concentrations, then extrapolate the results to the limit of zero concentration. Here are some data[*] for the protein hemoglobin dissolved in water at $3°C$:

Concentration (grams/liter)	Δh (cm)
5.6	2.0
16.6	6.5
32.5	12.8
43.4	17.6
54.0	22.6

The quantity Δh is the equilibrium difference in fluid level between the solution and the pure solvent, as shown in Figure 5.39. From these measurements, determine the approximate molecular weight of hemoglobin (in grams per mole).

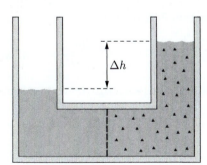

Figure 5.39. An experimental arrangement for measuring osmotic pressure. Solvent flows across the membrane from left to right until the difference in fluid level, Δh, is just enough to supply the osmotic pressure.

[*]From H. B. Bull, *An Introduction to Physical Biochemistry*, second edition (F. A. Davis, Philadelphia, 1971), p. 182. The measurements were made by H. Gutfreund.

Problem 5.78. Because osmotic pressures can be quite large, you may wonder whether the approximation made in equation 5.74 is valid in practice: Is μ_0 really a linear function of P to the required accuracy? Answer this question by discussing whether the derivative of this function changes significantly, over the relevant pressure range, in realistic examples.

Boiling and Freezing Points

In Section 5.4 we saw how impurities can shift the boiling and freezing points of a substance. For dilute solutions, we are now in a position to compute this shift quantitatively.

Consider first the case of a dilute solution at its boiling point, when it is in equilibrium with its gas phase (see Figure 5.40). Assume for simplicity that the solute does not evaporate at all—this is an excellent approximation for salt in water, for instance. Then the gas contains no solute, so we need only consider the equilibrium condition for the solvent:

$$\mu_{A,\text{liq}}(T,P) = \mu_{A,\text{gas}}(T,P). \tag{5.81}$$

Using equation 5.69 to rewrite the left-hand side, this condition becomes

$$\mu_0(T,P) - \frac{N_B kT}{N_A} = \mu_{\text{gas}}(T,P), \tag{5.82}$$

where μ_0 is the chemical potential of the pure solvent.

Now, as in the osmotic pressure derivation above, the procedure is to expand each μ function about the nearby point where the pure solvent would be in equilibrium. Because μ depends on both temperature and pressure, we can hold either fixed while allowing the other to vary. Let us first vary the pressure. Let P_0 be the vapor pressure of the pure solvent at temperature T, so that

$$\mu_0(T,P_0) = \mu_{\text{gas}}(T,P_0). \tag{5.83}$$

In terms of the chemical potentials at P_0, equation 5.82 becomes

$$\mu_0(T,P_0) + (P-P_0)\frac{\partial \mu_0}{\partial P} - \frac{N_B kT}{N_A} = \mu_{\text{gas}}(T,P_0) + (P-P_0)\frac{\partial \mu_{\text{gas}}}{\partial P}. \tag{5.84}$$

The first term on each side cancels by equation 5.83, and each $\partial\mu/\partial P$ is the volume

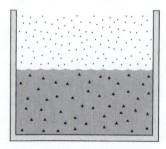

Figure 5.40. The presence of a solute reduces the tendency of a solvent to evaporate.

per particle for that phase, so

$$(P - P_0)\left(\frac{V}{N}\right)_{\text{liq}} - \frac{N_B kT}{N_A} = (P - P_0)\left(\frac{V}{N}\right)_{\text{gas}}. \tag{5.85}$$

The volume per particle in the gas phase is just kT/P_0, while the volume per particle in the liquid is negligible in comparison. This equation therefore reduces to

$$P - P_0 = \frac{-N_B}{N_A}P_0 \qquad \text{or} \qquad \frac{P}{P_0} = 1 - \frac{N_B}{N_A}. \tag{5.86}$$

The vapor pressure is reduced by a fraction equal to the ratio of the numbers of solute and solvent molecules. This result is known as **Raoult's law**. At the molecular level, the reduction in vapor pressure happens because the addition of solute reduces the number of solvent molecules at the surface of the liquid—hence they escape into the vapor less frequently.

Alternatively, we could hold the pressure fixed and solve for the shift in temperature needed to maintain equilibrium in the presence of the solute. Let T_0 be the boiling point of the pure solvent at pressure P, so that

$$\mu_0(T_0, P) = \mu_{\text{gas}}(T_0, P). \tag{5.87}$$

In terms of the chemical potentials at T_0, equation 5.82 becomes

$$\mu_0(T_0, P) + (T - T_0)\frac{\partial \mu_0}{\partial T} - \frac{N_B kT}{N_A} = \mu_{\text{gas}}(T_0, P) + (T - T_0)\frac{\partial \mu_{\text{gas}}}{\partial T}. \tag{5.88}$$

Again the first term on each side cancels. Each $\partial \mu/\partial T$ is just minus the entropy per particle for that phase (because $\partial G/\partial T = -S$), so

$$-(T - T_0)\left(\frac{S}{N}\right)_{\text{liq}} - \frac{N_B kT}{N_A} = -(T - T_0)\left(\frac{S}{N}\right)_{\text{gas}}. \tag{5.89}$$

It's simplest to set the N under each S equal to N_A, remembering that each S now refers to N_A molecules of solvent. The *difference* in entropy between the gas and the liquid is L/T_0, where L is the latent heat of vaporization. Therefore the temperature shift is

$$T - T_0 = \frac{N_B kT_0^2}{L} = \frac{n_B RT_0^2}{L}, \tag{5.90}$$

where I've approximated $T \approx T_0$ on the right-hand side.

As an example let's compute the boiling temperature of seawater. A convenient quantity to consider is one kilogram; then L is 2260 kJ. A kilogram of seawater contains 35 g of dissolved salts, mostly NaCl. The average atomic mass of Na and Cl is about 29, so 35 g of salt dissolves into $35/29 = 1.2$ moles of ions. Therefore the boiling temperature is shifted (relative to fresh water) by

$$T - T_0 = \frac{(1.2 \text{ mol})(8.3 \text{ J/mol·K})(373 \text{ K})^2}{2260 \text{ kJ}} = 0.6 \text{ K}. \tag{5.91}$$

To compute the shift in the vapor pressure at a given temperature, we need to know that a kilogram of water contains $1000/18 = 56$ moles of water molecules. Hence, by Raoult's law,

$$\frac{\Delta P}{P_0} = -\frac{1.2 \text{ mol}}{56 \text{ mol}} = -0.022. \tag{5.92}$$

Both of these effects are quite small: Seawater evaporates almost as easily as fresh water. Ironically, the shift in boiling temperature becomes large only for a nondilute solution, when the formulas of this section become inaccurate (though they can still give a rough estimate).

A formula essentially identical to equation 5.90 gives the shift in the *freezing* temperature of a dilute solution. Because the proof is so similar, I'll let you do it (see Problem 5.81). I'll also let you think about why the freezing temperature is *reduced* rather than increased. For water and most other solvents the shift in freezing temperature is somewhat larger than the shift in boiling temperature, due to the smaller value of L.

Together with osmotic pressure, the shifts in the vapor pressure, boiling temperature, and freezing temperature are all known as **colligative properties** of dilute solutions. All of these effects depend only on the *amount* of solute, not on *what* the solute is.

> **Problem 5.79.** Most pasta recipes instruct you to add a teaspoon of salt to a pot of boiling water. Does this have a significant effect on the boiling temperature? Justify your answer with a rough numerical estimate.

> **Problem 5.80.** Use the Clausius-Clapeyron relation to derive equation 5.90 directly from Raoult's law. Be sure to explain the logic carefully.

> **Problem 5.81.** Derive a formula, similar to equation 5.90, for the shift in the freezing temperature of a dilute solution. Assume that the solid phase is pure solvent, no solute. You should find that the shift is negative: The freezing temperature of a solution is *less* than that of the pure solvent. Explain in general terms why the shift should be negative.

> **Problem 5.82.** Use the result of the previous problem to calculate the freezing temperature of seawater.

5.6 Chemical Equilibrium

One interesting fact about chemical reactions is that they hardly ever go to completion. Consider, for example, the dissociation of water into H^+ and OH^- ions:

$$H_2O \longleftrightarrow H^+ + OH^-. \tag{5.93}$$

Under ordinary conditions, this reaction tends strongly to go to the left; an ordinary glass of water at equilibrium contains about 500 million water molecules for every pair of H^+ and OH^- ions. Naively, we tend to think of the water molecule as being "more stable" than the ions. But this can't be the whole story—otherwise there would be *no* ions in a glass full of water, when in fact there are quadrillions of them.

One way to understand why there are always some "unstable" ions even at equilibrium is to visualize the collisions at the molecular level. At room temperature, the water molecules are constantly colliding with each other at rather high speed. Every once in a while, one of these collisions is violent enough to break a molecule apart into two ions. The ions then tend to become separated, and do not recombine until they chance to meet new partners in the very dilute solution. Eventually an equilibrium is reached between the breaking apart and recombining, both of which occur rather rarely.

At a more abstract level, we can think about equilibrium in terms of the Gibbs free energy. At room temperature and atmospheric pressure, the concentration of each species at equilibrium is determined by the condition that the total Gibbs free energy,

$$G = U - TS + PV, \tag{5.94}$$

be minimized. We might expect that the minimum would occur when there are only water molecules, with no ions present. Indeed, it is true that a glass of water has much less Gibbs free energy than a glass full of H^+ and OH^- ions, simply because it has so much less energy. However, breaking just a *few* molecules apart into ions can lower the Gibbs free energy still further, because the *entropy* increases substantially. At higher temperature this entropy will contribute more to G, so there will be more ions.

It is instructive to plot the Gibbs free energy as a function of the **extent** of the reaction, which in this case is the fraction x of the water molecules that are split into ions. If every water molecule is intact then $x = 0$, while if every molecule were dissociated then x would equal 1. If we were to keep the dissociated ions separate from the intact water, then a graph of G vs. x would be a straight line with a large positive slope (see Figure 5.41). When we let the ions mix with the molecules, however, the entropy of mixing introduces an additional concave-upward term in G. As discussed in Section 5.4, the derivative of this term is infinite at the endpoints $x = 0$ and $x = 1$. Therefore, no matter how great the energy difference between the reactants and the products, the equilibrium point—the minimum of G—will lie

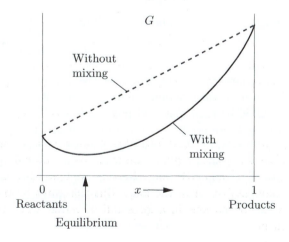

Figure 5.41. If reactants and products remained separate, the free energy would be a linear function of the extent of the reaction. With mixing, however, G has a minimum somewhere between $x = 0$ and $x = 1$.

at least a little bit inward from the lower of the two endpoints. (In practice it may be closer than one part in Avogadro's number; in such cases the reaction effectively *does* go to completion.)

We can characterize the equilibrium point by the condition that the slope of the Gibbs free energy graph is zero. This means that if one more H_2O molecule dissociates, G is unchanged:

$$0 = dG = \sum_i \mu_i \, dN_i. \tag{5.95}$$

In the last expression I've used the thermodynamic identity for G, plus the assumption that the temperature and pressure are fixed. The sum runs over all three species: H_2O, H^+, and OH^-. But the changes in the three N_i's are not independent: An increase of one H^+ is always accompanied by an increase of one OH^- and a decrease of one H_2O. One set of possible changes is

$$dN_{H_2O} = -1, \qquad dN_{H^+} = 1, \qquad dN_{OH^-} = 1. \tag{5.96}$$

Plugging these numbers into equation 5.95 yields

$$\mu_{H_2O} = \mu_{H^+} + \mu_{OH^-}. \tag{5.97}$$

This relation among the chemical potentials must be satisfied at equilibrium. Since each chemical potential is a function of the concentration of that species (a higher concentration implying a higher chemical potential), this condition determines the various concentrations at equilibrium.

Before generalizing this result to an arbitrary chemical reaction, let's consider another example, the combination of nitrogen and hydrogen to form ammonia:

$$N_2 + 3H_2 \longleftrightarrow 2NH_3. \tag{5.98}$$

Again, the reaction is at equilibrium when $\sum_i \mu_i \, dN_i = 0$. One possible set of dN's is

$$dN_{N_2} = -1, \qquad dN_{H_2} = -3, \qquad dN_{NH_3} = +2, \tag{5.99}$$

resulting in the equilibrium condition

$$\mu_{N_2} + 3\mu_{H_2} = 2\mu_{NH_3}. \tag{5.100}$$

By now you can probably see the pattern: The equilibrium condition is always the same as the reaction equation itself, but with the names of the chemical species replaced by their chemical potentials and $\longleftrightarrow$ replaced by $=$. To write this rule as a formula we need some notation. Let X_i represent the chemical name of the ith species involved in a reaction, and let ν_i represent the stoichiometric coefficient of this species in the reaction equation, that is, the number of i molecules that participate each time the reaction happens. (For instance, $\nu_{H_2} = 3$ in the previous example.) An arbitrary reaction equation then looks like this:

$$\nu_1 X_1 + \nu_2 X_2 + \cdots \longleftrightarrow \nu_3 X_3 + \nu_4 X_4 + \cdots. \tag{5.101}$$

In the corresponding equilibrium condition, we simply replace each species name with its chemical potential:

$$\nu_1 \mu_1 + \nu_2 \mu_2 + \cdots = \nu_3 \mu_3 + \nu_4 \mu_4 + \cdots. \tag{5.102}$$

The next step in understanding chemical equilibrium is to write each chemical potential μ_i in terms of the concentration of that species; then one can solve for the equilibrium concentrations. I could try to explain how to do this in general, but because gases, solutes, solvents, and pure substances must all be treated differently, I think it's easier (and more interesting) to demonstrate the procedure through the four worked examples that make up the rest of this section.

Problem 5.83. Write down the equilibrium condition for each of the following reactions:

    (a) $2H \leftrightarrow H_2$

    (b) $2CO + O_2 \leftrightarrow 2CO_2$

    (c) $CH_4 + 2O_2 \leftrightarrow 2H_2O + CO_2$

    (d) $H_2SO_4 \leftrightarrow 2H^+ + SO_4^{2-}$

    (e) $2p + 2n \leftrightarrow {}^4He$

Nitrogen Fixation

First consider the gaseous reaction 5.98, in which N_2 and H_2 combine to form ammonia (NH_3). This reaction is called nitrogen "fixation" because it puts the nitrogen into a form that can be used by plants to synthesize amino acids and other important molecules.

The equilibrium condition for this reaction is written in equation 5.100. If we assume that each species is an ideal gas, we can use equation 5.40 for each chemical potential to obtain

$$\mu^\circ_{N_2} + kT \ln\left(\frac{P_{N_2}}{P^\circ}\right) + 3\mu^\circ_{H_2} + 3kT \ln\left(\frac{P_{H_2}}{P^\circ}\right) = 2\mu^\circ_{NH_3} + 2kT \ln\left(\frac{P_{NH_3}}{P^\circ}\right). \tag{5.103}$$

Here each μ° represents the chemical potential of that species in its "standard state," when its partial pressure is P°. Normally we take P° to be 1 bar. Gathering all the μ°'s on the right and all the logarithms on the left gives

$$kT \ln\left(\frac{P_{N_2}}{P^\circ}\right) + 3kT \ln\left(\frac{P_{H_2}}{P^\circ}\right) - 2kT \ln\left(\frac{P_{NH_3}}{P^\circ}\right) = 2\mu^\circ_{NH_3} - \mu^\circ_{N_2} - 3\mu^\circ_{H_2}. \tag{5.104}$$

Now if we multiply through by Avogadro's number, what's on the right is the "standard" Gibbs free energy of the reaction, written ΔG°. This quantity is the hypothetical change in G when one mole of pure N_2 reacts with three moles of pure H_2 to form two moles of pure ammonia, all at 1 bar. The important thing about ΔG° is that you can often look it up in reference tables. Meanwhile, we can combine the logarithms on the left into one big logarithm; thus,

$$RT \ln\left(\frac{P_{N_2} P_{H_2}^3}{P_{NH_3}^2 (P^\circ)^2}\right) = \Delta G^\circ, \tag{5.105}$$

or with a bit more rearranging,

$$\frac{P_{NH_3}^2 (P^\circ)^2}{P_{N_2} P_{H_2}^3} = e^{-\Delta G^\circ / RT}. \tag{5.106}$$

Equation 5.106 is our final result. On the left-hand side are the equilibrium partial pressures of the three gases, raised to the powers of their stoichiometric coefficients, with reactants in the denominator and products in the numerator. There are also enough powers of the reference pressure P° to make the whole expression dimensionless. The quantity on the right-hand side is called the **equilibrium constant**, K:

$$K \equiv e^{-\Delta G^\circ / RT}. \tag{5.107}$$

It is a function of temperature (both through ΔG° and through the explicit T) but not of the amounts of the gases that are present. Often we compute K once and for all (at a given T) and then simply write

$$\frac{P_{NH_3}^2 (P^\circ)^2}{P_{N_2} P_{H_2}^3} = K. \tag{5.108}$$

This equation is called the **law of mass action** (don't ask me why).

Even if you don't know the value of K, equation 5.108 tells you quite a bit about this reaction. If the gases are initially in equilibrium and you add more nitrogen or hydrogen, some of what you add will have to react to form ammonia in order to maintain equilibrium. If you add more ammonia, some will have to convert to nitrogen and hydrogen. If you double the partial pressure of both the hydrogen and the nitrogen, the partial pressure of the ammonia must *quadruple* in order to maintain equilibrium. Increasing the *total* pressure therefore favors the production of more ammonia. One way to remember how a system in equilibrium will respond to these kinds of changes, at least qualitatively, is **Le Chatelier's principle**:

> When you disturb a system in equilibrium, it will respond in a way that partially offsets the disturbance.

So for instance, when you *increase* the total pressure, more nitrogen and hydrogen will react to form ammonia, decreasing the total number of molecules and thus *reducing* the total pressure.

To be more quantitative, we need a numerical value for the equilibrium constant K. Sometimes you can find tabulated values of equilibrium constants, but more often you need to compute them from ΔG° values using equation 5.107. For the production of two moles of ammonia at 298 K, standard tables give the value $\Delta G^\circ = -32.9$ kJ. The equilibrium constant at room temperature is therefore

$$K = \exp\left(\frac{+32,900 \text{ J}}{(8.31 \text{ J/K})(298 \text{ K})}\right) = 5.9 \times 10^5, \tag{5.109}$$

so this reaction tends strongly to the right, favoring the production of ammonia from nitrogen and hydrogen.

At higher temperatures, K becomes much smaller (see Problem 5.86), so you might think that industrial production of ammonia would be carried out at relatively low temperature. However, the equilibrium condition tells us absolutely nothing about the *rate* of the reaction. It turns out that, unless a good catalyst is present, this reaction proceeds negligibly slowly at temperatures below about 700°C. Certain bacteria do contain excellent catalysts (enzymes) that can fix nitrogen at room temperature. For industrial production, though, the best known catalyst still requires a temperature of about 500°C to achieve an acceptable production rate. At this temperature the equilibrium constant is only 6.9×10^{-5}, so very high pressures are needed to produce a reasonable amount of ammonia. The industrial nitrogen-fixation process used today, employing an iron-molybdenum catalyst, temperatures around 500°C, and total pressures around 400 atm, was developed in the early 20th century by the German chemist Fritz Haber. This process has revolutionized the production of chemical fertilizers, and also, unfortunately, facilitated the manufacture of explosives.

Problem 5.84. A mixture of one part nitrogen and three parts hydrogen is heated, in the presence of a suitable catalyst, to a temperature of 500°C. What fraction of the nitrogen (atom for atom) is converted to ammonia, if the final total pressure is 400 atm? Pretend for simplicity that the gases behave ideally despite the very high pressure. The equilibrium constant at 500°C is 6.9×10^{-5}. (Hint: You'll have to solve a quadratic equation.)

Problem 5.85. Derive the **van't Hoff equation**,

$$\frac{d \ln K}{dT} = \frac{\Delta H^\circ}{RT^2},$$

which gives the dependence of the equilibrium constant on temperature.[*] Here ΔH° is the enthalpy change of the reaction, for pure substances in their standard states (1 bar pressure for gases). Notice that if ΔH° is positive (loosely speaking, if the reaction requires the absorption of heat), then higher temperature makes the reaction tend more to the right, as you might expect. Often you can neglect the temperature dependence of ΔH°; solve the equation in this case to obtain

$$\ln K(T_2) - \ln K(T_1) = \frac{\Delta H^\circ}{R}\left(\frac{1}{T_1} - \frac{1}{T_2}\right).$$

Problem 5.86. Use the result of the previous problem to estimate the equilibrium constant of the reaction $N_2 + 3H_2 \leftrightarrow 2NH_3$ at 500°C, using only the room-temperature data at the back of this book. Compare your result to the actual value of K at 500°C quoted in the text.

---

[*]Van't Hoff's *equation* is not to be confused with van't Hoff's *formula* for osmotic pressure. Same person, different physical principle.

Dissociation of Water

As a second example of chemical equilibrium, consider again the dissociation of water into H^+ and OH^- ions, discussed briefly at the beginning of this section:

$$H_2O \longleftrightarrow H^+ + OH^-. \tag{5.110}$$

At equilibrium the chemical potentials of the three species satisfy

$$\mu_{H_2O} = \mu_{H^+} + \mu_{OH^-}. \tag{5.111}$$

Assuming that the solution is dilute (a very good approximation under normal conditions), the chemical potentials are given by equations 5.69 (for H_2O) and 5.72 (for H^+ and OH^-). Furthermore, the deviation of μ_{H_2O} from its value for pure water is negligible. The equilibrium condition is therefore

$$\mu^{\circ}_{H_2O} = \mu^{\circ}_{H^+} + kT \ln m_{H^+} + \mu^{\circ}_{OH^-} + kT \ln m_{OH^-}, \tag{5.112}$$

where each μ° represents the chemical potential of the substance in its "standard state"—pure liquid for H_2O and a concentration of one mole per kilogram solvent for the ions. The m's are molalities, understood to be measured in units of one mole solute per kilogram of solvent.

As in the previous example, the next step is to gather the μ°'s on the right, the logarithms on the left, and multiply through by Avogadro's number:

$$RT \ln(m_{H^+} m_{OH^-}) = -N_A(\mu^{\circ}_{H^+} + \mu^{\circ}_{OH^-} - \mu^{\circ}_{H_2O}) = -\Delta G^{\circ}, \tag{5.113}$$

where ΔG° is the standard change in G for the reaction, again a value that can be looked up in tables. A bit of rearrangement gives

$$m_{H^+} m_{OH^-} = e^{-\Delta G^{\circ}/RT}, \tag{5.114}$$

the equilibrium condition for the ion molalities.

Before plugging in numbers, it's worthwhile to pause and compare this result to the equilibrium condition in the previous example, equation 5.106. In both cases the right-hand side is called the equilibrium constant,

$$K = e^{-\Delta G^{\circ}/RT}, \tag{5.115}$$

and is given by the same exponential function of the standard change in the Gibbs free energy. But the "standard" states are now completely different: pure liquid for the solvent and 1 molal for the solutes instead of 1 bar partial pressure for the gases of the previous example. Correspondingly, the left-hand side of equation 5.114 involves molalities instead of partial pressures (but still raised to the powers of their stoichiometric coefficients, in this case both equal to 1). Most significantly, the amount or concentration of water does not appear at all on the left-hand side of equation 5.114. This is because in a dilute solution there is always plenty of solvent available for the reaction, no matter how much has already reacted. (The same would be true of a pure liquid or solid that reacts only at its surface.)

A final difference between ideal gas reactions and reactions in solution is that in the latter case, the equilibrium constant can in principle depend on the total pressure. In practice, however, this dependence is usually negligible except at very high (e.g., geological) pressures (see Problem 5.88).

The value of ΔG° for the dissociation of one mole of water at room temperature and atmospheric pressure is 79.9 kJ, so the equilibrium constant for this reaction is

$$K = \exp\left(-\frac{79,900 \text{ J}}{(8.31 \text{ J/K})(298 \text{ K})}\right) = 1.0 \times 10^{-14}. \tag{5.116}$$

If all the H^+ and OH^- ions come from dissociation of water molecules, then they must be equally abundant, so in this case

$$m_{H^+} = m_{OH^-} = 1.0 \times 10^{-7}. \tag{5.117}$$

The 7 in this result is called the **pH** of pure water. More generally, the pH is defined as minus the base-10 logarithm of the molality of H^+ ions:

$$\text{pH} \equiv -\log_{10} m_{H^+}. \tag{5.118}$$

If other substances are dissolved in water, the pH can shift significantly. When the pH is less than 7 (indicating a higher H^+ concentration) we say the solution is **acidic**, while when the pH is greater than 7 (indicating a lower H^+ concentration) we say the solution is **basic**.

Problem 5.87. Sulfuric acid, H_2SO_4, readily dissociates into H^+ and HSO_4^- ions:

$$H_2SO_4 \longrightarrow H^+ + HSO_4^-.$$

The hydrogen sulfate ion, in turn, can dissociate again:

$$HSO_4^- \longleftrightarrow H^+ + SO_4^{2-}.$$

The equilibrium constants for these reactions, in aqueous solutions at 298 K, are approximately 10^2 and $10^{-1.9}$, respectively. (For dissociation of acids it is usually more convenient to look up K than ΔG°. By the way, the negative base-10 logarithm of K for such a reaction is called **pK**, in analogy to pH. So for the first reaction pK $= -2$, while for the second reaction pK $= 1.9$.)

(a) Argue that the first reaction tends so strongly to the right that we might as well consider it to have gone to completion, in any solution that could possibly be considered dilute. At what pH values would a significant fraction of the sulfuric acid *not* be dissociated?

(b) In industrialized regions where lots of coal is burned, the concentration of sulfate in rainwater is typically 5×10^{-5} mol/kg. The sulfate can take any of the chemical forms mentioned above. Show that, at this concentration, the second reaction will also have gone essentially to completion, so all the sulfate is in the form of SO_4^{2-}. What is the pH of this rainwater?

(c) Explain why you can neglect dissociation of water into H^+ and OH^- in answering the previous question.

(d) At what pH would dissolved sulfate be equally distributed between HSO_4^- and SO_4^{2-}?

Problem 5.88. Express $\partial(\Delta G^{\circ})/\partial P$ in terms of the volumes of solutions of reactants and products, for a chemical reaction of dilute solutes. Plug in some reasonable numbers, to show that a pressure increase of 1 atm has only a negligible effect on the equilibrium constant.

Oxygen Dissolving in Water

When oxygen (O_2) gas dissolves in water (see Figure 5.42), there is no chemical reaction per se, but we can still apply the techniques of this section to find out how much O_2 will dissolve. The "reaction" equation and tabulated ΔG° value are

$$O_2(g) \longleftrightarrow O_2(aq), \qquad \Delta G^{\circ} = 16.4 \text{ kJ}, \qquad (5.119)$$

where g is for "gas" and aq is for "aqueous" (i.e., dissolved in water). The ΔG° value is for one mole of oxygen at 1 bar pressure dissolving in 1 kg of water (to give a solution with molality 1), all at 298 K.

When the dissolved oxygen is in equilibrium with the oxygen in the adjacent gas, their chemical potentials must be equal:

$$\mu_{\text{gas}} = \mu_{\text{solute}}. \qquad (5.120)$$

Using equation 5.40 for μ_{gas} and equation 5.72 for μ_{solute}, we can write both chemical potentials in terms of standard-state values and the respective concentrations:

$$\mu_{\text{gas}}^{\circ} + kT\ln(P/P^{\circ}) = \mu_{\text{solute}}^{\circ} + kT\ln m. \qquad (5.121)$$

Here P is the partial pressure of O_2 in the gas, P° is the standard pressure of 1 bar, and m is the molality of the dissolved oxygen in moles per kilogram of water. Once again, the procedure is to gather the μ°'s on the right and the logarithms on the left, then multiply through by Avogadro's number:

$$RT\ln\left(\frac{P/P^{\circ}}{m}\right) = N_A(\mu_{\text{solute}}^{\circ} - \mu_{\text{gas}}^{\circ}) = \Delta G^{\circ}; \qquad (5.122)$$

or equivalently,

$$\frac{m}{P/P^{\circ}} = e^{-\Delta G^{\circ}/RT}. \qquad (5.123)$$

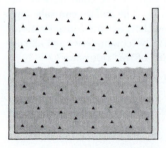

Figure 5.42. The dissolution of a gas in a liquid, such as oxygen in water, can be treated as a chemical reaction with its own equilibrium constant.

Equation 5.123 says that the ratio of the amount of dissolved oxygen to the amount in the adjacent gas is a constant, at any given temperature and total pressure. This result is known as **Henry's law**. As in the previous example, the dependence of $\Delta G°$ on the total pressure is usually negligible unless the pressure is very large. The constant on the right-hand side of the equation is sometimes called a "Henry's law constant," but one often finds these constants tabulated in very different ways—as reciprocals and/or in terms of mole fraction rather than molality.

For oxygen in water at room temperature the right-hand side of equation 5.123 is

$$\exp\left(-\frac{16,400 \text{ J}}{(8.31 \text{ J/K})(298 \text{ K})}\right) = 0.00133 = \frac{1}{750}, \tag{5.124}$$

meaning that if the partial pressure of oxygen is 1 bar, about 1/750 of a mole of oxygen will dissolve in each kilogram of water. In our atmosphere at sea level the partial pressure of oxygen is only about 1/5 as much, and the amount of dissolved oxygen in water is proportionally less. Still, each liter of water contains the equivalent of about 7 cm^3 of pure oxygen (if it were a gas at atmospheric pressure), enough for fish to breathe.

Problem 5.89. The standard enthalpy change upon dissolving one mole of oxygen at 25°C is −11.7 kJ. Use this number and the van't Hoff equation (Problem 5.85) to calculate the equilibrium (Henry's law) constant for oxygen in water at 0°C and at 100°C. Discuss the results briefly.

Problem 5.90. When solid quartz "dissolves" in water, it combines with water molecules in the reaction

$$\text{SiO}_2(\text{s}) + 2\text{H}_2\text{O}(\text{l}) \longleftrightarrow \text{H}_4\text{SiO}_4(\text{aq}).$$

(a) Use this data in the back of this book to compute the amount of silica dissolved in water in equilibrium with solid quartz, at 25°C.

(b) Use the van't Hoff equation (Problem 5.85) to compute the amount of silica dissolved in water in equilibrium with solid quartz at 100°C.

Problem 5.91. When carbon dioxide "dissolves" in water, essentially all of it reacts to form carbonic acid, H_2CO_3:

$$\text{CO}_2(\text{g}) + \text{H}_2\text{O}(\text{l}) \longleftrightarrow \text{H}_2\text{CO}_3(\text{aq}).$$

The carbonic acid can then dissociate into H^+ and bicarbonate ions,

$$\text{H}_2\text{CO}_3(\text{aq}) \longleftrightarrow \text{H}^+(\text{aq}) + \text{HCO}_3^-(\text{aq}).$$

(The table at the back of this book gives thermodynamic data for both of these reactions.) Consider a body of otherwise pure water (or perhaps a raindrop) that is in equilibrium with the atmosphere near sea level, where the partial pressure of carbon dioxide is 3.4×10^{-4} bar (or 340 parts per million). Calculate the molality of carbonic acid and of bicarbonate ions in the water, and determine the pH of the solution. Note that even "natural" precipitation is somewhat acidic.

Ionization of Hydrogen

As a final example of chemical equilibrium, let's consider the ionization of atomic hydrogen into a proton and an electron,

$$ \text{H} \longleftrightarrow p + e, \tag{5.125} $$

an important reaction in stars such as our sun. This reaction is so simple that we can compute the equilibrium constant from first principles, without looking up anything in a table.

Following the same steps as in the previous examples, we can write the equilibrium condition for the partial pressures as

$$ kT \ln\left(\frac{P_{\text{H}} P^\circ}{P_p P_e}\right) = \mu_p^\circ + \mu_e^\circ - \mu_{\text{H}}^\circ, \tag{5.126} $$

where each μ° is the chemical potential of that species at 1 bar pressure. Under most conditions we can treat all three species as structureless monatomic gases, for which we derived an explicit formula for μ in Section 3.5:

$$ \mu = -kT \ln\left[\frac{V}{N}\left(\frac{2\pi m kT}{h^2}\right)^{3/2}\right] = -kT \ln\left[\frac{kT}{P}\left(\frac{2\pi m kT}{h^2}\right)^{3/2}\right]. \tag{5.127} $$

(Here m is the mass of the particle, not molality.) The only subtlety is that this formula includes only kinetic energy, taking the energy zero-point to be the state where all the particles are at rest. In computing the *difference* of the μ°'s we also need to include the ionization energy, $I = 13.6$ eV, that you need to put in to convert H to $p + e$ even if there is no kinetic energy. I'll put this in by subtracting I from μ_{H}:

$$ \mu_{\text{H}}^\circ = -kT \ln\left[\frac{kT}{P^\circ}\left(\frac{2\pi m_{\text{H}} kT}{h^2}\right)^{3/2}\right] - I. \tag{5.128} $$

For p and e, the formulas for μ° are identical but with different masses and without the final $-I$.

Plugging all three μ°'s into equation 5.126 gives a big mess, but since $m_p \approx m_{\text{H}}$, everything except the $-I$ in the chemical potentials of these two species cancels. Dividing through by $-kT$, we're left with

$$ -\ln\left(\frac{P_{\text{H}} P^\circ}{P_p P_e}\right) = \ln\left[\frac{kT}{P^\circ}\left(\frac{2\pi m_e kT}{h^2}\right)^{3/2}\right] - \frac{I}{kT}. \tag{5.129} $$

A bit of algebra then yields the following result:

$$ \frac{P_p}{P_{\text{H}}} = \frac{kT}{P_e}\left(\frac{2\pi m_e kT}{h^2}\right)^{3/2} e^{-I/kT}. \tag{5.130} $$

This formula is called the **Saha equation**. It gives the ratio of the amount of ionized hydrogen (that is, protons) to the amount of un-ionized hydrogen as a function of temperature and the concentration of electrons. (Note that the combination P_e/kT is the same as N_e/V, the number of electrons per unit volume.) At the surface of the sun the temperature is about 5800 K, so the exponential factor is only $e^{-I/kT} = 1.5 \times 10^{-12}$. Meanwhile the electron concentration is roughly 2×10^{19} m^{-3}; the Saha equation thus predicts a ratio of

$$\frac{P_p}{P_H} = \frac{(1.07 \times 10^{27} \text{ m}^{-3})(1.5 \times 10^{-12})}{2 \times 10^{19} \text{ m}^{-3}} = 8 \times 10^{-5}. \qquad (5.131)$$

Even at the surface of the sun, less than one hydrogen atom in 10,000 is ionized.

Problem 5.92. Suppose you have a box of atomic hydrogen, initially at room temperature and atmospheric pressure. You then raise the temperature, keeping the volume fixed.

(a) Find an expression for the *fraction* of the hydrogen that is ionized as a function of temperature. (You'll have to solve a quadratic equation.) Check that your expression has the expected behavior at very low and very high temperatures.

(b) At what temperature is exactly half of the hydrogen ionized?

(c) Would raising the initial pressure cause the temperature you found in part (b) to increase or decrease? Explain.

(d) Plot the expression you found in part (a) as a function of the dimensionless variable $t = kT/I$. Choose the range of t values to clearly show the interesting part of the graph.

Thermodynamics has something to say about everything but does not tell us everything about anything.

—Martin Goldstein and Inge F. Goldstein,
 The Refrigerator and the Universe. Copyright 1993 by the President and Fellows of Harvard College. Reprinted by permission of Harvard University Press.

6 Boltzmann Statistics

Most of this book so far has dealt with the second law of thermodynamics: its origin in the statistical behavior of large numbers of particles, and its applications in physics, chemistry, earth science, and engineering. However, the second law by itself usually doesn't tell us all we would like to know. In the last two chapters especially, we have often had to rely on experimental measurements (of enthalpies, entropies, and so on) before we could extract any predictions from the second law. This approach to thermodynamics can be extremely powerful, provided that the needed measurements can be made to the required precision.

Ideally, though, we would like to be able to calculate *all* thermodynamic quantities from first principles, starting from microscopic models of various systems of interest. In this book we have already worked with three important microscopic models: the two-state paramagnet, the Einstein solid, and the monatomic ideal gas. For each of these models we were able to write down an explicit combinatoric formula for the multiplicity, Ω, and from there go on to compute the entropy, temperature, and other thermodynamic properties. In this chapter and the next two, we will study a number of more complicated models, representing a much greater variety of physical systems. For these more complicated models the direct combinatoric approach used in Chapters 2 and 3 would be too difficult mathematically. We therefore need to develop some new theoretical tools.

6.1 The Boltzmann Factor

In this section I will introduce the most powerful tool in all of statistical mechanics: an amazingly simple formula for the *probability* of finding a system in any particular microstate, when that system is in thermal equilibrium with a "reservoir" at a specified temperature (see Figure 6.1).

The system can be almost anything, but for definiteness, let's say it's a single atom. The microstates of the system then correspond to the various energy levels of

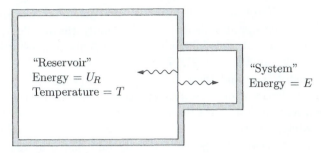

Figure 6.1. A "system" in thermal contact with a much larger "reservoir" at some well-defined temperature.

the atom, although for a given energy level there is often more than one independent state. For instance, a hydrogen atom has only one ground state (neglecting spin), with energy -13.6 eV. But it has *four* independent states with energy -3.4 eV, *nine* states with energy -1.5 eV, and so on (see Figure 6.2). Each of these independent states counts as a separate microstate. When an energy level corresponds to more than one independent state, we say that level is **degenerate**. (For a more precise definition of degeneracy, and a more thorough discussion of the hydrogen atom, see Appendix A.)

If our atom were completely isolated from the rest of the universe, then its energy would be fixed, and all microstates with that energy would be equally probable. Now, however, we're interested in the situation where the atom is not isolated, but instead is exchanging energy with lots of other atoms, which form a large "reservoir" with a fixed temperature. In this case the atom could conceivably be found in any of its microstates, but some will be more likely than others, depending on their energies. (Microstates with the *same* energy will still have the same probability.)

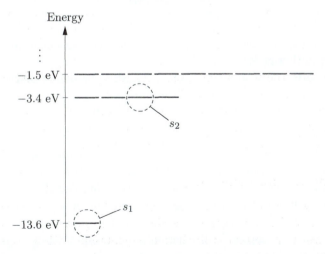

Figure 6.2. Energy level diagram for a hydrogen atom, showing the three lowest energy levels. There are four independent states with energy -3.4 eV, and nine independent states with energy -1.5 eV.

Since the probability of finding the atom in any particular microstate depends on how many other microstates there are, I'll simplify the problem, at first, by looking only at the *ratio* of probabilities for two particular microstates of interest (such as those circled in Figure 6.2). Let me call these states s_1 and s_2, their energies $E(s_1)$ and $E(s_2)$, and their probabilities $\mathcal{P}(s_1)$ and $\mathcal{P}(s_2)$. How can I find a formula for the ratio of these probabilities? Let's go all the way back to the fundamental assumption of statistical mechanics: For an *isolated* system, all accessible microstates are equally probable. Our atom is not an isolated system, but the atom and the reservoir together do make an isolated system, and we are equally likely to find this combined system in any of its accessible microstates.

Now we don't care what the state of the reservoir is; we just want to know what state the atom is in. But if the atom is in state s_1, then the reservoir will have some very large number of accessible states, all equally probable. I'll call this number $\Omega_R(s_1)$: the multiplicity of the *reservoir* when the *atom* is in state s_1. Similarly, I'll use $\Omega_R(s_2)$ to denote the multiplicity of the reservoir when the atom is in state s_2. These two multiplicities will generally be different, because when the atom is in a lower-energy state, more energy is left for the reservoir.

Suppose, for instance, that state s_1 has the lower energy, so that $\Omega_R(s_1) > \Omega_R(s_2)$. As a specific example, let's say $\Omega_R(s_1) = 100$ and $\Omega_R(s_2) = 50$ (though in a true reservoir the multiplicities would be *much* larger). Now fundamentally, all microstates of the combined system are equally probable. Therefore since there are twice as many states of the combined system when the atom is in state s_1 than when it is in state s_2, the former state must be twice as probable as the latter. More generally, the probability of finding the atom in any particular state is directly proportional to the number of microstates that are accessible to the reservoir. Thus the ratio of probabilities for any two states is

$$\frac{\mathcal{P}(s_2)}{\mathcal{P}(s_1)} = \frac{\Omega_R(s_2)}{\Omega_R(s_1)}. \tag{6.1}$$

Now we just have to get this expression into a more convenient form, using some math and a bit of thermodynamics.

First I'll rewrite each Ω in terms of entropy, using the definition $S = k \ln \Omega$:

$$\frac{\mathcal{P}(s_2)}{\mathcal{P}(s_1)} = \frac{e^{S_R(s_2)/k}}{e^{S_R(s_1)/k}} = e^{[S_R(s_2) - S_R(s_1)]/k}. \tag{6.2}$$

The exponent now contains the *change* in the entropy of the reservoir, when the atom undergoes a transition from state 1 to state 2. This change will be tiny, since the atom is so small compared to the reservoir. We can therefore invoke the thermodynamic identity:

$$dS_R = \frac{1}{T}\left(dU_R + P\, dV_R - \mu\, dN_R\right). \tag{6.3}$$

The right-hand side involves the changes in the reservoir's energy, volume, and number of particles. But anything gained by the reservoir is lost by the atom, so we can write each of these changes as minus the change in the same quantity for the atom.

I'm going to throw away the $P\,dV$ and $\mu\,dN$ terms, but for different reasons. The quantity $P\,dV_R$ is usually nonzero, but much smaller than dU_R and therefore negligible. For instance, when an atom goes into an excited state, its effective volume might increase by a cubic ångstrom, so at atmospheric pressure, the term $P\,dV$ is of order 10^{-25} J. This is a million times less than the typical change in the atom's energy of a few electron-volts. Meanwhile, dN really is zero, at least when the small system is a single atom, and also in the other cases that we'll consider in this chapter. (In the following chapter I'll put the dN term back, in order to deal with other types of systems.)

So the difference of entropies in equation 6.2 can be rewritten

$$S_R(s_2) - S_R(s_1) = \frac{1}{T}[U_R(s_2) - U_R(s_1)] = -\frac{1}{T}[E(s_2) - E(s_1)], \qquad (6.4)$$

where E is the energy of the atom. Plugging this expression back into equation 6.2, we obtain

$$\frac{\mathcal{P}(s_2)}{\mathcal{P}(s_1)} = e^{-[E(s_2)-E(s_1)]/kT} = \frac{e^{-E(s_2)/kT}}{e^{-E(s_1)/kT}}. \qquad (6.5)$$

The ratio of probabilities is equal to a ratio of simple exponential factors, each of which is a function of the energy of the corresponding microstate and the temperature of the reservoir. Each of these exponential factors is called a **Boltzmann factor**:

$$\text{Boltzmann factor} = e^{-E(s)/kT}. \qquad (6.6)$$

It would be nice if we could just say that the probability of each state is *equal* to the corresponding Boltzmann factor. Unfortunately, this isn't true. To arrive at the correct statement, let's manipulate equation 6.5 to get everything involving s_1 on one side and everything involving s_2 on the other side:

$$\frac{\mathcal{P}(s_2)}{e^{-E(s_2)/kT}} = \frac{\mathcal{P}(s_1)}{e^{-E(s_1)/kT}}. \qquad (6.7)$$

Notice that the left side of this equation is independent of s_1; therefore the right side must be as well. Similarly, since the right side is independent of s_2, so is the left side. But if both sides are independent of both s_1 and s_2, then in fact both sides must be equal to a constant, the same for all states. The constant is called $1/Z$; it is the constant of proportionality that converts a Boltzmann factor into a probability. In conclusion, for any state s,

$$\mathcal{P}(s) = \frac{1}{Z}\,e^{-E(s)/kT}. \qquad (6.8)$$

This is the most useful formula in all of statistical mechanics. Memorize it.[*]

---

[*]Equation 6.8 is sometimes called the **Boltzmann distribution** or the **canonical distribution**.

To interpret equation 6.8, let's suppose for a moment that the ground state energy of our atom is $E_0 = 0$, while the excited states all have positive energies. Then the probability of the ground state is $1/Z$, and all other states have smaller probabilities. States with energies much less than kT have probabilities only slightly less than $1/Z$, while states with energies much greater than kT have negligible probabilities, suppressed by the exponential Boltzmann factor. Figure 6.3 shows a bar graph of the probabilities for the states of a hypothetical system.

But what if the ground state energy is not zero? Physically, we should expect that shifting all the energies by a constant has no effect on the probabilities, and indeed, the probabilities *are* unchanged. It's true that all the Boltzmann factors get multiplied by an additional factor of $e^{-E_0/kT}$, but we'll see in a moment that Z gets multiplied by this same factor, so it cancels out in equation 6.8. Thus, the ground state still has the highest probability, and the remaining states have probabilities that are either a little less or a lot less, depending on how their energies, as measured from the ground state, compare to kT.

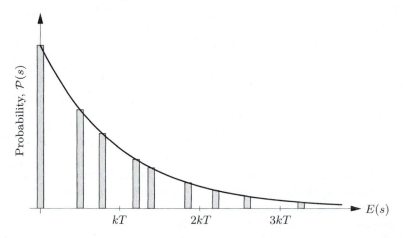

Figure 6.3. Bar graph of the relative probabilities of the states of a hypothetical system. The horizontal axis is energy. The smooth curve represents the Boltzmann distribution, equation 6.8, for one particular temperature. At lower temperatures it would fall off more suddenly, while at higher temperatures it would fall off more gradually.

Problem 6.1. Consider a system of two Einstein solids, where the first "solid" contains just a single oscillator, while the second solid contains 100 oscillators. The total number of energy units in the combined system is fixed at 500. Use a computer to make a table of the multiplicity of the combined system, for each possible value of the energy of the first solid from 0 units to 20. Make a graph of the total multiplicity vs. the energy of the first solid, and discuss, in some detail, whether the shape of the graph is what you would expect. Also plot the logarithm of the total multiplicity, and discuss the shape of this graph.

Problem 6.2. Prove that the probability of finding an atom in any particular energy *level* is $\mathcal{P}(E) = (1/Z)e^{-F/kT}$, where $F = E - TS$ and the "entropy" of a level is k times the logarithm of the number of degenerate states for that level.

The Partition Function

By now you're probably wondering how to actually *calculate* Z. The trick is to remember that the total probability of finding the atom in *some* state or other must be 1:

$$1 = \sum_s \mathcal{P}(s) = \sum_s \frac{1}{Z} e^{-E(s)/kT} = \frac{1}{Z} \sum_s e^{-E(s)/kT}. \tag{6.9}$$

Solving for Z therefore gives

$$Z = \sum_s e^{-E(s)/kT} = \text{sum of } all \text{ Boltzmann factors.} \tag{6.10}$$

This sum isn't always easy to carry out, since there may be an infinite number of states s and you may not have a simple formula for their energies. But the terms in the sum get smaller and smaller as the energies E_s get larger, so often you can just compute the first several terms numerically, neglecting states with energies much greater than kT.

The quantity Z is called the **partition function**, and turns out to be far more useful than I would have suspected. It is a "constant" in that it does not depend on any particular state s, but it *does* depend on temperature. To interpret it further, suppose once again that the ground state has energy zero. Then the Boltzmann factor for the ground state is 1, and the rest of the Boltzmann factors are less than 1, by a little or a lot, in proportion to the probabilities of the associated states. Thus the partition function essentially *counts* how many states are accessible to the atom, weighting each one in proportion to its probability. At very low temperature, $Z \approx 1$, since all the excited states have very small Boltzmann factors. At high temperature, Z will be much larger. And if we shift all the energies by a constant E_0, the whole partition function just gets multiplied by an uninteresting factor of $e^{-E_0/kT}$, which cancels when we compute probabilities.

Problem 6.3. Consider a hypothetical atom that has just two states: a ground state with energy zero and an excited state with energy 2 eV. Draw a graph of the partition function for this system as a function of temperature, and evaluate the partition function numerically at $T = 300$ K, 3000 K, 30,000 K, and 300,000 K.

Problem 6.4. Estimate the partition function for the hypothetical system represented in Figure 6.3. Then estimate the probability of this system being in its ground state.

Problem 6.5. Imagine a particle that can be in only three states, with energies -0.05 eV, 0, and 0.05 eV. This particle is in equilibrium with a reservoir at 300 K.

  (a) Calculate the partition function for this particle.

  (b) Calculate the probability for this particle to be in each of the three states.

  (c) Because the zero point for measuring energies is arbitrary, we could just as well say that the energies of the three states are 0, +0.05 eV, and +0.10 eV, respectively. Repeat parts (a) and (b) using these numbers. Explain what changes and what doesn't.

Thermal Excitation of Atoms

As a simple application of Boltzmann factors, let's consider a hydrogen atom in the atmosphere of the sun, where the temperature is about 5800 K. (We'll see in the following chapter how this temperature can be measured from earth.) I'd like to compare the probability of finding the atom in one of its first excited states (s_2) to the probability of finding it in the ground state (s_1). The ratio of probabilities is the ratio of Boltzmann factors, so

$$\frac{\mathcal{P}(s_2)}{\mathcal{P}(s_1)} = \frac{e^{-E_2/kT}}{e^{-E_1/kT}} = e^{-(E_2-E_1)/kT}. \tag{6.11}$$

The difference in energy is 10.2 eV, while kT is $(8.62 \times 10^{-5} \text{ eV/K})(5800 \text{ K}) = 0.50$ eV. So the ratio of probabilities is approximately $e^{-20.4} = 1.4 \times 10^{-9}$. For every billion atoms in the ground state, roughly 1.4 (on average) will be in any one of the first excited states. Since there are four such excited states, all with the same energy, the *total* number of atoms in these states will be four times as large, about 5.6 (for every billion in the ground state).

Atoms in the atmosphere of the sun can absorb sunlight on its way toward earth, but only at wavelengths that can induce transitions of the atoms into higher excited states. A hydrogen atom in its first excited state can absorb wavelengths in the Balmer series: 656 nm, 486 nm, 434 nm, and so on. These wavelengths are therefore missing, in part, from the sunlight we receive. If you put a narrow beam of sunlight through a good diffraction grating, you can see dark lines at the missing wavelengths (see Figure 6.4). There are also other prominent dark lines, created by other types of atoms in the solar atmosphere: iron, magnesium, sodium, calcium, and so on. The weird thing is, all these other wavelengths are absorbed by atoms

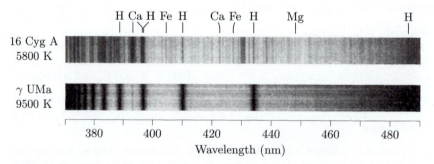

Figure 6.4. Photographs of the spectra of two stars. The upper spectrum is of a sunlike star (in the constellation Cygnus) with a surface temperature of about 5800 K; notice that the hydrogen absorption lines are clearly visible among a number of lines from other elements. The lower spectrum is of a hotter star (in Ursa Major, the Big Dipper), with a surface temperature of 9500 K. At this temperature a much larger fraction of the hydrogen atoms are in their first excited states, so the hydrogen lines are much more prominent than any others. Reproduced with permission from Helmut A. Abt et al., *An Atlas of Low-Dispersion Grating Stellar Spectra* (Kitt Peak National Observatory, Tucson, AZ, 1968).

(or ions) that start out either in their ground states or in very low-energy excited states (less than 3 eV above the ground state). The Balmer lines, by contrast, come only from the very rare hydrogen atoms that are excited more than 10 eV above the ground state. (A hydrogen atom in its ground state does not absorb any visible wavelengths.) Since the Balmer lines are quite prominent among the others, we can only conclude that hydrogen atoms are *much* more abundant in the sun's atmosphere than any of these other types.*

Problem 6.6. Estimate the probability that a hydrogen atom at room temperature is in one of its first excited states (relative to the probability of being in the ground state). Don't forget to take degeneracy into account. Then repeat the calculation for a hydrogen atom in the atmosphere of the star γ UMa, whose surface temperature is approximately 9500 K.

Problem 6.7. Each of the hydrogen atom states shown in Figure 6.2 is actually twofold degenerate, because the electron can be in two independent spin states, both with essentially the same energy. Repeat the calculation given in the text for the relative probability of being in a first excited state, taking spin degeneracy into account. Show that the results are unaffected.

Problem 6.8. The energy required to ionize a hydrogen atom is 13.6 eV, so you might expect that the number of ionized hydrogen atoms in the sun's atmosphere would be even less than the number in the first excited state. Yet at the end of Chapter 5 I showed that the fraction of ionized hydrogen is much larger, nearly one atom in 10,000. Explain why this result is not a contradiction, and why it would be incorrect to try to calculate the fraction of ionized hydrogen using the methods of this section.

Problem 6.9. In the numerical example in the text, I calculated only the *ratio* of the probabilities of a hydrogen atom being in two different states. At such a low temperature the *absolute* probability of being in a first excited state is essentially the same as the relative probability compared to the ground state. Proving this rigorously, however, is a bit problematic, because a hydrogen atom has infinitely many states.

(a) Estimate the partition function for a hydrogen atom at 5800 K, by adding the Boltzmann factors for all the states shown explicitly in Figure 6.2. (For simplicity you may wish to take the ground state energy to be zero, and shift the other energies accordingly.)

(b) Show that if *all* bound states are included in the sum, then the partition function of a hydrogen atom is infinite, at any nonzero temperature. (See Appendix A for the full energy level structure of a hydrogen atom.)

(c) When a hydrogen atom is in energy level n, the approximate radius of the electron wavefunction is $a_0 n^2$, where a_0 is the Bohr radius, about 5×10^{-11} m. Going back to equation 6.3, argue that the $P\,dV$ term is *not* negligible for the very high-n states, and therefore that the result of part (a), not that of part (b), gives the physically relevant partition function for this problem. Discuss.

---

*The recipe of the stars was first worked out by Cecilia Payne in 1924. The story is beautifully told by Philip and Phylis Morrison in *The Ring of Truth* (Random House, New York, 1987).

Problem 6.10. A water molecule can vibrate in various ways, but the easiest type of vibration to excite is the "flexing" mode in which the hydrogen atoms move toward and away from each other but the HO bonds do not stretch. The oscillations of this mode are approximately harmonic, with a frequency of 4.8×10^{13} Hz. As for any quantum harmonic oscillator, the energy levels are $\frac{1}{2}hf$, $\frac{3}{2}hf$, $\frac{5}{2}hf$, and so on. None of these levels are degenerate.

(a) Calculate the probability of a water molecule being in its flexing ground state and in each of the first two excited states, assuming that it is in equilibrium with a reservoir (say the atmosphere) at 300 K. (Hint: Calculate Z by adding up the first few Boltzmann factors, until the rest are negligible.)

(b) Repeat the calculation for a water molecule in equilibrium with a reservoir at 700 K (perhaps in a steam turbine).

Problem 6.11. A lithium nucleus has four independent spin orientations, conventionally labeled by the quantum number $m = -3/2, -1/2, 1/2, 3/2$. In a magnetic field B, the energies of these four states are $E = -m\mu B$, where the constant μ is 1.03×10^{-7} eV/T. In the Purcell-Pound experiment described in Section 3.3, the maximum magnetic field strength was 0.63 T and the temperature was 300 K. Calculate the probability of a lithium nucleus being in each of its four spin states under these conditions. Then show that, if the field is suddenly reversed, the probabilities of the four states obey the Boltzmann distribution for $T = -300$ K.

Problem 6.12. Cold interstellar molecular clouds often contain the molecule cyanogen (CN), whose first rotational excited states have an energy of 4.7×10^{-4} eV (above the ground state). There are actually three such excited states, all with the same energy. In 1941, studies of the absorption spectrum of starlight that passes through these molecular clouds showed that for every ten CN molecules that are in the ground state, approximately three others are in the three first excited states (that is, an average of one in each of these states). To account for this data, astronomers suggested that the molecules might be in thermal equilibrium with some "reservoir" with a well-defined temperature. What is that temperature?[*]

Problem 6.13. At *very* high temperatures (as in the very early universe), the proton and the neutron can be thought of as two different states of the same particle, called the "nucleon." (The reactions that convert a proton to a neutron or vice versa require the absorption of an electron or a positron or a neutrino, but all of these particles tend to be very abundant at sufficiently high temperatures.) Since the neutron's mass is higher than the proton's by 2.3×10^{-30} kg, its energy is higher by this amount times c^2. Suppose, then, that at some very early time, the nucleons were in thermal equilibrium with the rest of the universe at 10^{11} K. What fraction of the nucleons at that time were protons, and what fraction were neutrons?

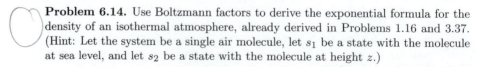

Problem 6.14. Use Boltzmann factors to derive the exponential formula for the density of an isothermal atmosphere, already derived in Problems 1.16 and 3.37. (Hint: Let the system be a single air molecule, let s_1 be a state with the molecule at sea level, and let s_2 be a state with the molecule at height z.)

---

[*]For a review of these measurements and calculations, see Patrick Thaddeus, *Annual Reviews of Astronomy and Astrophysics* **10**, 305–334 (1972).

6.2 Average Values

In the previous section we saw how to calculate the probability that a system is in any particular one of its microstates s, given that it is in equilibrium with a reservoir at temperature T:

$$\mathcal{P}(s) = \frac{1}{Z}\, e^{-\beta E(s)}, \tag{6.12}$$

where β is an abbreviation for $1/kT$. The exponential factor is called the **Boltzmann factor**, while Z is the **partition function**,

$$Z = \sum_s e^{-\beta E(s)}, \tag{6.13}$$

that is, the sum of the Boltzmann factors for all possible states.

Suppose, though, that we're not interested in knowing all the probabilities of all the various states our system could be in—suppose we just want to know the *average* value of some property of the system, such as its energy. Is there an easy way to compute this average, and if so, how?

Let me give a simple example. Suppose my system is an atom that has just three possible states: The ground state with energy 0 eV, a state with energy 4 eV, and a state with energy 7 eV. Actually, though, I have five such atoms, and at the moment, two of them are in the ground state, two are in the 4-eV state, and one is in the 7-eV state (see Figure 6.5). What is the average energy of all my atoms? Just add 'em up and divide by 5:

$$\overline{E} = \frac{(0\text{ eV})\cdot 2 + (4\text{ eV})\cdot 2 + (7\text{ eV})\cdot 1}{5} = 3\text{ eV}. \tag{6.14}$$

But there's another way to think about this computation. Instead of computing the numerator first and then dividing by 5, we can group the 1/5 with the factors of 2, 2, and 1, which represent the numbers of atoms in each state:

$$\overline{E} = (0\text{ eV})\cdot\frac{2}{5} + (4\text{ eV})\cdot\frac{2}{5} + (7\text{ eV})\cdot\frac{1}{5} = 3\text{ eV}. \tag{6.15}$$

In this expression, the energy of each state is multiplied by the *probability* of that state occurring (in any particular atom chosen from among my sample of 5); those probabilities are just 2/5, 2/5, and 1/5, respectively.

Figure 6.5. Five hypothetical atoms distributed among three different states.

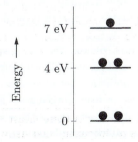

It's not hard to generalize this example into a formula. If I have a large sample of N atoms, and $N(s)$ is the number of atoms in any particular state s, then the average value of the energy is

$$\overline{E} = \frac{\sum_s E(s)N(s)}{N} = \sum_s E(s)\frac{N(s)}{N} = \sum_s E(s)\mathcal{P}(s), \qquad (6.16)$$

where $\mathcal{P}(s)$ is the probability of finding an atom in state s. So the average energy is just the sum of all the energies, weighted by their probabilities.

In the statistical mechanical systems that we're considering, each probability is given by equation 6.12, so

$$\overline{E} = \frac{1}{Z}\sum_s E(s)\,e^{-\beta E(s)}. \qquad (6.17)$$

Notice that the sum is similar to the partition function (6.13), but with an extra factor of $E(s)$ in each term.*

The average value of any other variable of interest can be computed in exactly the same way. If the variable is called X, and has the value $X(s)$ in state s, then

$$\overline{X} = \sum_s X(s)\mathcal{P}(s) = \frac{1}{Z}\sum_s X(s)\,e^{-\beta E(s)}. \qquad (6.18)$$

One nice feature of average values is that they are additive; for example, the average total energy of two objects is the sum of their individual average energies. This means that if you have a collection of many identical, independent particles, you can compute their total (average) energy from the average energy of just one, simply by multiplying by how many there are:

$$U = N\overline{E}. \qquad (6.19)$$

(Now you see why I've been using the symbol E for the energy of the atom; I've reserved U for the total energy of the much larger system that contains it.) So when I divided things into an "atom" and a "reservoir" in the previous section, it was partly just a trick. Even if you want to know the total energy of the *whole* system, you can often find it by concentrating first on one particle in the system, treating the rest as the reservoir. Once you know the average value of the quantity of interest for your particle, just multiply by N to get the total.

Technically, the U in equation 6.19 is merely the *average* energy of the entire system. If even this large system is in thermal contact with other objects, then the instantaneous value of U will fluctuate away from the average. However, if N is large, these fluctuations will almost always be negligible. Problem 6.17 shows how to calculate the size of typical fluctuations.

---

*In this chapter, the set of systems that we average *over* will be a hypothetical set whose members are assigned to states according to the Boltzmann probability distribution. This hypothetical set of systems is often called a **canonical ensemble**. In Chapters 2 and 3 we instead worked with isolated systems, where all allowed states had the same probability; a set of hypothetical systems with that (trivial) probability distribution is called a **microcanonical ensemble**.

Problem 6.15. Suppose you have 10 atoms of weberium: 4 with energy 0 eV, 3 with energy 1 eV, 2 with energy 4 eV, and 1 with energy 6 eV.

  (a) Compute the average energy of all your atoms, by adding up all their energies and dividing by 10.

  (b) Compute the probability that one of your atoms chosen at random would have energy E, for each of the four values of E that occur.

  (c) Compute the average energy again, using the formula $\overline{E} = \sum_s E(s)\mathcal{P}(s)$.

Problem 6.16. Prove that, for any system in equilibrium with a reservoir at temperature T, the average value of the energy is

$$\overline{E} = -\frac{1}{Z}\frac{\partial Z}{\partial \beta} = -\frac{\partial}{\partial \beta}\ln Z,$$

where $\beta = 1/kT$. These formulas can be extremely useful when you have an explicit formula for the partition function.

Problem 6.17. The most common measure of the *fluctuations* of a set of numbers away from the average is the **standard deviation**, defined as follows.

  (a) For each atom in the five-atom toy model of Figure 6.5, compute the *deviation* of the energy from the average energy, that is, $E_i - \overline{E}$, for $i = 1$ to 5. Call these deviations ΔE_i.

  (b) Compute the average of the *squares* of the five deviations, that is, $\overline{(\Delta E_i)^2}$. Then compute the square root of this quantity, which is the root-mean-square (rms) deviation, or standard deviation. Call this number σ_E. Does σ_E give a reasonable measure of how far the individual values tend to stray from the average?

  (c) Prove in general that

$$\sigma_E^2 = \overline{E^2} - (\overline{E})^2,$$

that is, the standard deviation squared is the average of the squares minus the square of the average. This formula usually gives the easier way of computing a standard deviation.

  (d) Check the preceding formula for the five-atom toy model of Figure 6.5.

Problem 6.18. Prove that, for any system in equilibrium with a reservoir at temperature T, the average value of E^2 is

$$\overline{E^2} = \frac{1}{Z}\frac{\partial^2 Z}{\partial \beta^2}.$$

Then use this result and the results of the previous two problems to derive a formula for σ_E in terms of the heat capacity, $C = \partial \overline{E}/\partial T$. You should find

$$\sigma_E = kT\sqrt{C/k}.$$

Problem 6.19. Apply the result of Problem 6.18 to obtain a formula for the standard deviation of the energy of a system of N identical harmonic oscillators (such as in an Einstein solid), in the high-temperature limit. Divide by the average energy to obtain a measure of the *fractional* fluctuation in energy. Evaluate this fraction numerically for $N = 1$, 10^4, and 10^{20}. Discuss the results briefly.

Paramagnetism

As a first application of these tools, I'd like to rederive some of our earlier results (see Section 3.3) for the ideal two-state paramagnet.

Recall that each elementary dipole in an ideal two-state paramagnet has just two possible states: an "up" state with energy $-\mu B$, and a "down" state with energy $+\mu B$. (Here B is the strength of the externally applied magnetic field, while the component of the dipole's magnetic moment in the direction of the field is $\pm\mu$.) The partition function for a single dipole is therefore

$$Z = \sum_s e^{-\beta E(s)} = e^{+\beta\mu B} + e^{-\beta\mu B} = 2\cosh(\beta\mu B). \tag{6.20}$$

The probability of finding the dipole in the "up" state is

$$\mathcal{P}_\uparrow = \frac{e^{+\beta\mu B}}{Z} = \frac{e^{+\beta\mu B}}{2\cosh(\beta\mu B)}, \tag{6.21}$$

while the probability of finding it in the "down" state is

$$\mathcal{P}_\downarrow = \frac{e^{-\beta\mu B}}{Z} = \frac{e^{-\beta\mu B}}{2\cosh(\beta\mu B)}. \tag{6.22}$$

You can easily check that these two probabilities add up to 1.

The average energy of our dipole is

$$\overline{E} = \sum_s E(s)\mathcal{P}(s) = (-\mu B)\mathcal{P}_\uparrow + (+\mu B)\mathcal{P}_\downarrow = -\mu B(\mathcal{P}_\uparrow - \mathcal{P}_\downarrow)$$

$$= -\mu B \frac{e^{\beta\mu B} - e^{-\beta\mu B}}{2\cosh(\beta\mu B)} = -\mu B\tanh(\beta\mu B). \tag{6.23}$$

If we have a collection of N such dipoles, the total energy is

$$U = -N\mu B\tanh(\beta\mu B), \tag{6.24}$$

in agreement with equation 3.31. In Section 3.3, however, we had to work much harder to derive this result: We started with the exact combinatoric formula for the multiplicity, then applied Stirling's approximation to simplify the entropy, then took a derivative and did lots of algebra to finally get U as a function of T. Here all we needed was Boltzmann factors.

According to the result of Problem 6.16, we can also compute the average energy by differentiating Z with respect to β, then multiplying by $-1/Z$:

$$\overline{E} = -\frac{1}{Z}\frac{\partial Z}{\partial \beta}. \tag{6.25}$$

Let's check this formula for the two-state paramagnet:

$$\overline{E} = -\frac{1}{Z}\frac{\partial}{\partial \beta}2\cosh(\beta\mu B) = -\frac{1}{Z}(2\mu B)\sinh(\beta\mu B) = -\mu B\tanh(\beta\mu B). \tag{6.26}$$

Yep, it works.

Finally, we can compute the average value of a dipole's magnetic moment along the direction of B:

$$\overline{\mu_z} = \sum_s \mu_z(s)\mathcal{P}(s) = (+\mu)\mathcal{P}_\uparrow + (-\mu)\mathcal{P}_\downarrow = \mu \tanh(\beta\mu B). \qquad (6.27)$$

Thus the total magnetization of the sample is

$$M = N\overline{\mu_z} = N\mu \tanh(\beta\mu B), \qquad (6.28)$$

in agreement with equation 3.32.

Problem 6.20. This problem concerns a collection of N identical harmonic oscillators (perhaps an Einstein solid or the internal vibrations of gas molecules) at temperature T. As in Section 2.2, the allowed energies of each oscillator are 0, hf, $2hf$, and so on.

(a) Prove by long division that

$$\frac{1}{1-x} = 1 + x + x^2 + x^3 + \cdots.$$

For what values of x does this series have a finite sum?

(b) Evaluate the partition function for a single harmonic oscillator. Use the result of part (a) to simplify your answer as much as possible.

(c) Use formula 6.25 to find an expression for the average energy of a single oscillator at temperature T. Simplify your answer as much as possible.

(d) What is the total energy of the system of N oscillators at temperature T? Your result should agree with what you found in Problem 3.25.

(e) If you haven't already done so in Problem 3.25, compute the heat capacity of this system and check that it has the expected limits as $T \to 0$ and $T \to \infty$.

Problem 6.21. In the real world, most oscillators are not perfectly harmonic. For a quantum oscillator, this means that the spacing between energy levels is not exactly uniform. The vibrational levels of an H_2 molecule, for example, are more accurately described by the approximate formula

$$E_n \approx \epsilon(1.03n - 0.03n^2), \qquad n = 0,\ 1,\ 2,\ \ldots,$$

where ϵ is the spacing between the two lowest levels. Thus, the levels get closer together with increasing energy. (This formula is reasonably accurate only up to about $n = 15$; for slightly higher n it would say that E_n decreases with increasing n. In fact, the molecule dissociates and there are no more discrete levels beyond $n \approx 15$.) Use a computer to calculate the partition function, average energy, and heat capacity of a system with this set of energy levels. Include all levels through $n = 15$, but check to see how the results change when you include fewer levels. Plot the heat capacity as a function of kT/ϵ. Compare to the case of a perfectly harmonic oscillator with evenly spaced levels, and also to the vibrational portion of the graph in Figure 1.13.

Problem 6.22. In most paramagnetic materials, the individual magnetic particles have more than two independent states (orientations). The number of independent states depends on the particle's angular momentum "quantum number" j, which must be a multiple of 1/2. For $j = 1/2$ there are just two independent states, as discussed in the text above and in Section 3.3. More generally, the allowed values of the z component of a particle's magnetic moment are

$$\mu_z = -j\delta_\mu, \; (-j+1)\delta_\mu, \; \ldots, \; (j-1)\delta_\mu, \; j\delta_\mu,$$

where δ_μ is a constant, equal to the difference in μ_z between one state and the next. (When the particle's angular momentum comes entirely from electron spins, δ_μ equals twice the Bohr magneton. When orbital angular momentum also contributes, δ_μ is somewhat different but comparable in magnitude. For an atomic nucleus, δ_μ is roughly a thousand times smaller.) Thus the number of states is $2j + 1$. In the presence of a magnetic field B pointing in the z direction, the particle's magnetic energy (neglecting interactions between dipoles) is $-\mu_z B$.

(a) Prove the following identity for the sum of a finite geometric series:

$$1 + x + x^2 + \cdots + x^n = \frac{1 - x^{n+1}}{1 - x}.$$

(Hint: Either prove this formula by induction on n, or write the series as a difference between two infinite series and use the result of Problem 6.20(a).)

(b) Show that the partition function of a single magnetic particle is

$$Z = \frac{\sinh[b(j + \frac{1}{2})]}{\sinh \frac{b}{2}},$$

where $b = \beta\delta_\mu B$.

(c) Show that the total magnetization of a system of N such particles is

$$M = N\delta_\mu \left[(j + \tfrac{1}{2}) \coth[b(j + \tfrac{1}{2})] - \tfrac{1}{2} \coth \tfrac{b}{2}\right],$$

where $\coth x$ is the hyperbolic cotangent, equal to $\cosh x / \sinh x$. Plot the quantity $M/N\delta_\mu$ vs. b, for a few different values of j.

(d) Show that the magnetization has the expected behavior as $T \to 0$.

(e) Show that the magnetization is proportional to $1/T$ (Curie's law) in the limit $T \to \infty$. (Hint: First show that $\coth x \approx \frac{1}{x} + \frac{x}{3}$ when $x \ll 1$.)

(f) Show that for $j = 1/2$, the result of part (c) reduces to the formula derived in the text for a two-state paramagnet.

Rotation of Diatomic Molecules

Now let's consider a more intricate application of Boltzmann factors and average values: the rotational motion of a diatomic molecule (assumed to be isolated, as in a low-density gas).

Rotational energies are quantized. (For details, see Appendix A.) For a diatomic molecule like CO or HCl, the allowed rotational energies are

$$E(j) = j(j+1)\epsilon, \tag{6.29}$$

where j can be 0, 1, 2, etc., and ϵ is a constant that is inversely proportional to the molecule's moment of inertia. The number of degenerate states for level j is $2j+1$,

Energy

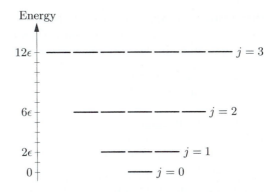

Figure 6.6. Energy level diagram for the rotational states of a diatomic molecule.

12ϵ — — — — — — — — $j = 3$

6ϵ — — — — — $j = 2$

2ϵ — — — $j = 1$

0 — $j = 0$

as shown in Figure 6.6. (I'm assuming, for now, that the two atoms making up the molecule are of different types. For molecules made of identical atoms, like H_2 or N_2, there is a subtlety that I'll deal with later.)

Given this energy level structure, we can write the partition function as a sum over j:

$$Z_{\text{rot}} = \sum_{j=0}^{\infty}(2j + 1)e^{-E(j)/kT} = \sum_{j=0}^{\infty}(2j + 1)e^{-j(j+1)\epsilon/kT}. \qquad (6.30)$$

Figure 6.7 shows a pictorial representation of this sum as the area under a bar graph. Unfortunately, there is no way to evaluate the sum exactly in closed form. But it's not hard to evaluate the sum numerically, for any particular temperature. Even better, in most cases of interest we can approximate the sum as an integral that yields a very simple result.

Let's look at some numbers. The constant ϵ, which sets the energy scale for rotational excitations, is never more than a small fraction of an electron-volt. For a CO molecule, for instance, $\epsilon = 0.00024$ eV, so that $\epsilon/k = 2.8$ K. Ordinarily we are interested only in temperatures much higher than ϵ/k, so the quantity kT/ϵ will be much greater than 1. In this case the number of terms that contribute significantly

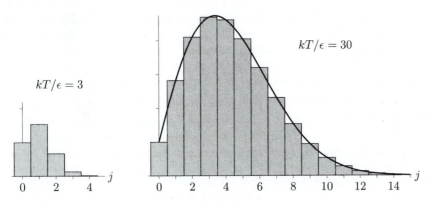

Figure 6.7. Bar-graph representations of the partition sum 6.30, for two different temperatures. At high temperatures the sum can be approximated as the area under a smooth curve.

to the partition function will be quite large, so we can, to a good approximation, replace the bar graph in Figure 6.7 with the smooth curve. The partition function is then approximately the area under this curve:

$$Z_{\text{rot}} \approx \int_0^\infty (2j+1) e^{-j(j+1)\epsilon/kT} \, dj = \frac{kT}{\epsilon} \qquad \text{(when } kT \gg \epsilon\text{)}. \qquad (6.31)$$

(To evaluate the integral, make the substitution $x = j(j+1)\epsilon/kT$.) This result should be accurate in the high-temperature limit where $Z_{\text{rot}} \gg 1$. As expected, the partition function increases with increasing temperature. For CO at room temperature, Z_{rot} is slightly greater than 100 (see Problem 6.23).

Still working in the high-temperature approximation, we can calculate the average rotational energy of a molecule using the magical formula 6.25:

$$\overline{E}_{\text{rot}} = -\frac{1}{Z} \frac{\partial Z}{\partial \beta} = -(\beta \epsilon) \frac{\partial}{\partial \beta} \frac{1}{\beta \epsilon} = \frac{1}{\beta} = kT \qquad \text{(when } kT \gg \epsilon\text{)}. \qquad (6.32)$$

This is just the prediction of the equipartition theorem, since a diatomic molecule has two rotational degrees of freedom. Differentiating $\overline{E}$ with respect to T gives the contribution of this energy to the heat capacity, simply k (for each molecule), again in agreement with the equipartition theorem. At low temperature, however, the third law tells us that the heat capacity must go to *zero*; and indeed it does, as you can confirm from the exact expression 6.30 (see Problem 6.26).

So much for diatomic molecules made of distinguishable atoms. Now, what about the case of identical atoms, such as the important molecules N_2 and O_2? The subtlety here is that turning the molecule by 180° does not change its spatial configuration, so the molecule actually has only *half* as many states as it otherwise would. In the high-temperature limit, when $Z \gg 1$, we can account for this symmetry by inserting a factor of $1/2$ into the partition function:

$$Z_{\text{rot}} \approx \frac{kT}{2\epsilon} \qquad \text{(identical atoms, } kT \gg \epsilon\text{)}. \qquad (6.33)$$

The factor of $1/2$ cancels out of the average energy (equation 6.32), so it has no effect on the heat capacity. At lower temperatures, however, things become more complicated: One must figure out exactly which terms should be omitted from the partition function (equation 6.30). At ordinary pressures, all diatomic gases except hydrogen will liquefy long before such low temperatures are reached. The behavior of hydrogen at low temperature is the subject of Problem 6.30.

Problem 6.23. For a CO molecule, the constant ϵ is approximately 0.00024 eV. (This number is measured using microwave spectroscopy, that is, by measuring the microwave frequencies needed to excite the molecules into higher rotational states.) Calculate the rotational partition function for a CO molecule at room temperature (300 K), first using the exact formula 6.30 and then using the approximate formula 6.31.

Problem 6.24. For an O_2 molecule, the constant ϵ is approximately 0.00018 eV. Estimate the rotational partition function for an O_2 molecule at room temperature.

Problem 6.25. The analysis of this section applies also to linear polyatomic molecules, for which no rotation about the axis of symmetry is possible. An example is CO_2, with $\epsilon = 0.000049$ eV. Estimate the rotational partition function for a CO_2 molecule at room temperature. (Note that the arrangement of the atoms is OCO, and the two oxygen atoms are identical.)

Problem 6.26. In the *low*-temperature limit ($kT \ll \epsilon$), each term in the rotational partition function (equation 6.30) is much smaller than the one before. Since the first term is independent of T, cut off the sum after the second term and compute the average energy and the heat capacity in this approximation. Keep only the largest T-dependent term at each stage of the calculation. Is your result consistent with the third law of thermodynamics? Sketch the behavior of the heat capacity at all temperatures, interpolating between the high-temperature and low-temperature expressions.

Problem 6.27. Use a computer to sum the exact rotational partition function (equation 6.30) numerically, and plot the result as a function of kT/ϵ. Keep enough terms in the sum to be confident that the series has converged. Show that the approximation in equation 6.31 is a bit low, and estimate by how much. Explain the discrepancy by referring to Figure 6.7.

Problem 6.28. Use a computer to sum the rotational partition function (equation 6.30) algebraically, keeping terms through $j = 6$. Then calculate the average energy and the heat capacity. Plot the heat capacity for values of kT/ϵ ranging from 0 to 3. Have you kept enough terms in Z to give accurate results within this temperature range?

Problem 6.29. Although an ordinary H_2 molecule consists of two identical atoms, this is not the case for the molecule HD, with one atom of deuterium (i.e., heavy hydrogen, ^2H). Because of its small moment of inertia, the HD molecule has a relatively large value of ϵ: 0.0057 eV. At approximately what temperature would you expect the rotational heat capacity of a gas of HD molecules to "freeze out," that is, to fall significantly below the constant value predicted by the equipartition theorem?

Problem 6.30. In this problem you will investigate the behavior of ordinary hydrogen, H_2, at low temperatures. The constant ϵ is 0.0076 eV. As noted in the text, only half of the terms in the rotational partition function, equation 6.30, contribute for any given molecule. More precisely, the set of allowed j values is determined by the *spin* configuration of the two atomic nuclei. There are four independent spin configurations, classified as a single "singlet" state and three "triplet" states. The time required for a molecule to convert between the singlet and triplet configurations is ordinarily quite long, so the properties of the two types of molecules can be studied independently. The singlet molecules are known as **parahydrogen** while the triplet molecules are known as **orthohydrogen**.

  (a) For parahydrogen, only the rotational states with even values of j are allowed.* Use a computer (as in Problem 6.28) to calculate the rotational

---

*For those who have studied quantum mechanics, here's why: Even-j wavefunctions are symmetric (unchanged) under the operation of replacing $\vec{r}$ with $-\vec{r}$, which is equivalent to interchanging the two nuclei; odd-j wavefunctions are antisymmetric under this operation. The two hydrogen nuclei (protons) are fermions, so their overall wavefunction must be antisymmetric under interchange. The singlet state ($\uparrow\downarrow - \downarrow\uparrow$) is already antisymmetric in

partition function, average energy, and heat capacity of a parahydrogen molecule. Plot the heat capacity as a function of kT/ϵ.[*]

(b) For orthohydrogen, only the rotational states with odd values of j are allowed. Repeat part (a) for orthohydrogen.

(c) At high temperature, where the number of accessible even-j states is essentially the same as the number of accessible odd-j states, a sample of hydrogen gas will ordinarily consist of a mixture of 1/4 parahydrogen and 3/4 orthohydrogen. A mixture with these proportions is called **normal hydrogen**. Suppose that normal hydrogen is cooled to low temperature without allowing the spin configurations of the molecules to change. Plot the rotational heat capacity of this mixture as a function of temperature. At what temperature does the rotational heat capacity fall to half its high-temperature value (i.e., to $k/2$ per molecule)?

(d) Suppose now that some hydrogen is cooled in the presence of a catalyst that allows the nuclear spins to frequently change alignment. In this case *all* terms in the original partition function are allowed, but the odd-j terms should be counted three times each because of the nuclear spin degeneracy. Calculate the rotational partition function, average energy, and heat capacity of this system, and plot the heat capacity as a function of kT/ϵ.

(e) A deuterium molecule, D_2, has nine independent nuclear spin configurations, of which six are "symmetric" and three are "antisymmetric." The rule for nomenclature is that the variety with more independent states gets called "ortho-," while the other gets called "para-." For orthodeuterium only even-j rotational states are allowed, while for paradeuterium only odd-j states are allowed.[†] Suppose, then, that a sample of D_2 gas, consisting of a normal equilibrium mixture of 2/3 ortho and 1/3 para, is cooled without allowing the nuclear spin configurations to change. Calculate and plot the rotational heat capacity of this system as a function of temperature.[‡]

6.3 The Equipartition Theorem

I've been invoking the equipartition theorem throughout this book, and we've verified that it is true in a number of particular cases, but so far I haven't shown you an actual proof. The proof is quite easy, if you use Boltzmann factors.

The equipartition theorem doesn't apply to all systems. It applies only to systems whose energy is in the form of quadratic "degrees of freedom," of the form

$$E(q) = cq^2, \tag{6.34}$$

where c is a constant coefficient and q is any coordinate or momentum variable, like x, or p_x, or L_x (angular momentum). I'm going to treat just this single degree

---

spin, so its spatial wavefunction must be symmetric, while the triplet states ($\uparrow\uparrow$, $\downarrow\downarrow$, and $\uparrow\downarrow + \downarrow\uparrow$) are symmetric in spin, so their spatial wavefunctions must be antisymmetric.

[*]For a molecule such as O_2 with spin-0 nuclei, this graph is the whole story; the only nuclear spin configuration is a singlet and only the even-j states are allowed.

[†]Deuterium nuclei are bosons, so the overall wavefunction must be symmetric under interchange.

[‡]For a good discussion of hydrogen at low temperature, with references to experiments, see Gopal (1966).

Figure 6.8. To count states over a continuous variable q, pretend that they're discretely spaced, separated by Δq.

of freedom as my "system," assume that it's in equilibrium with a reservoir at temperature T, and calculate its average energy, $\overline{E}$.

I'll analyze this system in classical mechanics, where each value of q corresponds to a separate, independent state. To *count* the states, I'll pretend that they're discretely spaced, separated by small intervals Δq, as shown in Figure 6.8. As long as Δq is extremely small, we expect it to cancel out of the final result for $\overline{E}$.

The partition function for this system is

$$Z = \sum_q e^{-\beta E(q)} = \sum_q e^{-\beta c q^2}. \tag{6.35}$$

To evaluate the sum, I'll multiply by Δq inside the sum and divide by Δq outside the sum:

$$Z = \frac{1}{\Delta q} \sum_q e^{-\beta c q^2} \, \Delta q. \tag{6.36}$$

Now the sum can be interpreted as the area under a bar graph whose height is determined by the Boltzmann factor (see Figure 6.9). Since Δq is very small, we can approximate the bar graph by the smooth curve, changing the sum into an integral:

$$Z = \frac{1}{\Delta q} \int_{-\infty}^{\infty} e^{-\beta c q^2} \, dq. \tag{6.37}$$

Before trying to evaluate the integral, let's change variables to $x = \sqrt{\beta c}\, q$, so that $dq = dx/\sqrt{\beta c}$. Then

$$Z = \frac{1}{\Delta q} \frac{1}{\sqrt{\beta c}} \int_{-\infty}^{\infty} e^{-x^2} \, dx. \tag{6.38}$$

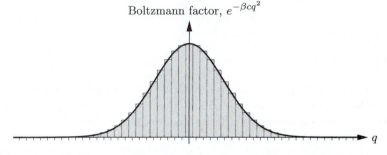

Figure 6.9. The partition function is the area under a bar graph whose height is the Boltzmann factor, $e^{-\beta c q^2}$. To calculate this area, we pretend that the bar graph is a smooth curve.

The integral over x is now just some number, whose value isn't very important as far as the physics is concerned. However, the integral is rather interesting mathematically. The function e^{-x^2} is called a **Gaussian**, and unfortunately its antiderivative cannot be written in terms of elementary functions. But there is a clever trick (described in Appendix B) for evaluating the *definite* integral from $-\infty$ to ∞, and the result is simply $\sqrt{\pi}$. So our final result for the partition function is

$$Z = \frac{1}{\Delta q}\sqrt{\frac{\pi}{\beta c}} = C\beta^{-1/2}, \tag{6.39}$$

where C is just an abbreviation for $\sqrt{\pi/c}/\Delta q$.

Once you have an explicit formula for the partition function, it's easy to calculate the average energy, using the magical formula 6.25:

$$\begin{aligned} \overline{E} &= -\frac{1}{Z}\frac{\partial Z}{\partial \beta} = -\frac{1}{C\beta^{-1/2}}\frac{\partial}{\partial \beta}C\beta^{-1/2} \\ &= -\frac{1}{C\beta^{-1/2}}(-\tfrac{1}{2})C\beta^{-3/2} = \tfrac{1}{2}\beta^{-1} = \tfrac{1}{2}kT. \end{aligned} \tag{6.40}$$

This is just the equipartition theorem. Notice that the constants c, Δq, and $\sqrt{\pi}$ have all canceled out.

The most important fact about this proof is that it does not carry over to quantum-mechanical systems. You can sort of see this from Figure 6.9: If the number of distinct states that have significant probabilities is too small, then the smooth Gaussian curve will not be a good approximation to the bar graph. And indeed, as we've seen in the case of an Einstein solid, the equipartition theorem is true only in the high-temperature limit, where many distinct states contribute and therefore the spacing between the states is unimportant. In general, the equipartition theorem applies only when the spacing between energy levels is much less than kT.

Problem 6.31. Consider a classical "degree of freedom" that is linear rather than quadratic: $E = c|q|$ for some constant c. (An example would be the kinetic energy of a highly relativistic particle in one dimension, written in terms of its momentum.) Repeat the derivation of the equipartition theorem for this system, and show that the average energy is $\overline{E} = kT$.

Problem 6.32. Consider a classical particle moving in a one-dimensional potential well $u(x)$, as shown in Figure 6.10. The particle is in thermal equilibrium with a reservoir at temperature T, so the probabilities of its various states are determined by Boltzmann statistics.

(a) Show that the average position of the particle is given by

$$\overline{x} = \frac{\int xe^{-\beta u(x)}\,dx}{\int e^{-\beta u(x)}\,dx},$$

where each integral is over the entire x axis.

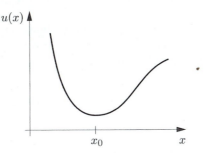

Figure 6.10. A one-dimensional potential well. The higher the temperature, the farther the particle will
stray from the equilibrium point.

(b) If the temperature is reasonably low (but still high enough for classical mechanics to apply), the particle will spend most of its time near the bottom
of the potential well. In that case we can expand $u(x)$ in a Taylor series
about the equilibrium point x_0:

$$u(x) = u(x_0) + (x - x_0) \left.\frac{du}{dx}\right|_{x_0} + \frac{1}{2}(x - x_0)^2 \left.\frac{d^2 u}{dx^2}\right|_{x_0}$$

$$+ \frac{1}{3!}(x - x_0)^3 \left.\frac{d^3 u}{dx^3}\right|_{x_0} + \cdots.$$

Show that the linear term must be zero, and that truncating the series after
the quadratic term results in the trivial prediction $\bar{x} = x_0$.

(c) If we keep the cubic term in the Taylor series as well, the integrals in
the formula for $\bar{x}$ become difficult. To simplify them, assume that the
cubic term is small, so its exponential can be expanded in a Taylor series
(leaving the quadratic term in the exponent). Keeping only the smallest
temperature-dependent term, show that in this limit $\bar{x}$ differs from x_0 by
a term proportional to kT. Express the coefficient of this term in terms of
the coefficients of the Taylor series for $u(x)$.

(d) The interaction of noble gas atoms can be modeled using the **Lennard-
Jones potential**,

$$u(x) = u_0 \left[\left(\frac{x_0}{x}\right)^{12} - 2\left(\frac{x_0}{x}\right)^{6} \right].$$

Sketch this function, and show that the minimum of the potential well is
at $x = x_0$, with depth u_0. For argon, $x_0 = 3.9$ Å and $u_0 = 0.010$ eV. Expand the Lennard-Jones potential in a Taylor series about the equilibrium
point, and use the result of part (c) to predict the linear thermal expansion
coefficient (see Problem 1.8) of a noble gas crystal in terms of u_0. Evaluate the result numerically for argon, and compare to the measured value
$\alpha = 0.0007$ K^{-1} (at 80 K).

6.4 The Maxwell Speed Distribution

For our next application of Boltzmann factors, I'd like to take a detailed look at the motion of molecules in an ideal gas. We already know (from the equipartition theorem) that the root-mean-square speed of the molecules is given by the formula

$$v_{\text{rms}} = \sqrt{\frac{3kT}{m}}. \tag{6.41}$$

But this is just a sort of average. Some of the molecules will be moving faster than this, others slower. In practice, we might want to know exactly how many molecules are moving at any given speed. Equivalently, let's ask what is the probability of some particular molecule moving at a given speed.

Technically, the probability that a molecule is moving *at* any given speed v is *zero*. Since speed can vary continuously, there are infinitely many possible speeds, and therefore each of them has infinitesimal probability (which is essentially the same as zero). However, some speeds are less probable than others, and we can still represent the *relative* probabilities of various speeds by a graph, which turns out to look like Figure 6.11. The most probable speed is where the graph is the highest, and other speeds are less probable, in proportion to the height of the graph. Furthermore, if we normalize the graph (that is, adjust the vertical scale) in the right way, it has a more precise interpretation: The *area* under the graph between any two speeds v_1 and v_2 equals the probability that the molecule's speed is between v_1 and v_2:

$$\text{Probability}(v_1 \ldots v_2) = \int_{v_1}^{v_2} \mathcal{D}(v)\, dv, \tag{6.42}$$

where $\mathcal{D}(v)$ is the height of the graph. If the interval between v_1 and v_2 is infinitesimal, then $\mathcal{D}(v)$ doesn't change significantly within the interval and we can write simply

$$\text{Probability}(v \ldots v + dv) = \mathcal{D}(v)\, dv. \tag{6.43}$$

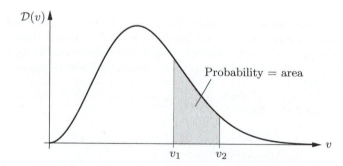

Figure 6.11. A graph of the relative probabilities for a gas molecule to have various speeds. More precisely, the vertical scale is defined so that the area under the graph within any interval equals the probability of the molecule having a speed in that interval.

The function $\mathcal{D}(v)$ is called a **distribution function**. Its actual *value* at any point isn't very meaningful by itself. Instead, $\mathcal{D}(v)$ is a function *whose purpose in life is to be integrated.* To turn $\mathcal{D}(v)$ into a probability you *must* integrate over some interval of v's (or, if the interval is small, just multiply by the width of the interval). The function $\mathcal{D}(v)$ itself doesn't even have the right *units* (namely, none) for a probability; instead, it has units of $1/v$, or $(\text{m/s})^{-1}$.

Now that we know how to interpret the answer, I'd like to derive a formula for the function $\mathcal{D}(v)$. The most important ingredient in the derivation is the Boltzmann factor. But another important element is the fact that space is three dimensional, which implies that for any given *speed*, there are many possible velocity *vectors*. In fact, we can write the function $\mathcal{D}(v)$ schematically as

$$\mathcal{D}(v) \propto \begin{pmatrix} \text{probability of a molecule} \\ \text{having velocity } \vec{v} \end{pmatrix} \times \begin{pmatrix} \text{number of vectors } \vec{v} \\ \text{corresponding to speed } v \end{pmatrix}. \qquad (6.44)$$

There's also a constant of proportionality, which we'll worry about later.

The first factor in equation 6.44 is just the Boltzmann factor. Each velocity vector corresponds to a distinct molecular state, and the probability of a molecule being in any given state s is proportional to the Boltzmann factor $e^{-E(s)/kT}$. In this case the energy is just the translational kinetic energy, $\frac{1}{2}mv^2$ (where $v = |\vec{v}|$), so

$$\begin{pmatrix} \text{probability of a molecule} \\ \text{having velocity } \vec{v} \end{pmatrix} \propto e^{-mv^2/2kT}. \qquad (6.45)$$

I've neglected any variables besides velocity that might affect the state of the molecule, such as its position in space or its internal motion. This simplification is valid for an ideal gas, where the translational motion is independent of all other variables.

Equation 6.45 says that the most likely velocity *vector* for a molecule in an ideal gas is *zero*. Given what we know about Boltzmann factors, this result should hardly be surprising: Low-energy states are always more probable than high-energy states, for any system at finite (positive) temperature. However, the most likely velocity *vector* does *not* correspond to the most likely *speed*, because for some speeds there are *more* distinct velocity vectors than for others.

So let us turn to the second factor in equation 6.44. To evaluate this factor, imagine a three-dimensional "velocity space" in which each point represents a velocity vector (see Figure 6.12). The set of velocity vectors corresponding to any given speed v lives on the surface of a sphere with radius v. The larger v is, the bigger the sphere, and the more possible velocity vectors there are. So I claim that the second factor in equation 6.44 is the surface area of the sphere in velocity space:

$$\begin{pmatrix} \text{number of vectors } \vec{v} \\ \text{corresponding to speed } v \end{pmatrix} \propto 4\pi v^2. \qquad (6.46)$$

Putting this "degeneracy" factor together with the Boltzmann factor (6.45), we obtain

$$\mathcal{D}(v) = C \cdot 4\pi v^2\, e^{-mv^2/2kT}, \qquad (6.47)$$

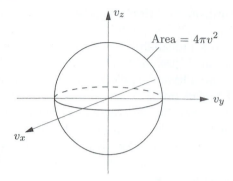

Figure 6.12. In "velocity space" each point represents a possible velocity vector. The set of all vectors for a given speed v lies on the surface of a sphere with radius v.

where C is a constant of proportionality. To determine C, note that the total probability of finding the molecule at *some* speed must equal 1:

$$1 = \int_0^\infty \mathcal{D}(v)\, dv = 4\pi C \int_0^\infty v^2\, e^{-mv^2/2kT}\, dv. \tag{6.48}$$

Changing variables to $x = v\sqrt{m/2kT}$ puts this integral into the form

$$1 = 4\pi C \left(\frac{2kT}{m}\right)^{3/2} \int_0^\infty x^2\, e^{-x^2}\, dx. \tag{6.49}$$

Like the pure Gaussian e^{-x^2}, the function $x^2 e^{-x^2}$ cannot be anti-differentiated in terms of elementary functions. Again, however, there are tricks (explained in Appendix B) for evaluating the *definite* integral from 0 to ∞; in this case the answer is $\sqrt{\pi}/4$. The 4 cancels, leaving us with $C = (m/2\pi kT)^{3/2}$.

Our final result for the distribution function $\mathcal{D}(v)$ is therefore

$$\mathcal{D}(v) = \left(\frac{m}{2\pi kT}\right)^{3/2} 4\pi v^2\, e^{-mv^2/2kT}. \tag{6.50}$$

This result is called the **Maxwell distribution** (after James Clerk Maxwell) for the speeds of molecules in an ideal gas. It's a complicated formula, but I hope you won't find it hard to remember the important parts: the Boltzmann factor involving the translational kinetic energy, and the geometrical factor of the surface area of the sphere in velocity space.

Figure 6.13 shows another plot of the Maxwell distribution. At very small v, the Boltzmann factor is approximately 1 so the curve is a parabola; in particular, the distribution goes to zero at $v = 0$. This result does not contradict the fact that zero is the *most* likely velocity vector, because now we're talking about *speeds*, and there are simply too few velocity vectors corresponding to very small speeds.

Meanwhile, the Maxwell distribution also goes to zero at very high speeds (much greater than $\sqrt{kT/m}$), because of the exponential fall-off of the Boltzmann factor.

In between $v = 0$ and $v = \infty$, the Maxwell distribution rises and falls. By setting the derivative of equation 6.50 equal to zero, you can show that the maximum value of $\mathcal{D}(v)$ occurs at $v_{\max} = \sqrt{2kT/m}$. As you would expect, the peak shifts to the

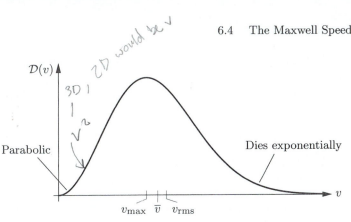

(handwritten annotations:) 2D would be ✓   3D ✓ v^2 ✓

$v_{RMS} = \sqrt{\overline{v^2}}$

square, the average, then square root

$v_{RMS} = \sqrt{\dfrac{3kT}{m}}$

Figure 6.13. The Maxwell speed distribution falls off as $v \to 0$ and as $v \to \infty$. The average speed is slightly larger than the most likely speed, while the rms speed is a bit larger still.

right as the temperature is increased. The most likely speed is not the same as the rms speed; referring to equation 6.41, we see that the rms speed is the greater of the two, by about 22%. The *average* speed is different still; to compute it, add up all the possible speeds, weighted by their probabilities:

$$\overline{v} = \sum_{\text{all } v} v \, \mathcal{D}(v) \, dv. \tag{6.51}$$

In this equation I'm imagining the various speeds to be discretely spaced, separated by dv. If you turn the sum into an integral and evaluate it, you get

$$\overline{v} = \sqrt{\frac{8kT}{\pi m}}, \tag{6.52}$$

which lies in between v_{max} and v_{rms}.

As an example, consider the nitrogen molecules in air at room temperature. You can easily calculate the most probable speed, which turns out to be 422 m/s at 300 K. But some of the molecules are moving much faster than this, while others are moving much slower. What is the probability that a particular molecule is moving faster than 1000 m/s?

First let's make a graphical estimate. The speed 1000 m/s exceeds v_{max} by a factor of

$$\frac{1000 \text{ m/s}}{422 \text{ m/s}} = 2.37. \tag{6.53}$$

Looking at Figure 6.13, you can see that at this point the Maxwell distribution is rapidly dying out but not yet dead. The area under the graph beyond $2.37v_{\text{max}}$ looks like only one or two percent of the total area under the graph.

Quantitatively, the probability is given by the integral of the Maxwell distribution from 1000 m/s to infinity:

$$\text{Probability}(v > 1000 \text{ m/s}) = 4\pi \left(\frac{m}{2\pi kT} \right)^{3/2} \int_{1000 \text{ m/s}}^{\infty} v^2 \, e^{-mv^2/2kT} \, dv. \tag{6.54}$$

With this nontrivial lower limit, the integral *cannot* be carried out analytically; the best option is to do it numerically, by calculator or computer. You *could* go ahead and plug in numbers at this point, instructing the computer to work in units of m/s. But it's much cleaner to first change variables to $x = v\sqrt{m/2kT} = v/v_{max}$, just as in equation 6.49. The integral then becomes

$$4\pi\left(\frac{m}{2\pi kT}\right)^{3/2}\left(\frac{2kT}{m}\right)^{3/2}\int_{x_{min}}^{\infty} x^2\, e^{-x^2}\, dx = \frac{4}{\sqrt{\pi}}\int_{x_{min}}^{\infty} x^2\, e^{-x^2}\, dx, \qquad (6.55)$$

where the lower limit is the value of x when $v = 1000$ m/s, that is, $x_{min} = (1000 \text{ m/s})/(422 \text{ m/s}) = 2.37$. Now it's *easy* to type the integral into a computer. I did so and got an answer of 0.0105 for the probability. Only about 1% of the nitrogen molecules are moving faster than 1000 m/s.

Problem 6.33. Calculate the most probable speed, average speed, and rms speed for oxygen (O_2) molecules at room temperature.

Problem 6.34. Carefully plot the Maxwell speed distribution for nitrogen molecules at $T = 300$ K and at $T = 600$ K. Plot both graphs on the same axes, and label the axes with numbers.

Problem 6.35. Verify from the Maxwell speed distribution that the most likely speed of a molecule is $\sqrt{2kT/m}$.

Problem 6.36. Fill in the steps between equations 6.51 and 6.52, to determine the average speed of the molecules in an ideal gas.

Problem 6.37. Use the Maxwell distribution to calculate the average value of v^2 for the molecules in an ideal gas. Check that your answer agrees with equation 6.41.

Problem 6.38. At room temperature, what fraction of the nitrogen molecules in the air are moving at less than 300 m/s?

Problem 6.39. A particle near earth's surface traveling faster than about 11 km/s has enough kinetic energy to completely escape from the earth, despite earth's gravitational pull. Molecules in the upper atmosphere that are moving faster than this will therefore escape if they do not suffer any collisions on the way out.

(a) The temperature of earth's upper atmosphere is actually quite high, around 1000 K. Calculate the probability of a nitrogen molecule at this temperature moving faster than 11 km/s, and comment on the result.

(b) Repeat the calculation for a hydrogen molecule (H_2) and for a helium atom, and discuss the implications.

(c) Escape speed from the moon's surface is only about 2.4 km/s. Explain why the moon has no atmosphere.

Problem 6.40. You might wonder why all the molecules in a gas in thermal equilibrium don't have exactly the *same* speed. After all, when two molecules collide, doesn't the faster one always lose energy and the slower one gain energy? And if so, wouldn't repeated collisions eventually bring all the molecules to some common speed? Describe an example of a billiard-ball collision in which this is *not* the case: The faster ball *gains* energy and the slower ball *loses* energy. Include numbers, and be sure that your collision conserves both energy and momentum.

Problem 6.41. Imagine a world in which space is two-dimensional, but the laws of physics are otherwise the same. Derive the speed distribution formula for an ideal gas of nonrelativistic particles in this fictitious world, and sketch this distribution. Carefully explain the similarities and differences between the two-dimensional and three-dimensional cases. What is the most likely velocity vector? What is the most likely speed?

6.5 Partition Functions and Free Energy

For an isolated system with fixed energy U, the most fundamental statistical quantity is the multiplicity, $\Omega(U)$—the number of available microstates. The logarithm of the multiplicity gives the entropy, which tends to increase.

For a system in equilibrium with a reservoir at temperature T (see Figure 6.14), the quantity most analogous to Ω is the partition function, $Z(T)$. Like $\Omega(U)$, the partition function is more or less equal to the number of microstates available to the system (but at fixed temperature, not fixed energy). We might therefore expect its logarithm to be a quantity that tends to increase under these conditions. But we already know a quantity that tends to *decrease* under these conditions: the Helmholtz free energy, F. The quantity that tends to *increase* would be $-F$, or, if we want a dimensionless quantity, $-F/kT$. Taking a giant intuitive leap, we might therefore guess the formula

$$F = -kT \ln Z \qquad \text{or} \qquad Z = e^{-F/kT}. \qquad (6.56)$$

Indeed, this formula turns out to be true. Let me prove it.

First recall the definition of F:

$$F \equiv U - TS. \qquad (6.57)$$

Also, recall the partial derivative relation

$$\left(\frac{\partial F}{\partial T}\right)_{V,N} = -S. \qquad (6.58)$$

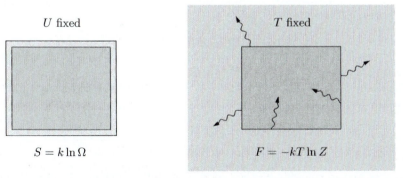

Figure 6.14. For an isolated system (left), S tends to increase. For a system at constant temperature (right), F tends to decrease. Like S, F can be written as the logarithm of a statistical quantity, in this case Z.

Solving equation 6.57 for S and plugging into equation 6.58 gives

$$\left(\frac{\partial F}{\partial T}\right)_{V,N} = \frac{F - U}{T}.$$
(6.59)

This is a differential equation for the function $F(T)$, for any fixed V and N. To prove equation 6.56, I'll show that the quantity $-kT \ln Z$ obeys the same differential equation, with the same "initial" condition at $T = 0$.

Let me define the symbol $\widetilde{F}$ to stand for the quantity $-kT \ln Z$. Holding V and N fixed, I want to find a formula for the derivative of this quantity with respect to T:

$$\frac{\partial \widetilde{F}}{\partial T} = \frac{\partial}{\partial T}(-kT \ln Z) = -k \ln Z - kT \frac{\partial}{\partial T} \ln Z.$$
(6.60)

In the second term I'll use the chain rule to rewrite the derivative in terms of $\beta = 1/kT$:

$$\frac{\partial}{\partial T} \ln Z = \frac{\partial \beta}{\partial T} \frac{\partial}{\partial \beta} \ln Z = \frac{-1}{kT^2} \frac{1}{Z} \frac{\partial Z}{\partial \beta} = \frac{U}{kT^2}.$$
(6.61)

(Here I'm using U instead of $\overline{E}$ for the system's average energy, because these ideas are most useful when applied to fairly large systems.) Plugging this result back into equation 6.60, we obtain

$$\frac{\partial \widetilde{F}}{\partial T} = -k \ln Z - kT \frac{U}{kT^2} = \frac{\widetilde{F}}{T} - \frac{U}{T},$$
(6.62)

that is, $\widetilde{F}$ obeys exactly the same differential equation as F.

A first-order differential equation has an infinite family of solutions, corresponding to different "initial" conditions. So to complete the proof that $\widetilde{F} = F$, I need to show that they're the same for at least one particular value of T, say $T = 0$. At $T = 0$, the original F is simply equal to U, the energy of the system when it is at zero temperature. This energy must be the lowest possible energy, U_0, since the Boltzmann factors $e^{-U(s)/kT}$ for all excited states will be infinitely suppressed in comparison to the ground state. Meanwhile, the partition function at $T = 0$ is simply $e^{-U_0/kT}$, again since all other Boltzmann factors are infinitely suppressed in comparison. Therefore

$$\widetilde{F}(0) = -kT \ln Z(0) = U_0 = F(0),$$
(6.63)

completing the proof that $\widetilde{F} = F$ for all T.

The usefulness of the formula $F = -kT \ln Z$ is that from F we can compute the entropy, pressure, and chemical potential, using the partial-derivative formulas

$$S = -\left(\frac{\partial F}{\partial T}\right)_{V,N}, \qquad P = -\left(\frac{\partial F}{\partial V}\right)_{T,N}, \qquad \mu = +\left(\frac{\partial F}{\partial N}\right)_{T,V}.$$
(6.64)

In this way we can compute all the thermodynamic properties of a system, once we know its partition function. In Section 6.7 I'll apply this technique to analyze an ideal gas.

Problem 6.42. In Problem 6.20 you computed the partition function for a quantum harmonic oscillator: $Z_{\text{h.o.}} = 1/(1-e^{-\beta\epsilon})$, where $\epsilon = hf$ is the spacing between energy levels.

  (a) Find an expression for the Helmholtz free energy of a system of N harmonic oscillators.

  (b) Find an expression for the entropy of this system as a function of temperature. (Don't worry, the result is fairly complicated.)

Problem 6.43. Some advanced textbooks define entropy by the formula

$$S = -k \sum_s \mathcal{P}(s) \ln \mathcal{P}(s),$$

where the sum runs over all microstates accessible to the system and $\mathcal{P}(s)$ is the probability of the system being in microstate s.

  (a) For an isolated system, $\mathcal{P}(s) = 1/\Omega$ for all accessible states s. Show that in this case the preceding formula reduces to our familiar definition of entropy.

  (b) For a system in thermal equilibrium with a reservoir at temperature T, $\mathcal{P}(s) = e^{-E(s)/kT}/Z$. Show that in this case as well, the preceding formula agrees with what we already know about entropy.

6.6 Partition Functions for Composite Systems

Before trying to write down the partition function for an ideal gas, it is useful to ask in general how the partition function for a system of several particles is related to the partition function for each individual particle. For instance, consider a system of just two particles, 1 and 2. If these particles do not interact with each other, so their total energy is simply $E_1 + E_2$, then

$$Z_{\text{total}} = \sum_s e^{-\beta[E_1(s)+E_2(s)]} = \sum_s e^{-\beta E_1(s)} e^{-\beta E_2(s)}, \qquad (6.65)$$

where the sum runs over all states s for the composite system. If, in addition, the two particles are distinguishable (either by their fixed positions or by some intrinsic properties), then the set of states for the composite system is equivalent to the set of all possible pairs of states, (s_1, s_2), for the two particles individually. In this case,

$$Z_{\text{total}} = \sum_{s_1} \sum_{s_2} e^{-\beta E_1(s_1)} e^{-\beta E_2(s_2)}, \qquad (6.66)$$

where s_1 represents the state of particle 1 and s_2 represents the state of particle 2. The first Boltzmann factor can be moved outside the sum over s_2. This sum, now of just the second Boltzmann factor, is simply the partition function for particle 2 alone, Z_2. This partition function is independent of s_1, and can therefore be taken out of the remaining sum. Finally, the sum over s_1 gives simply Z_1, leaving us with

$$Z_{\text{total}} = Z_1 Z_2 \qquad \text{(noninteracting, distinguishable particles)}. \qquad (6.67)$$

If the particles are *in*distinguishable, however, the step going from equation 6.65 to equation 6.66 is not valid. The problem is exactly the same as the one

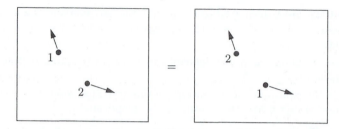

Figure 6.15. Interchanging the states of two indistinguishable particles leaves the system in the same state as before.

we encountered when computing the multiplicity of an ideal gas in Section 2.5: Putting particle 1 in state A and particle 2 in state B is the same thing as putting particle 2 in state A and particle 1 in state B (see Figure 6.15). Equation 6.66 therefore counts nearly every state twice, and a more accurate formula would be

$$Z_{\text{total}} = \frac{1}{2} Z_1 Z_2 \qquad \text{(noninteracting, indistinguishable particles).} \qquad (6.68)$$

This formula still isn't precisely correct, because there are some terms in the double sum of equation 6.66 in which both particles are in the *same* state, that is, $s_1 = s_2$. These terms have *not* been double-counted, so we shouldn't divide their number by 2. But for ideal gases and many other familiar systems, the density is low enough that the chances of both particles being in the same state are negligible. The terms with $s_1 = s_2$ are therefore only a tiny fraction of all the terms in equation 6.66; and it doesn't much matter whether we count them correctly or not.[*]

The generalization of equations 6.67 and 6.68 to systems of more than two particles is straightforward. If the particles are distinguishable, the total partition function is the product of all the individual partition functions:

$$Z_{\text{total}} = Z_1 Z_2 Z_3 \cdots Z_N \qquad \text{(noninteracting, distinguishable systems).} \qquad (6.69)$$

This equation also applies to the total partition function of a *single* particle that can store energy in several ways; for instance, Z_1 could be the partition function for its motion in the x direction, Z_2 for its motion in the y direction, and so on.

For a not-too-dense system of N indistinguishable particles, the general formula is

$$Z_{\text{total}} = \frac{1}{N!} Z_1^N \qquad \text{(noninteracting, indistinguishable particles),} \qquad (6.70)$$

where Z_1 is the partition function for any one of the particles individually. The number of ways of interchanging N particles with each other is $N!$, hence the prefactor.

When we deal with multiparticle systems, one point of terminology can be confusing. It is important to distinguish the "state" of an individual particle from

---

[*]The following chapter deals with very dense systems for which this issue is important. Until then, don't worry about it.

the "state" of the entire system. Unfortunately, I don't know of a good concise way to distinguish between these two concepts. When the context is ambiguous, I'll write **single-particle state** or **system state**, as appropriate. In the preceding discussion, s is the system state while s_1 and s_2 are single-particle states. In general, to specify the system state, you must specify the single-particle states of all the particles in the system.

> **Problem 6.44.** Consider a large system of N indistinguishable, noninteracting molecules (perhaps in an ideal gas or a dilute solution). Find an expression for the Helmholtz free energy of this system, in terms of Z_1, the partition function for a single molecule. (Use Stirling's approximation to eliminate the $N!$.) Then use your result to find the chemical potential, again in terms of Z_1.

6.7 Ideal Gas Revisited

The Partition Function

We now have all the tools needed to calculate the partition function, and hence all the other thermal quantities, of an ideal gas. An *ideal* gas, as before, means one in which the molecules are usually far enough apart that we can neglect any energy due to forces between them. If the gas contains N molecules (all identical), then its partition function has the form

$$Z = \frac{1}{N!}Z_1^N, \tag{6.71}$$

where Z_1 is the partition function for one individual molecule.

To calculate Z_1, we must add up the Boltzmann factors for all possible microstates of a single molecule. Each Boltzmann factor has the form

$$e^{-E(s)/kT} = e^{-E_{\mathrm{tr}}(s)/kT}e^{-E_{\mathrm{int}}(s)/kT}, \tag{6.72}$$

where E_{tr} is the molecule's translational kinetic energy and E_{int} is its **internal energy** (rotational, vibrational, or whatever), for the state s. The sum over all single-particle states can be written as a double sum over translational states and internal states, allowing us to factor the partition function as in the previous section. The result is simply

$$Z_1 = Z_{\mathrm{tr}}Z_{\mathrm{int}}, \tag{6.73}$$

where

$$Z_{\mathrm{tr}} = \sum_{\substack{\text{translational} \\ \text{states}}} e^{-E_{\mathrm{tr}}/kT} \quad \text{and} \quad Z_{\mathrm{int}} = \sum_{\substack{\text{internal} \\ \text{states}}} e^{-E_{\mathrm{int}}/kT}. \tag{6.74}$$

The internal partition functions for rotational and vibrational states are treated in Section 6.2. For a given rotational and vibrational state, a molecule can also have various *electronic* states, in which its electrons are in different independent wavefunctions. For most molecules at ordinary temperatures, electronic *excited* states have negligible Boltzmann factors, due to their rather high energies. The electronic *ground state*, however, can sometimes be degenerate. An oxygen molecule, for example, has a threefold-degenerate ground state, which contributes a factor of 3 to its internal partition function.

Now let us put aside the internal partition function and concentrate on the translational part, Z_{tr}. To compute Z_{tr}, we need to add up the Boltzmann factors for all possible translational states of a molecule. One way to enumerate these states is to count all the possible position and momentum vectors for a molecule, slipping in a factor of $1/h^3$ to account for quantum mechanics as in Section 2.5. Instead, however, I'd now like to use the more rigorous method of counting all the independent definite-energy wavefunctions, just as we've been doing with internal states of atoms and molecules. I'll start with the case of a molecule confined to a one-dimensional box, then generalize to three dimensions.

A few of the definite-energy wavefunctions for a molecule in a one-dimensional box are shown in Figure 6.16. Because the molecule is confined to the box, its wavefunction must go to zero at each end, and therefore the allowed standing-wave patterns are limited to wavelengths of

$$\lambda_n = \frac{2L}{n}, \qquad n = 1, 2, \ldots, \tag{6.75}$$

where L is the length of the box and n is the number of "bumps." Each of these standing waves can be thought of as a superposition of left- and right-moving traveling waves with equal and opposite momenta; the magnitude of the momentum is given by the de Broglie relation $p = h/\lambda$, that is,

$$p_n = \frac{h}{\lambda_n} = \frac{hn}{2L}. \tag{6.76}$$

Finally, the relation between energy and momentum for a nonrelativistic particle is $E = p^2/2m$, where m is its mass. So the allowed energies for a molecule in a one-dimensional box are

$$E_n = \frac{p_n^2}{2m} = \frac{h^2 n^2}{8mL^2}. \tag{6.77}$$

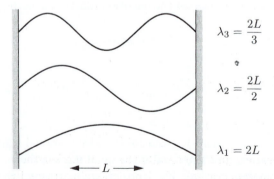

Figure 6.16. The three lowest-energy wavefunctions for a particle confined to a one-dimensional box.

Knowing the energies, we can immediately write down the translational partition function for this molecule (still in one dimension):

$$Z_{1d} = \sum_n e^{-E_n/kT} = \sum_n e^{-h^2 n^2/8mL^2 kT}. \tag{6.78}$$

Unless L and/or T is *extremely* small, the energy levels are extremely close together, so we may as well approximate the sum as an integral:

$$Z_{1d} = \int_0^\infty e^{-h^2 n^2/8mL^2 kT}\, dn = \frac{\sqrt{\pi}}{2}\sqrt{\frac{8mL^2 kT}{h^2}} = \sqrt{\frac{2\pi mkT}{h^2}}\, L \equiv \frac{L}{\ell_Q}, \tag{6.79}$$

where ℓ_Q is defined as the reciprocal of the square root in the previous expression:

$$\ell_Q \equiv \frac{h}{\sqrt{2\pi mkT}}. \tag{6.80}$$

I like to call ℓ_Q the **quantum length**; aside from the factor of π, it is the de Broglie wavelength of a particle of mass m whose kinetic energy is kT. For a nitrogen molecule at room temperature, the quantum length works out to 1.9×10^{-11} m. The ratio L/ℓ_Q is therefore quite large for any realistic box, meaning that *many* translational states are available to the molecule under these conditions: roughly the number of de Broglie wavelengths that would fit inside the box.

So much for a molecule moving in one dimension. For a molecule moving in three dimensions, the total kinetic energy is

$$E_{tr} = \frac{p_x^2}{2m} + \frac{p_y^2}{2m} + \frac{p_z^2}{2m}, \tag{6.81}$$

where each momentum component can take on infinitely many different values according to formula 6.76. Since the n's for the three momentum components can be chosen independently, we can again factor the partition function, into a piece for each of the three dimensions:

$$\begin{aligned}
Z_{tr} &= \sum_s e^{-E_{tr}/kT} = \sum_{n_x}\sum_{n_y}\sum_{n_z} e^{-h^2 n_x^2/8mL_x^2 kT}\, e^{-h^2 n_y^2/8mL_y^2 kT}\, e^{-h^2 n_z^2/8mL_z^2 kT} \\
&= \frac{L_x}{\ell_Q}\frac{L_y}{\ell_Q}\frac{L_z}{\ell_Q} = \frac{V}{v_Q},
\end{aligned} \tag{6.82}$$

where V is the total volume of the box and v_Q is the **quantum volume**,

$$v_Q = \ell_Q^3 = \left(\frac{h}{\sqrt{2\pi mkT}}\right)^3. \tag{6.83}$$

The quantum volume is just the cube of the quantum length, so it's very small for a molecule at room temperature. The translational partition function is essentially the number of de Broglie-wavelength cubes that would fit inside the entire volume of the box, and is again quite large under ordinary conditions.

Combining this result with equation 6.73, we obtain for the single-particle partition function

$$Z_1 = \frac{V}{v_Q} Z_{int},$$ (6.84)

where Z_{int} is a sum over all relevant internal states. The partition function for the entire gas of N molecules is then

$$Z = \frac{1}{N!} \left(\frac{V Z_{int}}{v_Q} \right)^N.$$ (6.85)

For future reference, the logarithm of the partition function is

$$\ln Z = N \big[\ln V + \ln Z_{int} - \ln N - \ln v_Q + 1 \big].$$ (6.86)

Predictions

At this point we can compute all of the thermal properties of an ideal gas. Let's start with the total (average) energy, using the formula derived in Problem 6.16:

$$U = -\frac{1}{Z} \frac{\partial Z}{\partial \beta} = -\frac{\partial}{\partial \beta} \ln Z.$$ (6.87)

The quantities in equation 6.86 that depend on β are Z_{int} and v_Q, so

$$U = -N \frac{\partial}{\partial \beta} \ln Z_{int} + N \frac{1}{v_Q} \frac{\partial v_Q}{\partial \beta} = N \overline{E}_{int} + N \cdot \frac{3}{2} \frac{1}{\beta} = U_{int} + \frac{3}{2} NkT.$$ (6.88)

Here $\overline{E}_{int}$ is the average internal energy of a molecule. The average translational kinetic energy is $\frac{3}{2}kT$, as we already knew from the equipartition theorem. Taking another derivative gives the heat capacity,

$$C_V = \frac{\partial U}{\partial T} = \frac{\partial U_{int}}{\partial T} + \frac{3}{2} Nk.$$ (6.89)

For a diatomic gas, the internal contribution to the heat capacity comes from rotation and vibration. As shown in Section 6.2, each of these contributions adds approximately Nk to the heat capacity at sufficiently high temperatures, but goes to zero at lower temperatures. The translational contribution could also freeze out in theory, but only at temperatures so low that ℓ_Q is of order L, so replacing the sum by an integral in equation 6.79 becomes invalid. We have now explained all the features in the graph of C_V for hydrogen shown in Figure 1.13.

To compute the remaining thermal properties of an ideal gas, we need the Helmholtz free energy,

$$F = -kT \ln Z = -NkT \big[\ln V + \ln Z_{int} - \ln N - \ln v_Q + 1 \big]$$
$$= -NkT \big[\ln V - \ln N - \ln v_Q + 1 \big] + F_{int},$$ (6.90)

where F_{int} is the internal contribution to F, namely $-NkT \ln Z_{int}$. From this expression it's easy to compute the pressure,

$$P = -\left(\frac{\partial F}{\partial V}\right)_{T,N} = \frac{NkT}{V}.$$

(6.91)

I'll let you work out the entropy and chemical potential. The results are

$$S = -\left(\frac{\partial F}{\partial T}\right)_{V,N} = Nk\left[\ln\left(\frac{V}{Nv_Q}\right) + \frac{5}{2}\right] - \frac{\partial F_{int}}{\partial T}$$

(6.92)

and

$$\mu = \left(\frac{\partial F}{\partial N}\right)_{T,V} = -kT \ln\left(\frac{V Z_{int}}{Nv_Q}\right).$$

(6.93)

If we neglect the internal contributions, both of these quantities reduce to our earlier results for a monatomic ideal gas.

Problem 6.45. Derive equations 6.92 and 6.93 for the entropy and chemical potential of an ideal gas.

Problem 6.46. Equations 6.92 and 6.93 for the entropy and chemical potential involve the logarithm of the quantity $V Z_{int}/Nv_Q$. Is this logarithm normally positive or negative? Plug in some numbers for an ordinary gas and discuss.

Problem 6.47. Estimate the temperature at which the translational motion of a nitrogen molecule would freeze out, in a box of width 1 cm.

Problem 6.48. For a diatomic gas near room temperature, the internal partition function is simply the rotational partition function computed in Section 6.2, multiplied by the degeneracy Z_e of the electronic ground state.

(a) Show that the entropy in this case is

$$S = Nk\left[\ln\left(\frac{V Z_e Z_{rot}}{Nv_Q}\right) + \frac{7}{2}\right].$$

Calculate the entropy of a mole of oxygen ($Z_e = 3$) at room temperature and atmospheric pressure, and compare to the measured value in the table at the back of this book.[*]

(b) Calculate the chemical potential of oxygen in earth's atmosphere near sea level, at room temperature. Express the answer in electron-volts.

Problem 6.49. For a mole of nitrogen (N_2) gas at room temperature and atmospheric pressure, compute the following: U, H, F, G, S, and μ. (The rotational constant ϵ for N_2 is 0.00025 eV. The electronic ground state is not degenerate.)

Problem 6.50. Show explicitly from the results of this section that $G = N\mu$ for an ideal gas.

---

[*]See Rock (1983) or Gopal (1966) for a discussion of the comparison of theoretical and experimental entropies.

Problem 6.51. In this section we computed the single-particle translational partition function, Z_{tr}, by summing over all definite-energy wavefunctions. An alternative approach, however, is to sum over all possible position and momentum vectors, as we did in Section 2.5. Because position and momentum are continuous variables, the sums are really integrals, and we need to slip in a factor of $1/h^3$ to get a unitless number that actually counts the independent wavefunctions. Thus, we might guess the formula

$$Z_{tr} = \frac{1}{h^3} \int d^3r \, d^3p \, e^{-E_{tr}/kT},$$

where the single integral sign actually represents six integrals, three over the position components (denoted d^3r) and three over the momentum components (denoted d^3p). The region of integration includes all momentum vectors, but only those position vectors that lie within a box of volume V. By evaluating the integrals explicitly, show that this expression yields the same result for the translational partition function as that obtained in the text. (The only time this formula would not be valid would be when the box is so small that we could not justify converting the sum in equation 6.78 to an integral.)

Problem 6.52. Consider an ideal gas of highly relativistic particles (such as photons or fast-moving electrons), whose energy-momentum relation is $E = pc$ instead of $E = p^2/2m$. Assume that these particles live in a one-dimensional universe. By following the same logic as above, derive a formula for the single-particle partition function, Z_1, for one particle in this gas.

Problem 6.53. The dissociation of molecular hydrogen into atomic hydrogen,

$$H_2 \longleftrightarrow 2H,$$

can be treated as an ideal gas reaction using the techniques of Section 5.6. The equilibrium constant K for this reaction is defined as

$$K = \frac{P_H^2}{P^\circ P_{H_2}},$$

where P° is a reference pressure conventionally taken to be 1 bar, and the other P's are the partial pressures of the two species at equilibrium. Now, using the methods of Boltzmann statistics developed in this chapter, you are ready to calculate K from first principles. Do so. That is, derive a formula for K in terms of more basic quantities such as the energy needed to dissociate one molecule (see Problem 1.53) and the internal partition function for molecular hydrogen. This internal partition function is a product of rotational and vibrational contributions, which you can estimate using the methods and data in Section 6.2. (An H_2 molecule doesn't have any electronic spin degeneracy, but an H atom does—the electron can be in two different spin states. Neglect electronic excited states, which are important only at very high temperatures. The degeneracy due to nuclear spin alignments cancels, but include it if you wish.) Calculate K numerically at $T = 300$ K, 1000 K, 3000 K, and 6000 K. Discuss the implications, working out a couple of numerical examples to show when hydrogen is mostly dissociated and when it is not.

7 Quantum Statistics

7.1 The Gibbs Factor

In deriving the Boltzmann factor in Section 6.1, I allowed the small system and the reservoir to exchange energy, but not particles. Often, however, it is useful to consider a system that *can* exchange particles with its environment (see Figure 7.1). Let me now modify the previous derivation to allow for this possibility.

As in Section 6.1, we can write the ratio of probabilities for two different microstates as

$$\frac{\mathcal{P}(s_2)}{\mathcal{P}(s_1)} = \frac{\Omega_R(s_2)}{\Omega_R(s_1)} = \frac{e^{S_R(s_2)/k}}{e^{S_R(s_1)/k}} = e^{[S_R(s_2) - S_R(s_1)]/k}. \tag{7.1}$$

The exponent now contains the change in the entropy of the reservoir as the system goes from state 1 to state 2. This is an infinitesimal change from the reservoir's viewpoint, so we can invoke the thermodynamic identity:

$$dS_R = \frac{1}{T}\left(dU_R + P\,dV_R - \mu\,dN_R\right). \tag{7.2}$$

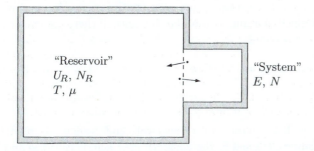

Figure 7.1. A system in thermal and diffusive contact with a much larger reservoir, whose temperature and chemical potential are effectively constant.

Since any energy, volume, or particles gained by the reservoir must be lost by the system, each of the changes on the right-hand side can be written as minus the same change for the system.

As in Section 6.1, I'll throw away the $P\,dV$ term; this term is often zero, or at least very small compared to the others. This time, however, I'll keep the $\mu\,dN$ term. Then the change in entropy can be written

$$S_R(s_2) - S_R(s_1) = -\frac{1}{T}\big[E(s_2) - E(s_1) - \mu N(s_2) + \mu N(s_1)\big]. \qquad (7.3)$$

On the right-hand side both E and N refer to the small system, hence the overall minus sign. Plugging this expression into equation 7.1 gives

$$\frac{\mathcal{P}(s_2)}{\mathcal{P}(s_1)} = \frac{e^{-[E(s_2)-\mu N(s_2)]/kT}}{e^{-[E(s_1)-\mu N(s_1)]/kT}}. \qquad (7.4)$$

As before, the ratio of probabilities is a ratio of simple exponential factors, each of which is a function of the temperature of the reservoir and the energy of the corresponding microstate. Now, however, the factor depends also on the number of particles in the system for state s. This new exponential factor is called a **Gibbs factor**:

$$\text{Gibbs factor} = e^{-[E(s)-\mu N(s)]/kT}. \qquad (7.5)$$

If we want an absolute probability instead of a ratio of probabilities, again we have to slip a constant of proportionality in front of the exponential:

$$\mathcal{P}(s) = \frac{1}{\mathcal{Z}}\, e^{-[E(s)-\mu N(s)]/kT}. \qquad (7.6)$$

The quantity $\mathcal{Z}$ is called the **grand partition function*** or the **Gibbs sum**. By requiring that the sum of the probabilities of all states equal 1, you can easily show that

$$\mathcal{Z} = \sum_s e^{-[E(s)-\mu N(s)]/kT}, \qquad (7.7)$$

where the sum runs over all possible states (including all possible values of N).

If more than one type of particle can be present in the system, then the $\mu\,dN$ term in equation 7.2 becomes a sum over species of $\mu_i\,dN_i$, and each subsequent equation is modified in a similar way. For instance, if there are two types of particles, the Gibbs factor becomes

$$\text{Gibbs factor} = e^{-[E(s)-\mu_A N_A(s)-\mu_B N_B(s)]/kT} \qquad \text{(two species).} \qquad (7.8)$$

---

*In analogy with the terms "microcanonical" and "canonical" used to describe the methods of Chapters 2–3 and 6, the approach used here is called **grand canonical**. A hypothetical set of systems with probabilities assigned according to equation 7.6 is called a **grand canonical ensemble**.

An Example: Carbon Monoxide Poisoning

A good example of a system to illustrate the use of Gibbs factors is an adsorption site on a hemoglobin molecule, which carries oxygen in the blood. A single hemoglobin molecule has four adsorption sites, each consisting of an Fe^{2+} ion surrounded by various other atoms. Each site can carry one O_2 molecule. For simplicity I'll take the system to be just one of the four sites, and pretend that it is completely independent of the other three.* Then if oxygen is the only molecule that can occupy the site, the system has just two possible states: unoccupied and occupied (see Figure 7.2). I'll take the energies of these two states to be 0 and ϵ, with $\epsilon = -0.7$ eV.[†]

The grand partition function for this single-site system has just two terms:

$$\mathcal{Z} = 1 + e^{-(\epsilon - \mu)/kT}. \tag{7.9}$$

The chemical potential μ is relatively high in the lungs, where oxygen is abundant, but is much lower in the cells where the oxygen is used. Let's consider the situation near the lungs. There the blood is in approximate diffusive equilibrium with the atmosphere, an ideal gas in which the partial pressure of oxygen is about 0.2 atm. The chemical potential can therefore be calculated from equation 6.93:

$$\mu = -kT \ln\left(\frac{V Z_{\text{int}}}{N v_Q}\right) \approx -0.6 \text{ eV} \tag{7.10}$$

at body temperature, 310 K. Plugging in these numbers gives for the second Gibbs factor

$$e^{-(\epsilon - \mu)/kT} \approx e^{(0.1\,\text{eV})/kT} \approx 40. \tag{7.11}$$

The probability of any given site being occupied is therefore

$$\mathcal{P}(\text{occupied by } O_2) = \frac{40}{1 + 40} = 98\%. \tag{7.12}$$

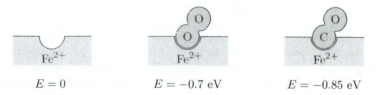

$$E = 0 \qquad\qquad E = -0.7 \text{ eV} \qquad\qquad E = -0.85 \text{ eV}$$

Figure 7.2. A single heme site can be unoccupied, occupied by oxygen, or occupied by carbon monoxide. (The energy values are only approximate.)

---

*The assumption of independent sites is quite accurate for myoglobin, a related protein that binds oxygen in muscles, which has only one adsorption site per molecule. A more accurate model of hemoglobin is presented in Problem 7.2.

[†]Biochemists *never* express energies in electron-volts. In fact, they rarely talk about individual bond energies at all (perhaps because these energies can vary so much under different conditions). I've chosen the ϵ values in this section to yield results that are in rough agreement with experimental measurements.

Suppose, however, that there is also some carbon monoxide present, which can also be adsorbed into the heme site. Now there are three states available to the site: unoccupied, occupied by O_2, and occupied by CO. The grand partition function is

$$\mathcal{Z} = 1 + e^{-(\epsilon - \mu)/kT} + e^{-(\epsilon' - \mu')/kT}, \qquad (7.13)$$

where ϵ' is the negative energy of a bound CO molecule and μ' is the chemical potential of CO in the environment. On the one hand, CO will never be as abundant as oxygen. If it is 100 times less abundant, then its chemical potential is lower by *roughly* $kT \ln 100 = 0.12$ eV, so μ' is roughly -0.72 eV. On the other hand, CO is more tightly bound to the site than oxygen, with $\epsilon' \approx -0.85$ eV. Plugging in these numbers gives for the third Gibbs factor

$$e^{-(\epsilon' - \mu')/kT} \approx e^{(0.13\,\mathrm{eV})/kT} \approx 120. \qquad (7.14)$$

The probability of the site being occupied by an *oxygen* molecule therefore drops to

$$\mathcal{P}(\text{occupied by } O_2) = \frac{40}{1 + 40 + 120} = 25\%. \qquad (7.15)$$

Problem 7.1. Near the cells where oxygen is used, its chemical potential is significantly lower than near the lungs. Even though there is no gaseous oxygen near these cells, it is customary to express the abundance of oxygen in terms of the partial pressure of gaseous oxygen that *would* be in equilibrium with the blood. Using the independent-site model just presented, with only oxygen present, calculate and plot the fraction of occupied heme sites as a function of the partial pressure of oxygen. This curve is called the **Langmuir adsorption isotherm** ("isotherm" because it's for a fixed temperature). Experiments show that adsorption by *myoglobin* follows the shape of this curve quite accurately.

Problem 7.2. In a real hemoglobin molecule, the tendency of oxygen to bind to a heme site increases as the other three heme sites become occupied. To model this effect in a simple way, imagine that a hemoglobin molecule has just two sites, either or both of which can be occupied. This system has four possible states (with only oxygen present). Take the energy of the unoccupied state to be zero, the energies of the two singly occupied states to be -0.55 eV, and the energy of the doubly occupied state to be -1.3 eV (so the change in energy upon binding the *second* oxygen is -0.75 eV). As in the previous problem, calculate and plot the fraction of occupied sites as a function of the effective partial pressure of oxygen. Compare to the graph from the previous problem (for independent sites). Can you think of why this behavior is preferable for the function of hemoglobin?

Problem 7.3. Consider a system consisting of a single hydrogen atom/ion, which has two possible states: unoccupied (i.e., no electron present) and occupied (i.e., one electron present, in the ground state). Calculate the ratio of the probabilities of these two states, to obtain the Saha equation, already derived in Section 5.6. Treat the electrons as a monatomic ideal gas, for the purpose of determining μ. Neglect the fact that an electron has two independent spin states.

Problem 7.4. Repeat the previous problem, taking into account the two independent spin states of the electron. Now the system has two "occupied" states, one with the electron in each spin configuration. However, the chemical potential of the electron gas is also slightly different. Show that the ratio of probabilities is the same as before: The spin degeneracy cancels out of the Saha equation.

Problem 7.5. Consider a system consisting of a single impurity atom/ion in a semiconductor. Suppose that the impurity atom has one "extra" electron compared to the neighboring atoms, as would a phosphorus atom occupying a lattice site in a silicon crystal. The extra electron is then easily removed, leaving behind a positively charged ion. The ionized electron is called a **conduction electron**, because it is free to move through the material; the impurity atom is called a **donor**, because it can "donate" a conduction electron. This system is analogous to the hydrogen atom considered in the previous two problems except that the ionization energy is much less, mainly due to the screening of the ionic charge by the dielectric behavior of the medium.

(a) Write down a formula for the probability of a single donor atom being ionized. Do not neglect the fact that the electron, if present, can have two independent spin states. Express your formula in terms of the temperature, the ionization energy I, and the chemical potential of the "gas" of ionized electrons.

(b) Assuming that the conduction electrons behave like an ordinary ideal gas (with two spin states per particle), write their chemical potential in terms of the number of conduction electrons per unit volume, N_c/V.

(c) Now assume that every conduction electron comes from an ionized donor atom. In this case the number of conduction electrons is equal to the number of donors that are ionized. Use this condition to derive a quadratic equation for N_c in terms of the number of donor atoms (N_d), eliminating μ. Solve for N_c using the quadratic formula. (Hint: It's helpful to introduce some abbreviations for dimensionless quantities. Try $x = N_c/N_d$, $t = kT/I$, and so on.)

(d) For phosphorus in silicon, the ionization energy is 0.044 eV. Suppose that there are 10^{17} P atoms per cubic centimeter. Using these numbers, calculate and plot the fraction of ionized donors as a function of temperature. Discuss the results.

Problem 7.6. Show that when a system is in thermal and diffusive equilibrium with a reservoir, the average number of particles in the system is

$$\overline{N} = \frac{kT}{\mathcal{Z}} \frac{\partial \mathcal{Z}}{\partial \mu},$$

where the partial derivative is taken at fixed temperature and volume. Show also that the mean *square* number of particles is

$$\overline{N^2} = \frac{(kT)^2}{\mathcal{Z}} \frac{\partial^2 \mathcal{Z}}{\partial \mu^2}.$$

Use these results to show that the standard deviation of N is

$$\sigma_N = \sqrt{kT(\partial \overline{N}/\partial \mu)},$$

in analogy with Problem 6.18. Finally, apply this formula to an ideal gas, to obtain a simple expression for σ_N in terms of $\overline{N}$. Discuss your result briefly.

Problem 7.7. In Section 6.5 I derived the useful relation $F = -kT \ln Z$ between the Helmholtz free energy and the ordinary partition function. Use an analogous argument to prove that

$$\Phi = -kT \ln \mathcal{Z},$$

where $\mathcal{Z}$ is the grand partition function and Φ is the grand free energy introduced in Problem 5.23.

7.2 Bosons and Fermions

The most important application of Gibbs factors is to **quantum statistics**, the study of dense systems in which two or more identical particles have a reasonable chance of wanting to occupy the same single-particle state. In this situation, my derivation (in Section 6.6) of the partition function for a system of N indistinguishable, noninteracting particles,

$$Z = \frac{1}{N!} Z_1^N, \tag{7.16}$$

breaks down. The problem is that the counting factor of $N!$, the number of ways of interchanging the particles among their various states, is correct only if the particles are always in *different* states. (In this section I'll use the word "state" to mean a single-particle state. For the state of the system as a whole I'll always say "system state.")

To better understand this issue, let's consider a very simple example: a system containing two noninteracting particles, either of which can occupy any of five states (see Figure 7.3). Imagine that all five of these states have energy zero, so every Boltzmann factor equals 1 (and therefore Z is the same as Ω).

If the two particles are *distinguishable*, then each has five available states and the total number of *system* states is $Z = 5 \times 5 = 25$. If the two particles are *indistinguishable*, equation 7.16 would predict $Z = 5^2/2 = 12.5$, and this can't be right, since Z must (for this system) be an integer.

So let's count the system states more carefully. Since the particles are indistinguishable, all that matters is the number of particles in any given state. I can therefore represent any system state by a sequence of five integers, each representing the number of particles in a particular state. For instance, 01100 would represent the system state in which the second and third states each contain one particle,

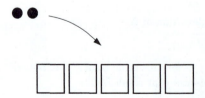

Figure 7.3. A simple model of five single-particle states, with two particles that can occupy these states.

while the rest contain none. Here, then, are all the allowed system states:

$$
\begin{array}{lll}
11000 & 01010 & 20000 \\
10100 & 01001 & 02000 \\
10010 & 00110 & 00200 \\
10001 & 00101 & 00020 \\
01100 & 00011 & 00002 \\
\end{array}
$$

(If you pretend that the states are harmonic oscillators and the particles are energy units, you can count the system states in the same way as for an Einstein solid.) There are 15 system states in all, of which 10 have the two particles in different states while 5 have the two particles in the same state. Each of the first 10 system states would actually be *two* different system states if the particles were distinguishable, since then they could be placed in either order. These 20 system states, plus the last 5 listed above, make the 25 counted in the previous paragraph. The factor of $1/N!$ in equation 7.16 correctly cuts the 20 down to 10, but also incorrectly cuts out half of the last five states.

Here I'm implicitly assuming that two identical particles *can* occupy the same state. It turns out that some types of particles can do this while others can't. Particles that *can* share a state with another of the same species are called **bosons**,* and include photons, pions, helium-4 atoms, and a variety of others. The number of identical bosons in a given state is unlimited. Experiments show, however, that many types of particles *cannot* share a state with another particle of the same type—not because they physically repel each other, but due to a quirk of quantum mechanics that I won't try to explain here (see Appendix A for some further discussion of this point). These particles are called **fermions**,[†] and include electrons, protons, neutrons, neutrinos, helium-3 atoms, and others. If the particles in the preceding example are identical fermions, then the five system states in the final column of the table are not allowed, so Z is only 10, not 15. (In formula 7.16, a system state with two particles in the same state is counted as half a system state, so this formula interpolates between the correct result for fermions and the correct result for bosons.) The rule that two identical fermions cannot occupy the same state is called the **Pauli exclusion principle**.

You can tell which particles are bosons and which are fermions by looking at their *spins*. Particles with integer spin (0, 1, 2, etc., in units of $h/2\pi$) are bosons, while particles with half-integer spin (1/2, 3/2, etc.) are fermions. This rule is *not* the *definition* of a boson or fermion, however; it is a nontrivial fact of nature, a deep consequence of the theories of relativity and quantum mechanics (as first derived by Wolfgang Pauli).

---

*After Satyendra Nath Bose, who in 1924 introduced the method of treating a photon gas presented in Section 7.4. The generalization to other bosons was provided by Einstein shortly thereafter.

[†]After Enrico Fermi, who in 1926 worked out the basic implications of the exclusion principle for statistical mechanics. Paul A. M. Dirac independently did the same thing, in the same year.

In many situations, however, it just doesn't matter whether the particles in a fluid are bosons or fermions. When the number of available single-particle states is much greater than the number of particles,

$$Z_1 \gg N, \tag{7.17}$$

the chance of any two particles wanting to occupy the same state is negligible. More precisely, only a tiny fraction of all system states have a significant number of states doubly occupied. For an ideal gas, the single-particle partition function is $Z_1 = V Z_{\text{int}}/v_Q$, where Z_{int} is some reasonably small number and v_Q is the quantum volume,

$$v_Q = \ell_Q^3 = \left(\frac{h}{\sqrt{2\pi m k T}} \right)^3, \tag{7.18}$$

roughly the cube of the average de Broglie wavelength. The condition (7.17) for the formula $Z = Z_1^N/N!$ to apply then translates to

$$\frac{V}{N} \gg v_Q, \tag{7.19}$$

which says that the average distance between particles must be much greater than the average de Broglie wavelength. For the air we breathe, the average distance between molecules is about 3 nm while the average de Broglie wavelength is less than 0.02 nm, so this condition is definitely satisfied. Notice, by the way, that this condition depends not only on the density of the system, but also on the temperature and the mass of the particles, both through v_Q.

It's hard to visualize what actually happens in a gas when condition 7.17 breaks down and multiple particles start trying to get into the same state. Figure 7.4, though imperfect, is about the best I can do. Picture each particle as being smeared out in a quantum wavefunction filling a volume equal to v_Q. (This is equivalent to putting the particles into wavefunctions that are as localized in space as possible. To squeeze them into narrower wavefunctions we would have to introduce uncer-

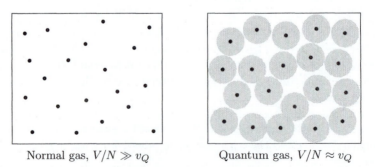

Normal gas, $V/N \gg v_Q$ Quantum gas, $V/N \approx v_Q$

Figure 7.4. In a normal gas, the space between particles is much greater than the typical size of a particle's wavefunction. When the wavefunctions begin to "touch" and overlap, we call it a **quantum gas**.

tainties in momentum that are large compared to the average momentum h/ℓ_Q, thus increasing the energy and temperature of the system.) In a normal gas, the effective volume thus occupied by all the particles will be much less than the volume of the container. (Often the quantum volume is less than the physical volume of a molecule.) But if the gas is sufficiently dense or v_Q is sufficiently large, then the wavefunctions will start trying to touch and overlap. At this point it starts to matter whether the particles are fermions or bosons; either way, the behavior will be much different from that of a normal gas.

There are plenty of systems that violate condition 7.17, either because they are very dense (like a neutron star), or very cold (like liquid helium), or composed of very light particles (like the electrons in a metal or the photons in a hot oven). The rest of this chapter is devoted to the study of these fascinating systems.

Problem 7.8. Suppose you have a "box" in which each particle may occupy any of 10 single-particle states. For simplicity, assume that each of these states has energy zero.

(a) What is the partition function of this system if the box contains only one particle?

(b) What is the partition function of this system if the box contains two distinguishable particles?

(c) What is the partition function if the box contains two identical bosons?

(d) What is the partition function if the box contains two identical fermions?

(e) What would be the partition function of this system according to equation 7.16?

(f) What is the probability of finding both particles in the same single-particle state, for the three cases of distinguishable particles, identical bosons, and identical fermions?

Problem 7.9. Compute the quantum volume for an N_2 molecule at room temperature, and argue that a gas of such molecules at atmospheric pressure can be treated using Boltzmann statistics. At about what temperature would quantum statistics become relevant for this system (keeping the density constant and pretending that the gas does not liquefy)?

Problem 7.10. Consider a system of five particles, inside a container where the allowed energy levels are nondegenerate and evenly spaced. For instance, the particles could be trapped in a one-dimensional harmonic oscillator potential. In this problem you will consider the allowed states for this system, depending on whether the particles are identical fermions, identical bosons, or distinguishable particles.

(a) Describe the ground state of this system, for each of these three cases.

(b) Suppose that the system has one unit of energy (above the ground state). Describe the allowed states of the system, for each of the three cases. How many possible system states are there in each case?

(c) Repeat part (b) for two units of energy and for three units of energy.

(d) Suppose that the temperature of this system is low, so that the total energy is low (though not necessarily zero). In what way will the behavior of the bosonic system differ from that of the system of distinguishable particles? Discuss.

The Distribution Functions

When a system violates the condition $Z_1 \gg N$, so that we cannot treat it using the methods of Chapter 6, we can use Gibbs factors instead. The idea is to first consider a "system" consisting of *one single-particle state*, rather than a particle itself. Thus the system will consist of a particular spatial wavefunction (and, for particles with spin, a particular spin orientation). This idea seems strange at first, because we normally work with wavefunctions of definite energy, and each of these wavefunctions shares its space with all the other wavefunctions. The "system" and the "reservoir" therefore occupy the same physical space, as in Figure 7.5. Fortunately, the mathematics that went into the derivation of the Gibbs factor couldn't care less whether the system is spatially distinct from the reservoir, so all those formulas still apply to a single-particle-state system.

So let's concentrate on just one single-particle state of a system (say, a particle in a box), whose energy when occupied by a single particle is ϵ. When the state is unoccupied, its energy is 0; if it can be occupied by n particles, then the energy will be $n\epsilon$. The probability of the state being occupied by n particles is

$$P(n) = \frac{1}{\mathcal{Z}}e^{-(n\epsilon - \mu n)/kT} = \frac{1}{\mathcal{Z}}e^{-n(\epsilon - \mu)/kT}, \tag{7.20}$$

where $\mathcal{Z}$ is the grand partition function, that is, the sum of the Gibbs factors for all possible n.

If the particles in question are fermions, then n can only be 0 or 1, so the grand partition function is

$$\mathcal{Z} = 1 + e^{-(\epsilon - \mu)/kT} \qquad \text{(fermions)}. \tag{7.21}$$

From this we can compute the probability of the state being occupied or unoccupied, as a function of ϵ, μ, and T. We can also compute the *average* number of particles in the state, also called the **occupancy** of the state:

$$\bar{n} = \sum_n nP(n) = 0 \cdot P(0) + 1 \cdot P(1) = \frac{e^{-(\epsilon - \mu)/kT}}{1 + e^{-(\epsilon - \mu)/kT}}$$

$$= \frac{1}{e^{(\epsilon - \mu)/kT} + 1} \qquad \text{(fermions)}. \tag{7.22}$$

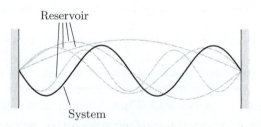

Figure 7.5. To treat a quantum gas using Gibbs factors, we consider a "system" consisting of one single-particle state (or wavefunction). The "reservoir" consists of all the other possible single-particle states.

This important formula is called the **Fermi-Dirac distribution**; I'll call it $\bar{n}_{\mathrm{FD}}$:

$$\bar{n}_{\mathrm{FD}} = \frac{1}{e^{(\epsilon - \mu)/kT} + 1}. \qquad (7.23)$$

The Fermi-Dirac distribution goes to zero when $\epsilon \gg \mu$, and goes to 1 when $\epsilon \ll \mu$. Thus, states with energy much less than μ tend to be occupied, while states with energy much greater than μ tend to be unoccupied. A state with energy exactly equal to μ has a 50% chance of being occupied, while the width of the fall-off from 1 to 0 is a few times kT. A graph of the Fermi-Dirac distribution vs. ϵ for three different temperatures is shown in Figure 7.6.

If instead the particles in question are bosons, then n can be any nonnegative integer, so the grand partition function is

$$
\begin{aligned}
\mathcal{Z} &= 1 + e^{-(\epsilon - \mu)/kT} + e^{-2(\epsilon - \mu)/kT} + \cdots \\
&= 1 + e^{-(\epsilon - \mu)/kT} + (e^{-(\epsilon - \mu)/kT})^2 + \cdots \\
&= \frac{1}{1 - e^{-(\epsilon - \mu)/kT}} \qquad \text{(bosons)}.
\end{aligned}
\qquad (7.24)
$$

(Since the Gibbs factors cannot keep growing without limit, μ must be less than ϵ and therefore the series must converge.) Meanwhile, the average number of particles in the state is

$$\bar{n} = \sum_n n\mathcal{P}(n) = 0 \cdot \mathcal{P}(0) + 1 \cdot \mathcal{P}(1) + 2 \cdot \mathcal{P}(2) + \cdots. \qquad (7.25)$$

To evaluate this sum let's abbreviate $x \equiv (\epsilon - \mu)/kT$. Then

$$\bar{n} = \sum_n n \frac{e^{-nx}}{\mathcal{Z}} = -\frac{1}{\mathcal{Z}} \sum_n \frac{\partial}{\partial x} e^{-nx} = -\frac{1}{\mathcal{Z}} \frac{\partial \mathcal{Z}}{\partial x}. \qquad (7.26)$$

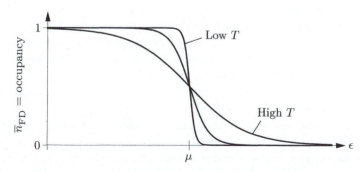

Figure 7.6. The Fermi-Dirac distribution goes to 1 for very low-energy states and to zero for very high-energy states. It equals $1/2$ for a state with energy μ, falling off suddenly for low T and gradually for high T. (Although μ is fixed on this graph, in the next section we'll see that μ normally varies with temperature.)

You can easily check that this formula works for fermions. For bosons, we have

$$\bar{n} = -(1 - e^{-x}) \frac{\partial}{\partial x} (1 - e^{-x})^{-1} = (1 - e^{-x})(1 - e^{-x})^{-2}(e^{-x})$$

$$= \frac{1}{e^{(\epsilon - \mu)/kT} - 1} \qquad \text{(bosons).}$$

(7.27)

This important formula is called the **Bose-Einstein distribution**; I'll call it $\bar{n}_{\text{BE}}$:

$$\bar{n}_{\text{BE}} = \frac{1}{e^{(\epsilon - \mu)/kT} - 1}.$$

(7.28)

Like the Fermi-Dirac distribution, the Bose-Einstein distribution goes to zero when $\epsilon \gg \mu$. Unlike the Fermi-Dirac distribution, however, it goes to infinity as ϵ approaches μ from above (see Figure 7.7). It would be negative if ϵ could be less than μ, but we've already seen that this cannot happen.

To better understand the Fermi-Dirac and Bose-Einstein distributions, it's useful to ask what $\bar{n}$ would be for particles obeying *Boltzmann* statistics. In this case, the probability of any single particle being in a certain state of energy ϵ is

$$\mathcal{P}(s) = \frac{1}{Z_1} e^{-\epsilon/kT} \qquad \text{(Boltzmann),}$$

(7.29)

so if there are N independent particles in total, the average number in this state is

$$\bar{n}_{\text{Boltzmann}} = N\mathcal{P}(s) = \frac{N}{Z_1} e^{-\epsilon/kT}.$$

(7.30)

But according to the result of Problem 6.44, the chemical potential for such a system is $\mu = -kT \ln(Z_1/N)$. Therefore the average occupancy can be written

$$\bar{n}_{\text{Boltzmann}} = e^{\mu/kT} e^{-\epsilon/kT} = e^{-(\epsilon - \mu)/kT}.$$

(7.31)

When ϵ is sufficiently greater than μ, so that this exponential is very small, we can neglect the 1 in the denominator of either the Fermi-Dirac distribution (7.23) or the Bose-Einstein distribution (7.28), and both reduce to the Boltzmann distribution (7.31). The equality of the three distribution functions in this limit is shown in Figure 7.7. The precise condition for the three distributions to agree is that the exponent $(\epsilon - \mu)/kT$ be much greater than 1. If we take the lowest-energy state to have $\epsilon \approx 0$, then this condition will be met for all states whenever $\mu \ll -kT$, that is, when $Z_1 \gg N$. This is the same condition that we arrived at through different reasoning at the beginning of this section.

We now know how to compute the average number of particles occupying a single-particle state, whether the particles are fermions or bosons, in terms of the energy of the state, the temperature, and the chemical potential. To apply these ideas to any particular system, we still need to know what the energies of all the states are. This is a problem in quantum mechanics, and can be extremely difficult in many cases. In this book we'll deal mostly with particles in a "box," where the quantum-mechanical wavefunctions are simple sine waves and the corresponding energies can

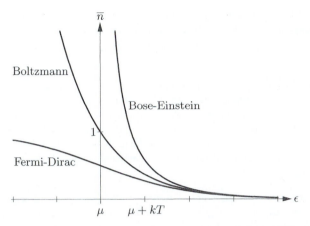

Figure 7.7. Comparison of the Fermi-Dirac, Bose-Einstein, and Boltzmann distributions, all for the same value of μ. When $(\epsilon - \mu)/kT \gg 1$, the three distributions become equal.

be determined straightforwardly. The particles could be electrons in a metal, neutrons in a neutron star, atoms in a fluid at very low temperature, photons inside a hot oven, or even "phonons," the quantized units of vibrational energy in a solid.

For any of these applications, before we can apply the Fermi-Dirac or Bose-Einstein distribution, we'll also have to figure out what the chemical potential is. In a few cases this is quite easy, but in other applications it will require considerable work. As we'll see, μ is usually determined indirectly by the total number of particles in the system.

Problem 7.11. For a system of fermions at room temperature, compute the probability of a single-particle state being occupied if its energy is

(a) 1 eV less than μ

(b) 0.01 eV less than μ

(c) equal to μ

(d) 0.01 eV greater than μ

(e) 1 eV greater than μ

Problem 7.12. Consider two single-particle states, A and B, in a system of fermions, where $\epsilon_A = \mu - x$ and $\epsilon_B = \mu + x$; that is, level A lies below μ by the same amount that level B lies above μ. Prove that the probability of level B being occupied is the same as the probability of level A being *unoccupied*. In other words, the Fermi-Dirac distribution is "symmetrical" about the point where $\epsilon = \mu$.

Problem 7.13. For a system of bosons at room temperature, compute the average occupancy of a single-particle state and the probability of the state containing 0, 1, 2, or 3 bosons, if the energy of the state is

(a) 0.001 eV greater than μ

(b) 0.01 eV greater than μ

(c) 0.1 eV greater than μ

(d) 1 eV greater than μ

Problem 7.14. For a system of particles at room temperature, how large must $\epsilon - \mu$ be before the Fermi-Dirac, Bose-Einstein, and Boltzmann distributions agree within 1%? Is this condition ever violated for the gases in our atmosphere? Explain.

Problem 7.15. For a system obeying Boltzmann statistics, we know what μ is from Chapter 6. Suppose, though, that you knew the distribution function (equation 7.31) but didn't know μ. You could still determine μ by requiring that the total number of particles, summed over all single-particle states, equal N. Carry out this calculation, to rederive the formula $\mu = -kT \ln(Z_1/N)$. (This is normally how μ is determined in quantum statistics, although the math is usually more difficult.)

Problem 7.16. Consider an isolated system of N identical fermions, inside a container where the allowed energy levels are nondegenerate and evenly spaced.* For instance, the fermions could be trapped in a one-dimensional harmonic oscillator potential. For simplicity, neglect the fact that fermions can have multiple spin orientations (or assume that they are all forced to have the same spin orientation). Then each energy level is either occupied or unoccupied, and any allowed system state can be represented by a column of dots, with a filled dot representing an occupied level and a hollow dot representing an unoccupied level. The lowest-energy system state has all levels below a certain point occupied, and all levels above that point unoccupied. Let η be the spacing between energy levels, and let q be the number of energy units (each of size η) in excess of the ground-state energy. Assume that $q < N$. Figure 7.8 shows all system states up to $q = 3$.

  (a) Draw dot diagrams, as in the figure, for all allowed system states with $q = 4$, $q = 5$, and $q = 6$.

  (b) According to the fundamental assumption, all allowed system states with a given value of q are equally probable. Compute the probability of each energy level being occupied, for $q = 6$. Draw a graph of this probability as a function of the energy of the level.

  (c) In the thermodynamic limit where q is large, the probability of a level being occupied should be given by the Fermi-Dirac distribution. Even though 6 is not a large number, estimate the values of μ and T that you would have to plug into the Fermi-Dirac distribution to best fit the graph you drew in part (b).

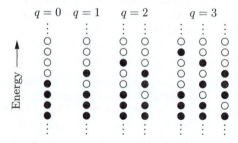

Figure 7.8. A representation of the system states of a fermionic system with evenly spaced, nondegenerate energy levels. A filled dot represents an occupied single-particle state, while a hollow dot represents an unoccupied single-particle state.

---

*This problem and Problem 7.27 are based on an article by J. Arnaud et al., *American Journal of Physics* **67**, 215 (1999).

(d) Calculate the entropy of this system for each value of q from 0 to 6, and draw a graph of entropy vs. energy. Make a rough estimate of the slope of this graph near $q = 6$, to obtain another estimate of the temperature of this system at that point. Check that it is in rough agreement with your answer to part (c).

Problem 7.17. In analogy with the previous problem, consider a system of identical spin-0 *bosons* trapped in a region where the energy levels are evenly spaced. Assume that N is a large number, and again let q be the number of energy units.

(a) Draw diagrams representing all allowed system states from $q = 0$ up to $q = 6$. Instead of using dots as in the previous problem, use numbers to indicate the number of bosons occupying each level.

(b) Compute the occupancy of each energy level, for $q = 6$. Draw a graph of the occupancy as a function of the energy of the level.

(c) Estimate the values of μ and T that you would have to plug into the Bose-Einstein distribution to best fit the graph of part (b).

(d) As in part (d) of the previous problem, draw a graph of entropy vs. energy and estimate the temperature at $q = 6$ from this graph.

Problem 7.18. Imagine that there exists a third type of particle, which can share a single-particle state with one other particle of the same type but no more. Thus the number of these particles in any state can be 0, 1, or 2. Derive the distribution function for the average occupancy of a state by particles of this type, and plot the occupancy as a function of the state's energy, for several different temperatures.

7.3   Degenerate Fermi Gases

As a first application of quantum statistics and the Fermi-Dirac distribution, I'd like to consider a "gas" of fermions at very low temperature. The fermions could be helium-3 atoms, or protons and neutrons in an atomic nucleus, or electrons in a white dwarf star, or neutrons in a neutron star. The most familiar example, though, is the conduction electrons inside a chunk of metal. In this section I'll say "electrons" to be specific, even though the results apply to other types of fermions as well.

By "very low temperature," I do *not* necessarily mean low compared to room temperature. What I mean is that the condition for Boltzmann statistics to apply to an ideal gas, $V/N \gg v_Q$, is badly violated, so that in fact $V/N \ll v_Q$. For an electron at room temperature, the quantum volume is

$$v_Q = \left(\frac{h}{\sqrt{2\pi mkT}} \right)^3 = (4.3 \text{ nm})^3. \tag{7.32}$$

But in a typical metal there is about one conduction electron per atom, so the volume per conduction electron is roughly the volume of an atom, $(0.2 \text{ nm})^3$. Thus, the temperature is *much* too low for Boltzmann statistics to apply. Instead, we are in the opposite limit, where for many purposes we can pretend that $T = 0$. Let us therefore first consider the properties of an electron gas *at* $T = 0$, and later ask what happens at small nonzero temperatures.

Zero Temperature

At $T = 0$ the Fermi-Dirac distribution becomes a step function (see Figure 7.9). All single-particle states with energy less than μ are occupied, while all states with energy greater than μ are unoccupied. In this context μ is also called the **Fermi energy**, denoted ϵ_F:

$$\epsilon_F \equiv \mu(T = 0). \tag{7.33}$$

When a gas of fermions is so cold that nearly all states below ϵ_F are occupied while nearly all states above ϵ_F are unoccupied, it is said to be **degenerate**. (This use of the word is completely unrelated to its other use to describe a set of quantum states that have the same energy.)

The *value* of ϵ_F is determined by the total number of electrons present. Imagine an empty box, to which you add electrons one at a time, with no excess energy. Each electron goes into the lowest available state, until the last electron goes into a state with energy just below ϵ_F. To add one more electron you would have to give it an energy essentially equal to $\epsilon_F = \mu$; in this context, the equation $\mu = (\partial U/\partial N)_{S,V}$ makes perfect physical sense, since $dU = \mu$ when $dN = 1$ (and S is fixed at zero when all the electrons are packed into the lowest-energy states).

In order to calculate ϵ_F, as well as other interesting quantities such as the total energy and the pressure of the electron gas, I'll make the approximation that the electrons are *free* particles, subject to no forces whatsoever except that they are confined inside a box of volume $V = L^3$. For the conduction electrons in a metal, this approximation is not especially accurate. Although it is reasonable to neglect long-range electrostatic forces in any electrically neutral material, each conduction electron still feels attractive forces from nearby ions in the crystal lattice, and I'm neglecting these forces.*

The definite-energy wavefunctions of a free electron inside a box are just sine waves, exactly as for the gas molecules treated in Section 6.7. For a one-dimensional

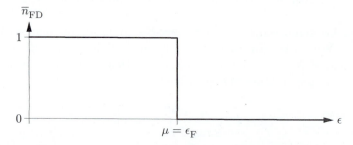

Figure 7.9. At $T = 0$, the Fermi-Dirac distribution equals 1 for all states with $\epsilon < \mu$ and equals 0 for all states with $\epsilon > \mu$.

---

*Problems 7.33 and 7.34 treat some of the effects of the crystal lattice on the conduction electrons. For much more detail, see a solid state physics textbook such as Kittel (1996) or Ashcroft and Mermin (1976).

box the allowed wavelengths and momenta are (as before)

$$\lambda_n = \frac{2L}{n}, \qquad p_n = \frac{h}{\lambda_n} = \frac{hn}{2L}, \tag{7.34}$$

where n is any positive integer. In a three-dimensional box these equations apply separately to the x, y, and z directions, so

$$p_x = \frac{hn_x}{2L}, \qquad p_y = \frac{hn_y}{2L}, \qquad p_z = \frac{hn_z}{2L}, \tag{7.35}$$

where (n_x, n_y, n_z) is a triplet of positive integers. The allowed energies are therefore

$$\epsilon = \frac{|\vec{p}|^2}{2m} = \frac{h^2}{8mL^2}(n_x^2 + n_y^2 + n_z^2). \tag{7.36}$$

To visualize the set of allowed states, I like to draw a picture of "n-space," the three-dimensional space whose axes are n_x, n_y, and n_z (see Figure 7.10). Each allowed $\vec{n}$ vector corresponds to a point in this space with positive integer coordinates; the set of all allowed states forms a huge lattice filling the first octant of n-space. Each lattice point actually represents *two* states, since for each spatial wavefunction there are two independent spin orientations.

In n-space, the energy of any state is proportional to the square of the *distance* from the origin, $n_x^2 + n_y^2 + n_z^2$. So as we add electrons to the box, they settle into states starting at the origin and gradually working outward. By the time we're done, the total number of occupied states is so huge that the occupied region of n-space is essentially an eighth of a sphere. (The roughness of the edges is insignificant, compared to the enormous size of the entire sphere.) I'll call the radius of this sphere $n_{\max}$.

It's now quite easy to relate the total number of electrons, N, to the chemical potential or Fermi energy, $\mu = \epsilon_{\mathrm{F}}$. On one hand, ϵ_{F} is the energy of a state that sits just on the surface of the sphere in n-space, so

$$\epsilon_{\mathrm{F}} = \frac{h^2 n_{\max}^2}{8mL^2}. \tag{7.37}$$

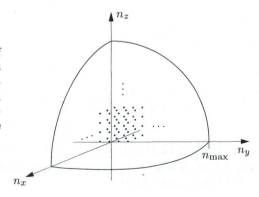

Figure 7.10. Each triplet of integers (n_x, n_y, n_z) represents a pair of definite-energy electron states (one with each spin orientation). The set of all independent states fills the positive octant of n-space.

On the other hand, the total volume of the eighth-sphere in n-space equals the number of lattice points enclosed, since the separation between lattice points is 1 in all three directions. Therefore the total number of occupied states is twice this volume (because of the two spin orientations):

$$N = 2 \times (\text{volume of eighth-sphere}) = 2 \cdot \frac{1}{8} \cdot \frac{4}{3}\pi n_{\text{max}}^3 = \frac{\pi n_{\text{max}}^3}{3}. \tag{7.38}$$

Combining these two equations gives the Fermi energy as a function of N and the volume $V = L^3$ of the box:

$$\epsilon_F = \frac{h^2}{8m}\left(\frac{3N}{\pi V}\right)^{2/3}. \tag{7.39}$$

Notice that this quantity is intensive, since it depends only on the number density of electrons, N/V. For a larger container with correspondingly more electrons, ϵ_F comes out the same. Although I have derived this result only for electrons in a cube-shaped box, it actually applies to macroscopic containers (or chunks of metal) of any shape.

The Fermi energy is the *highest* energy of all the electrons. On average, they'll have somewhat less energy, a little more than half ϵ_F. To be more precise, we have to do an integral, to find the *total* energy of all the electrons; the average is just the total divided by N.

To calculate the total energy of all the electrons, I'll add up the energies of the electrons in all occupied states. This entails a triple sum over n_x, n_y, and n_z:

$$U = 2 \sum_{n_x} \sum_{n_y} \sum_{n_z} \epsilon(\vec{n}) = 2 \iiint \epsilon(\vec{n})\, dn_x\, dn_y\, dn_z. \tag{7.40}$$

The factor of 2 is for the two spin orientations for each $\vec{n}$. I'm allowed to change the sum into an integral because the number of terms is so huge, it might as well be a continuous function. To evaluate the triple integral I'll use spherical coordinates, as illustrated in Figure 7.11. Note that the volume element $dn_x\, dn_y\, dn_z$ becomes

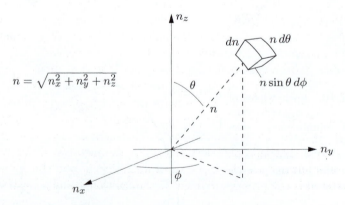

Figure 7.11. In spherical coordinates (n, θ, ϕ), the infinitesimal volume element is $(dn)(n\, d\theta)(n \sin\theta\, d\phi)$.

$n^2 \sin\theta\, dn\, d\theta\, d\phi$. The total energy of all the electrons is therefore

$$U = 2 \int_0^{n_{max}} dn \int_0^{\pi/2} d\theta \int_0^{\pi/2} d\phi\, n^2 \sin\theta\, \epsilon(n). \tag{7.41}$$

The angular integrals give $\pi/2$, one-eighth the surface area of a unit sphere. This leaves us with

$$U = \pi \int_0^{n_{max}} \epsilon(n)\, n^2\, dn = \frac{\pi h^2}{8mL^2} \int_0^{n_{max}} n^4\, dn = \frac{\pi h^2 n_{max}^5}{40mL^2} = \frac{3}{5} N \epsilon_F. \tag{7.42}$$

The *average* energy of the electrons is therefore 3/5 the Fermi energy.

If you plug in some numbers, you'll find that the Fermi energy for conduction electrons in a typical metal is a few electron-volts. This is *huge* compared to the average *thermal* energy of a particle at room temperature, roughly $kT \approx 1/40$ eV. In fact, comparing the Fermi energy to the average thermal energy is essentially the same as comparing the quantum volume to the average volume per particle, as I did at the beginning of this section:

$$\frac{V}{N} \ll v_Q \qquad \text{is the same as} \qquad kT \ll \epsilon_F. \tag{7.43}$$

When this condition is met, the approximation $T \approx 0$ is fairly accurate for many purposes, and the gas is said to be degenerate. The temperature that a Fermi gas would have to have in order for kT to equal ϵ_F is called the **Fermi temperature**: $T_F \equiv \epsilon_F/k$. This temperature is purely hypothetical for electrons in a metal, since metals liquefy and evaporate long before it is reached.

Using the formula $P = -(\partial U/\partial V)_{S,N}$, which you can derive from the thermodynamic identity or straight from classical mechanics, we can calculate the pressure of a degenerate electron gas:

$$P = -\frac{\partial}{\partial V}\left[\frac{3}{5}N\frac{h^2}{8m}\left(\frac{3N}{\pi}\right)^{2/3} V^{-2/3} \right] = \frac{2N\epsilon_F}{5V} = \frac{2}{3}\frac{U}{V}. \tag{7.44}$$

This quantity is called the **degeneracy pressure**. It is positive because when you compress a degenerate electron gas, the wavelengths of all the wavefunctions are reduced, hence the energies of all the wavefunctions increase. Degeneracy pressure is what keeps matter from collapsing under the huge electrostatic forces that try to pull electrons and protons together. Please note that degeneracy pressure has absolutely nothing to do with electrostatic repulsion between the electrons (which we've completely ignored); it arises purely by virtue of the exclusion principle.

Numerically, the degeneracy pressure comes out to a few *billion* N/m^2 for a typical metal. But this number is not directly measurable—it is canceled by the electrostatic forces that hold the electrons inside the metal in the first place. A more measurable quantity is the **bulk modulus**, that is, the change in pressure when the material is compressed, divided by the fractional change in volume:

$$B = -V\left(\frac{\partial P}{\partial V}\right)_T = \frac{10}{9}\frac{U}{V}. \tag{7.45}$$

This quantity is also quite large in SI units, but it is *not* completely canceled by the electrostatic forces; the formula actually agrees with experiment, within a factor of 3 or so, for most metals.

Problem 7.19. Each atom in a chunk of copper contributes one conduction electron. Look up the density and atomic mass of copper, and calculate the Fermi energy, the Fermi temperature, the degeneracy pressure, and the contribution of the degeneracy pressure to the bulk modulus. Is room temperature sufficiently low to treat this system as a degenerate electron gas?

Problem 7.20. At the center of the sun, the temperature is approximately 10^7 K and the concentration of electrons is approximately 10^{32} per cubic meter. Would it be (approximately) valid to treat these electrons as a "classical" ideal gas (using Boltzmann statistics), or as a degenerate Fermi gas (with $T \approx 0$), or neither?

Problem 7.21. An atomic nucleus can be crudely modeled as a gas of nucleons with a number density of 0.18 fm^{-3} (where 1 fm = 10^{-15} m). Because nucleons come in two different types (protons and neutrons), each with spin 1/2, each spatial wavefunction can hold *four* nucleons. Calculate the Fermi energy of this system, in MeV. Also calculate the Fermi temperature, and comment on the result.

Problem 7.22. Consider a degenerate electron gas in which essentially all of the electrons are highly relativistic ($\epsilon \gg mc^2$), so that their energies are $\epsilon = pc$ (where p is the magnitude of the momentum vector).

(a) Modify the derivation given above to show that for a relativistic electron gas at zero temperature, the chemical potential (or Fermi energy) is given by $\mu = hc(3N/8\pi V)^{1/3}$.

(b) Find a formula for the total energy of this system in terms of N and μ.

Problem 7.23. A **white dwarf** star (see Figure 7.12) is essentially a degenerate electron gas, with a bunch of nuclei mixed in to balance the charge and to provide the gravitational attraction that holds the star together. In this problem you will derive a relation between the mass and the radius of a white dwarf star, modeling the star as a uniform-density sphere. White dwarf stars tend to be extremely hot by our standards; nevertheless, it is an excellent approximation in this problem to set $T = 0$.

(a) Use dimensional analysis to argue that the gravitational potential energy of a uniform-density sphere (mass M, radius R) must equal

$$U_{\text{grav}} = -(\text{constant})\frac{GM^2}{R},$$

where (constant) is some numerical constant. Be sure to explain the minus sign. The constant turns out to equal 3/5; you can derive it by calculating the (negative) work needed to assemble the sphere, shell by shell, from the inside out.

(b) Assuming that the star contains one proton and one neutron for each electron, and that the electrons are nonrelativistic, show that the total (kinetic) energy of the degenerate electrons equals

$$U_{\text{kinetic}} = (0.0088)\frac{h^2 M^{5/3}}{m_e m_p^{5/3} R^2}.$$

Figure 7.12. The double star system Sirius A and B. Sirius A (greatly overexposed in the photo) is the brightest star in our night sky. Its companion, Sirius B, is hotter but very faint, indicating that it must be extremely small—a white dwarf. From the orbital motion of the pair we know that Sirius B has about the same mass as our sun. (UCO/Lick Observatory photo.)

The numerical factor can be expressed exactly in terms of π and cube roots and such, but it's not worth it.

(c) The equilibrium radius of the white dwarf is that which minimizes the total energy $U_{grav} + U_{kinetic}$. Sketch the total energy as a function of R, and find a formula for the equilibrium radius in terms of the mass. As the mass increases, does the radius increase or decrease? Does this make sense?

(d) Evaluate the equilibrium radius for $M = 2 \times 10^{30}$ kg, the mass of the sun. Also evaluate the density. How does the density compare to that of water?

(e) Calculate the Fermi energy and the Fermi temperature, for the case considered in part (d). Discuss whether the approximation $T = 0$ is valid.

(f) Suppose instead that the electrons in the white dwarf star are highly relativistic. Using the result of the previous problem, show that the total kinetic energy of the electrons is now proportional to $1/R$ instead of $1/R^2$. Argue that there is no stable equilibrium radius for such a star.

(g) The transition from the nonrelativistic regime to the ultrarelativistic regime occurs approximately where the average kinetic energy of an electron is equal to its rest energy, mc^2. Is the nonrelativistic approximation valid for a one-solar-mass white dwarf? Above what mass would you expect a white dwarf to become relativistic and hence unstable?

Problem 7.24. A star that is too heavy to stabilize as a white dwarf can collapse further to form a **neutron star**: a star made entirely of neutrons, supported against gravitational collapse by degenerate neutron pressure. Repeat the steps of the previous problem for a neutron star, to determine the following: the mass-radius relation; the radius, density, Fermi energy, and Fermi temperature of a one-solar-mass neutron star; and the critical mass above which a neutron star becomes relativistic and hence unstable to further collapse.

Small Nonzero Temperatures

One property of a Fermi gas that we *cannot* calculate using the approximation $T = 0$ is the heat capacity, since this is a measure of how the energy of the system depends on T. Let us therefore consider what happens when the temperature is very small but nonzero. Before doing any careful calculations, I'll explain what happens qualitatively and try to give some plausibility arguments.

At temperature T, all particles typically acquire a thermal energy of roughly kT. However, in a degenerate electron gas, most of the electrons *cannot* acquire such a small amount of energy, because all the states that they might jump into are already occupied (recall the shape of the Fermi-Dirac distribution, Figure 7.6).

The only electrons that *can* acquire some thermal energy are those that are already within about kT of the Fermi energy—these can jump up into unoccupied states above ϵ_F. (The spaces they leave behind allow *some*, but not many, of the lower-lying electrons to also gain energy.) Notice that the number of electrons that can be affected by the increase in T is proportional to T. This number must also be proportional to N, because it is an extensive quantity.

Thus, the additional energy that a degenerate electron gas acquires when its temperature is raised from zero to T is *doubly* proportional to T:

$$\text{additional energy} \propto (\text{number of affected electrons}) \times (\text{energy acquired by each})$$
$$\propto (NkT) \times (kT)$$
$$\propto N(kT)^2. \tag{7.46}$$

We can guess the constant of proportionality using dimensional analysis. The quantity $N(kT)^2$ has units of (energy)^2, so to get something with units of (energy)^1, we need to divide by some constant with units of energy. The only such constant available is ϵ_F, so the additional energy must be $N(kT)^2/\epsilon_F$, times some constant of order 1. In a few pages we'll see that this constant is $\pi^2/4$, so the total energy of a degenerate Fermi gas for $T \ll \epsilon_F/k$ is

$$U = \frac{3}{5}N\epsilon_F + \frac{\pi^2}{4}N\frac{(kT)^2}{\epsilon_F}. \tag{7.47}$$

From this result we can easily calculate the heat capacity:

$$C_V = \left(\frac{\partial U}{\partial T}\right)_V = \frac{\pi^2 N k^2 T}{2\epsilon_F}. \tag{7.48}$$

Notice that the heat capacity goes to zero at $T = 0$, as required by the third law of thermodynamics. The approach to zero is *linear* in T, and this prediction agrees well with experiments on metals at low temperatures. (Above a few kelvins, lattice vibrations also contribute significantly to the heat capacity of a metal.) The numerical coefficient of $\pi^2/2$ usually agrees with experiment to within 50% or better, but there are exceptions.

Problem 7.25. Use the results of this section to estimate the contribution of conduction electrons to the heat capacity of one mole of copper at room temperature. How does this contribution compare to that of lattice vibrations, assuming that these are not frozen out? (The electronic contribution has been measured at low temperatures, and turns out to be about 40% more than predicted by the free electron model used here.)

Problem 7.26. In this problem you will model helium-3 as a noninteracting Fermi gas. Although ^{3}He liquefies at low temperatures, the liquid has an unusually low density and behaves in many ways like a gas because the forces between the atoms are so weak. Helium-3 atoms are spin-1/2 fermions, because of the unpaired neutron in the nucleus.

(a) Pretending that liquid ^{3}He is a noninteracting Fermi gas, calculate the Fermi energy and the Fermi temperature. The molar volume (at low pressures) is 37 cm^3.

(b) Calculate the heat capacity for $T \ll T_F$, and compare to the experimental result $C_V = (2.8 \text{ K}^{-1})NkT$ (in the low-temperature limit). (Don't expect perfect agreement.)

(c) The entropy of *solid* ^{3}He below 1 K is almost entirely due to its multiplicity of nuclear spin alignments. Sketch a graph S vs. T for liquid and solid ^{3}He at low temperature, and estimate the temperature at which the liquid and solid have the same entropy. Discuss the shape of the solid-liquid phase boundary shown in Figure 5.13.

Problem 7.27. The argument given above for why $C_V \propto T$ does not depend on the details of the energy levels available to the fermions, so it should also apply to the model considered in Problem 7.16: a gas of fermions trapped in such a way that the energy levels are evenly spaced and nondegenerate.

(a) Show that, in this model, the number of possible system states for a given value of q is equal to the number of distinct ways of writing q as a sum of positive integers. (For example, there are three system states for $q = 3$, corresponding to the sums 3, $2 + 1$, and $1 + 1 + 1$. Note that $2 + 1$ and $1 + 2$ are not counted separately.) This combinatorial function is called the number of **unrestricted partitions** of q, denoted $p(q)$. For example, $p(3) = 3$.

(b) By enumerating the partitions explicitly, compute $p(7)$ and $p(8)$.

(c) Make a table of $p(q)$ for values of q up to 100, by either looking up the values in a mathematical reference book, or using a software package that can compute them, or writing your own program to compute them. From this table, compute the entropy, temperature, and heat capacity of this system, using the same methods as in Section 3.3. Plot the heat capacity as a function of temperature, and note that it is approximately linear.

(d) Ramanujan and Hardy (two famous mathematicians) have shown that when q is large, the number of unrestricted partitions of q is given approximately by

$$p(q) \approx \frac{e^{\pi\sqrt{2q/3}}}{4\sqrt{3}\, q}.$$

Check the accuracy of this formula for $q = 10$ and for $q = 100$. Working in this approximation, calculate the entropy, temperature, and heat capacity of this system. Express the heat capacity as a series in decreasing powers of kT/η, assuming that this ratio is large and keeping the two largest terms. Compare to the numerical results you obtained in part (c). Why is the heat capacity of this system independent of N, unlike that of the three-dimensional box of fermions discussed in the text?

The Density of States

To better visualize—and quantify—the behavior of a Fermi gas at small nonzero temperatures, I need to introduce a new concept. Let's go back to the energy integral (7.42), and change variables from n to the electron energy ϵ:

$$\epsilon = \frac{h^2}{8mL^2}n^2, \qquad n = \sqrt{\frac{8mL^2}{h^2}}\sqrt{\epsilon}, \qquad dn = \sqrt{\frac{8mL^2}{h^2}}\frac{1}{2\sqrt{\epsilon}}\,d\epsilon. \qquad (7.49)$$

With this substitution, you can show that the energy integral for a Fermi gas at zero temperature becomes

$$U = \int_0^{\epsilon_F} \epsilon \left[\frac{\pi}{2} \left(\frac{8mL^2}{h^2} \right)^{3/2} \sqrt{\epsilon} \right] d\epsilon \qquad (T = 0). \qquad (7.50)$$

The quantity in square brackets has a nice interpretation: It is the number of single-particle states per unit energy. To compute the total energy of the system we carry out a sum over all energies of the energy in question times the number of states with that energy.

The number of single-particle states per unit energy is called the **density of states**. The symbol for it is $g(\epsilon)$, and it can be written in various ways:

$$g(\epsilon) = \underbrace{\frac{\pi(8m)^{3/2}}{2h^3} V \sqrt{\epsilon}}_{} = \underbrace{\frac{3N}{2\epsilon_F^{3/2}} \sqrt{\epsilon}.}_{} \qquad (7.51)$$

The second expression is compact and handy, but perhaps rather confusing since it seems to imply that $g(\epsilon)$ depends on N, when in fact the N dependence is canceled by ϵ_F. I like the first expression better, since it shows explicitly that $g(\epsilon)$ is proportional to V and independent of N. But either way, the most important point is that $g(\epsilon)$, for a three-dimensional box of free particles, is proportional to $\sqrt{\epsilon}$. A graph of the function is a parabola opening to the right, as shown in Figure 7.13. If you want to know how many states there are between two energies ϵ_1 and ϵ_2, you just integrate this function over the desired range. The density of states is a function whose purpose in life is to be integrated.

The density-of-states idea can be applied to lots of other systems besides this one. Equation 7.51 and Figure 7.13 are for the specific case of a gas of "free" electrons, confined inside a fixed volume but not subject to any other forces. In more realistic models of metals we would want to take into account the attraction of the electrons toward the positive ions of the crystal lattice. Then the wavefunctions and their energies would be quite different, and therefore $g(\epsilon)$ would be a much more complicated function. The nice thing is that determining g is purely a problem of quantum mechanics, having nothing to do with thermal effects or temperature. And

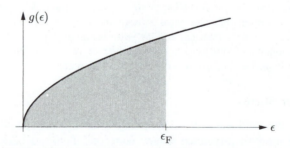

Figure 7.13. Density of states for a system of noninteracting, nonrelativistic particles in a three-dimensional box. The number of states within any energy interval is the area under the graph. For a Fermi gas at $T = 0$, all states with $\epsilon < \epsilon_F$ are occupied while all states with $\epsilon > \epsilon_F$ are unoccupied.

once you know g for some system, you can then forget about quantum mechanics and concentrate on the thermal physics.

For an electron gas at *zero* temperature, we can get the total number of electrons by just integrating the density of states up to the Fermi energy:

$$N = \int_0^{\epsilon_F} g(\epsilon)\, d\epsilon \qquad (T = 0). \qquad (7.52)$$

(For a free electron gas this is the same as equation 7.50 for the energy, but without the extra factor of ϵ.) But what if T is nonzero? Then we need to multiply $g(\epsilon)$ by the *probability* of a state with that energy being occupied, that is, by the Fermi-Dirac distribution function. Also we need to integrate all the way up to infinity, since any state could conceivably be occupied:

$$N = \int_0^{\infty} g(\epsilon)\, \bar{n}_{\mathrm{FD}}(\epsilon)\, d\epsilon = \int_0^{\infty} g(\epsilon)\, \frac{1}{e^{(\epsilon-\mu)/kT} + 1}\, d\epsilon \qquad (\text{any } T). \qquad (7.53)$$

And to get the total energy of all the electrons, just slip in an ϵ:

$$U = \int_0^{\infty} \epsilon\, g(\epsilon)\, \bar{n}_{\mathrm{FD}}(\epsilon)\, d\epsilon = \int_0^{\infty} \epsilon\, g(\epsilon)\, \frac{1}{e^{(\epsilon-\mu)/kT} + 1}\, d\epsilon \qquad (\text{any } T). \qquad (7.54)$$

Figure 7.14 shows a graph of the integrand of the N-integral (7.53), for a free electron gas at nonzero T. Instead of falling immediately to zero at $\epsilon = \epsilon_F$, the number of electrons per unit energy now drops more gradually, over a width of a few times kT. The chemical potential, μ, is the point where the probability of a state being occupied is exactly $1/2$, and it's important to note that this point is no longer the same as it was at zero temperature:

$$\mu(T) \neq \epsilon_F \qquad \text{except when } T = 0. \qquad (7.55)$$

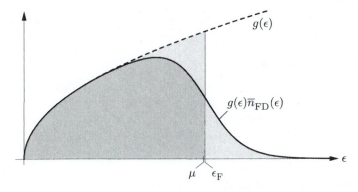

Figure 7.14. At nonzero T, the number of fermions per unit energy is given by the density of states times the Fermi-Dirac distribution. Because increasing the temperature does not change the total number of fermions, the two lightly shaded areas must be equal. Since $g(\epsilon)$ is greater above ϵ_F than below, this means that the chemical potential decreases as T increases. This graph is drawn for $T/T_F = 0.1$; at this temperature μ is about 1% less than ϵ_F.

Why not? Recall from Problem 7.12 that the Fermi-Dirac distribution function is symmetrical about $\epsilon = \mu$: The probability of a state above μ being occupied is the same as the probability of a state the same amount below μ being *unoccupied*. Now suppose that μ were to remain constant as T increases from zero. Then since the density of states is greater to the right of μ than to the left, the number of electrons we would be adding at $\epsilon > \mu$ would be greater than the number we are losing from $\epsilon < \mu$. In other words, we could increase the total number of electrons just by raising the temperature! To prevent such nonsense, the chemical potential has to decrease slightly, thus lowering all of the probabilities by a small amount.

The precise formula for $\mu(T)$ is determined implicitly by the integral for N, equation 7.53. *If* we could carry out this integral, we could take the resulting formula and solve it for $\mu(T)$ (since N is a fixed constant). Then we could plug our value of $\mu(T)$ into the energy integral (7.54), and try to carry out *that* integral to find $U(T)$ (and hence the heat capacity). The bad news is that these integrals cannot be evaluated exactly, even for the simple case of a free electron gas. The good news is that they *can* be evaluated approximately, in the limit $kT \ll \epsilon_F$. In this limit the answer for the energy integral is what I wrote in equation 7.47.

Problem 7.28. Consider a free Fermi gas in two dimensions, confined to a square area $A = L^2$.

(a) Find the Fermi energy (in terms of N and A), and show that the average energy of the particles is $\epsilon_F/2$.

(b) Derive a formula for the density of states. You should find that it is a constant, independent of ϵ.

(c) Explain how the chemical potential of this system should behave as a function of temperature, both when $kT \ll \epsilon_F$ and when T is much higher.

(d) Because $g(\epsilon)$ is a constant for this system, it is possible to carry out the integral 7.53 for the number of particles analytically. Do so, and solve for μ as a function of N. Show that the resulting formula has the expected qualitative behavior.

(e) Show that in the high-temperature limit, $kT \gg \epsilon_F$, the chemical potential of this system is the same as that of an ordinary ideal gas.

The Sommerfeld Expansion

After talking about the integrals 7.53 and 7.54 for so long, it's about time I explained how to evaluate them, to find the chemical potential and total energy of a free electron gas. The method for doing this in the limit $kT \ll \epsilon_F$ is due to Arnold Sommerfeld, and is therefore called the **Sommerfeld expansion**. None of the steps are particularly difficult, but taken as a whole the calculation is rather tricky and intricate. Hang on.

I'll start with the integral for N:

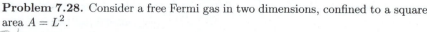

$$N = \int_0^\infty g(\epsilon)\, \overline{n}_{FD}(\epsilon)\, d\epsilon = g_0 \int_0^\infty \epsilon^{1/2}\, \overline{n}_{FD}(\epsilon)\, d\epsilon. \qquad (7.56)$$

(In the second expression I've introduced the abbreviation g_0 for the constant that multiplies $\sqrt{\epsilon}$ in equation 7.51 for the density of states.) Although this integral runs over all positive ϵ, the most interesting region is near $\epsilon = \mu$, where $\bar{n}_{FD}(\epsilon)$ falls off steeply (for $T \ll \epsilon_F$). So the first trick is to isolate this region, by integrating by parts:

$$N = \frac{2}{3}g_0 \epsilon^{3/2} \bar{n}_{FD}(\epsilon)\Big|_0^\infty + \frac{2}{3}g_0 \int_0^\infty \epsilon^{3/2}\left(-\frac{d\bar{n}_{FD}}{d\epsilon}\right) d\epsilon. \qquad (7.57)$$

The boundary term vanishes at both limits, leaving us with an integral that is much nicer, because $d\bar{n}_{FD}/d\epsilon$ is negligible everywhere except in a narrow region around $\epsilon = \mu$ (see Figure 7.15). Explicitly, we can compute

$$-\frac{d\bar{n}_{FD}}{d\epsilon} = -\frac{d}{d\epsilon}\left(e^{(\epsilon-\mu)/kT}+1\right)^{-1} = \frac{1}{kT}\frac{e^x}{(e^x+1)^2}, \qquad (7.58)$$

where $x = (\epsilon - \mu)/kT$. Thus the integral that we need to evaluate is

$$N = \frac{2}{3}g_0 \int_0^\infty \frac{1}{kT}\frac{e^x}{(e^x+1)^2}\epsilon^{3/2}\,d\epsilon = \frac{2}{3}g_0 \int_{-\mu/kT}^\infty \frac{e^x}{(e^x+1)^2}\epsilon^{3/2}\,dx, \qquad (7.59)$$

where in the last expression I've changed the integration variable to x.

Because the integrand in this expression dies out exponentially when $|\epsilon - \mu| \gg kT$, we can now make two approximations. First, we can extend the lower limit on the integral down to $-\infty$; this makes things look more symmetrical, and it's harmless because the integrand is utterly negligible at negative ϵ values anyway. Second, we can expand the function $\epsilon^{3/2}$ in a Taylor series about the point $\epsilon = \mu$, and keep only the first few terms:

$$\epsilon^{3/2} = \mu^{3/2} + (\epsilon - \mu)\frac{d}{d\epsilon}\epsilon^{3/2}\Big|_{\epsilon=\mu} + \frac{1}{2}(\epsilon-\mu)^2\frac{d^2}{d\epsilon^2}\epsilon^{3/2}\Big|_{\epsilon=\mu} + \cdots$$
$$= \mu^{3/2} + \frac{3}{2}(\epsilon-\mu)\mu^{1/2} + \frac{3}{8}(\epsilon-\mu)^2\mu^{-1/2} + \cdots. \qquad (7.60)$$

With these approximations our integral becomes

$$N = \frac{2}{3}g_0 \int_{-\infty}^\infty \frac{e^x}{(e^x+1)^2}\left[\mu^{3/2} + \frac{3}{2}xkT\mu^{1/2} + \frac{3}{8}(xkT)^2\mu^{-1/2} + \cdots\right]dx. \qquad (7.61)$$

Now, with only integer powers of x appearing, the integrals can actually be performed, term by term.

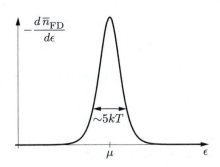

Figure 7.15. The derivative of the Fermi-Dirac distribution is negligible everywhere except within a few kT of μ.

The first term is easy:

$$\int_{-\infty}^{\infty} \frac{e^x}{(e^x + 1)^2}\, dx = \int_{-\infty}^{\infty} -\frac{d\bar{n}_{\mathrm{FD}}}{d\epsilon}\, d\epsilon = \bar{n}_{\mathrm{FD}}(-\infty) - \bar{n}_{\mathrm{FD}}(\infty) = 1 - 0 = 1. \quad (7.62)$$

The second term is also easy, since the integrand is an odd function of x:

$$\int_{-\infty}^{\infty} \frac{x\, e^x}{(e^x + 1)^2}\, dx = \int_{-\infty}^{\infty} \frac{x}{(e^x + 1)(1 + e^{-x})}\, dx = 0. \quad (7.63)$$

The third integral is the hard one. It *can* be evaluated analytically, as shown in Appendix B:

$$\int_{-\infty}^{\infty} \frac{x^2\, e^x}{(e^x + 1)^2}\, dx = \frac{\pi^2}{3}. \quad (7.64)$$

You can also look it up in tables, or evaluate it numerically.

Assembling the pieces of equation 7.61, we obtain for the number of electrons

$$\begin{aligned} N &= \frac{2}{3} g_0 \mu^{3/2} + \frac{1}{4} g_0 (kT)^2 \mu^{-1/2} \cdot \frac{\pi^2}{3} + \cdots \\ &= N\left(\frac{\mu}{\epsilon_{\mathrm{F}}}\right)^{3/2} + N \frac{\pi^2}{8} \frac{(kT)^2}{\epsilon_{\mathrm{F}}^{3/2} \mu^{1/2}} + \cdots. \end{aligned} \quad (7.65)$$

(In the second line I've plugged in $g_0 = 3N/2\epsilon_{\mathrm{F}}^{3/2}$, from equation 7.51.) Canceling the N's, we now see that $\mu/\epsilon_{\mathrm{F}}$ is approximately equal to 1, with a correction proportional to $(kT/\epsilon_{\mathrm{F}})^2$ (which we assume to be very small). Since the correction term is already quite small, we can approximate $\mu \approx \epsilon_{\mathrm{F}}$ in that term, then solve for $\mu/\epsilon_{\mathrm{F}}$ to obtain

$$\begin{aligned} \frac{\mu}{\epsilon_{\mathrm{F}}} &= \left[1 - \frac{\pi^2}{8}\left(\frac{kT}{\epsilon_{\mathrm{F}}}\right)^2 + \cdots\right]^{2/3} \\ &= 1 - \frac{\pi^2}{12}\left(\frac{kT}{\epsilon_{\mathrm{F}}}\right)^2 + \cdots. \end{aligned} \quad (7.66)$$

As predicted, the chemical potential gradually decreases as T is raised. The behavior of μ over a wide range of temperatures is shown in Figure 7.16.

The integral (7.54) for the total energy can be evaluated using exactly the same sequence of tricks. I'll leave it for you to do in Problem 7.29; the result is

$$U = \frac{3}{5} N \frac{\mu^{5/2}}{\epsilon_{\mathrm{F}}^{3/2}} + \frac{3\pi^2}{8} N \frac{(kT)^2}{\epsilon_{\mathrm{F}}} + \cdots. \quad (7.67)$$

Finally you can plug in formula 7.66 for μ and do just a bit more algebra to obtain

$$U = \frac{3}{5} N\epsilon_{\mathrm{F}} + \frac{\pi^2}{4} N \frac{(kT)^2}{\epsilon_{\mathrm{F}}} + \cdots, \quad (7.68)$$

as I claimed in equation 7.47.

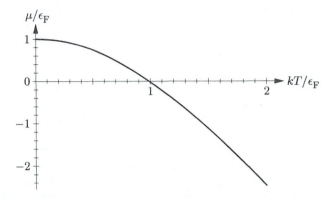

Figure 7.16. Chemical potential of a noninteracting, nonrelativistic Fermi gas in a three-dimensional box, calculated numerically as described in Problem 7.32. At low temperatures μ is given approximately by equation 7.66, while at high temperatures μ becomes negative and approaches the form for an ordinary gas obeying Boltzmann statistics.

Now admittedly, that was a lot of work just to get a factor of $\pi^2/4$ (since we had already guessed the rest by dimensional analysis). But I've presented this calculation in detail not so much because the *answer* is important, as because the *methods* are so typical of what professional physicists (and many other scientists and engineers) often do. Very few real-world problems can be solved exactly, so it's crucial for a scientist to learn when and how to make approximations. And more often than not, it's only *after* doing the hard calculation that one develops enough intuition to see how to guess most of the answer.

Problem 7.29. Carry out the Sommerfeld expansion for the energy integral (7.54), to obtain equation 7.67. Then plug in the expansion for μ to obtain the final answer, equation 7.68.

Problem 7.30. The Sommerfeld expansion is an expansion in powers of kT/ϵ_F, which is assumed to be small. In this section I kept all terms through order $(kT/\epsilon_F)^2$, omitting higher-order terms. Show at each relevant step that the term proportional to T^3 is zero, so that the next nonvanishing terms in the expansions for μ and U are proportional to T^4. (If you enjoy such things, you might try evaluating the T^4 terms, possibly with the aid of a computer algebra program.)

Problem 7.31. In Problem 7.28 you found the density of states and the chemical potential for a two-dimensional Fermi gas. Calculate the heat capacity of this gas in the limit $kT \ll \epsilon_F$. Also show that the heat capacity has the expected behavior when $kT \gg \epsilon_F$. Sketch the heat capacity as a function of temperature.

Problem 7.32. Although the integrals (7.53 and 7.54) for N and U cannot be carried out analytically for all T, it's not difficult to evaluate them numerically using a computer. This calculation has little relevance for electrons in metals (for which the limit $kT \ll \epsilon_F$ is always sufficient), but it is needed for liquid ^{3}He and for astrophysical systems like the electrons at the center of the sun.

(a) As a warm-up exercise, evaluate the N integral (7.53) for the case $kT = \epsilon_F$ and $\mu = 0$, and check that your answer is consistent with the graph shown above. (Hint: As always when solving a problem on a computer, it's best to first put everything in terms of dimensionless variables. So let $t = kT/\epsilon_F$, $c = \mu/\epsilon_F$, and $x = \epsilon/\epsilon_F$. Rewrite everything in terms of these variables, and *then* put it on the computer.)

(b) The next step is to vary μ, holding T fixed, until the integral works out to the desired value, N. Do this for values of kT/ϵ_F ranging from 0.1 up to 2, and plot the results to reproduce Figure 7.16. (It's probably not a good idea to try to use numerical methods when kT/ϵ_F is much smaller than 0.1, since you can start getting overflow errors from exponentiating large numbers. But this is the region where we've already solved the problem analytically.)

(c) Plug your calculated values of μ into the energy integral (7.54), and evaluate that integral numerically to obtain the energy as a function of temperature for kT up to $2\epsilon_F$. Plot the results, and evaluate the slope to obtain the heat capacity. Check that the heat capacity has the expected behavior at both low and high temperatures.

Problem 7.33. When the attractive forces of the ions in a crystal are taken into account, the allowed electron energies are no longer given by the simple formula 7.36; instead, the allowed energies are grouped into **bands**, separated by **gaps** where there are no allowed energies. In a **conductor** the Fermi energy lies within one of the bands; in this section we have treated the electrons in this band as "free" particles confined to a fixed volume. In an **insulator**, on the other hand, the Fermi energy lies within a gap, so that at $T = 0$ the band below the gap is completely occupied while the band above the gap is unoccupied. Because there are no empty states close in energy to those that are occupied, the electrons are "stuck in place" and the material does not conduct electricity. A **semiconductor** is an insulator in which the gap is narrow enough for a few electrons to jump across it at room temperature. Figure 7.17 shows the density of states in the vicinity of the Fermi energy for an idealized semiconductor, and defines some terminology and notation to be used in this problem.

(a) As a first approximation, let us model the density of states near the bottom of the conduction band using the same function as for a free Fermi gas, with an appropriate zero-point: $g(\epsilon) = g_0\sqrt{\epsilon - \epsilon_c}$, where g_0 is the same constant as in equation 7.51. Let us also model the density of states near the top

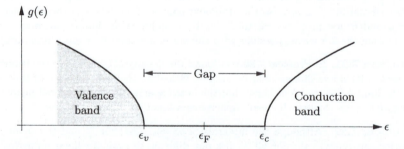

Figure 7.17. The periodic potential of a crystal lattice results in a density-of-states function consisting of "bands" (with many states) and "gaps" (with no states). For an insulator or a semiconductor, the Fermi energy lies in the middle of a gap so that at $T = 0$, the "valence band" is completely full while the "conduction band" is completely empty.

of the valence band as a mirror image of this function. Explain why, in this approximation, the chemical potential must always lie precisely in the middle of the gap, regardless of temperature.

(b) Normally the width of the gap is much greater than kT. Working in this limit, derive an expression for the number of conduction electrons per unit volume, in terms of the temperature and the width of the gap.

(c) For silicon near room temperature, the gap between the valence and conduction bands is approximately 1.11 eV. Roughly how many conduction electrons are there in a cubic centimeter of silicon at room temperature? How does this compare to the number of conduction electrons in a similar amount of copper?

(d) Explain why a semiconductor conducts electricity much better at higher temperatures. Back up your explanation with some numbers. (Ordinary conductors like copper, on the other hand, conduct better at *low* temperatures.)

(e) Very roughly, how wide would the gap between the valence and conduction bands have to be in order to consider a material an insulator rather than a semiconductor?

Problem 7.34. In a real semiconductor, the density of states at the bottom of the conduction band will differ from the model used in the previous problem by a numerical factor, which can be small or large depending on the material. Let us therefore write for the conduction band $g(\epsilon) = g_{0c}\sqrt{\epsilon - \epsilon_c}$, where g_{0c} is a new normalization constant that differs from g_0 by some fudge factor. Similarly, write $g(\epsilon)$ at the top of the valence band in terms of a new normalization constant g_{0v}.

(a) Explain why, if $g_{0v} \neq g_{0c}$, the chemical potential will now vary with temperature. When will it increase, and when will it decrease?

(b) Write down an expression for the number of conduction electrons, in terms of T, μ, ϵ_c, and g_{0c}. Simplify this expression as much as possible, assuming $\epsilon_c - \mu \gg kT$.

(c) An empty state in the valence band is called a **hole**. In analogy to part (b), write down an expression for the number of holes, and simplify it in the limit $\mu - \epsilon_v \gg kT$.

(d) Combine the results of parts (b) and (c) to find an expression for the chemical potential as a function of temperature.

(e) For silicon, $g_{0c}/g_0 = 1.09$ and $g_{0v}/g_0 = 0.44$.[*] Calculate the shift in μ for silicon at room temperature.

Problem 7.35. The previous two problems dealt with pure semiconductors, also called **intrinsic** semiconductors. Useful semiconductor devices are instead made from **doped** semiconductors, which contain substantial numbers of impurity atoms. One example of a doped semiconductor was treated in Problem 7.5. Let us now consider that system again. (Note that in Problem 7.5 we measured all energies relative to the bottom of the conduction band, ϵ_c. We also neglected the distinction between g_0 and g_{0c}; this simplification happens to be ok for conduction electrons in silicon.)

---

[*]These values can be calculated from the "effective masses" of electrons and holes. See, for example, S. M. Sze, *Physics of Semiconductor Devices*, second edition (Wiley, New York, 1981).

(a) Calculate and plot the chemical potential as a function of temperature, for silicon doped with 10^{17} phosphorus atoms per cm^3 (as in Problem 7.5). Continue to assume that the conduction electrons can be treated as an ordinary ideal gas.

(b) Discuss whether it is legitimate to assume for this system that the conduction electrons can be treated as an ordinary ideal gas, as opposed to a Fermi gas. Give some numerical examples.

(c) Estimate the temperature at which the number of valence electrons excited to the conduction band would become comparable to the number of conduction electrons from donor impurities. Which source of conduction electrons is more important at room temperature?

Problem 7.36. Most spin-1/2 fermions, including electrons and helium-3 atoms, have nonzero magnetic moments. A gas of such particles is therefore paramagnetic. Consider, for example, a gas of free electrons, confined inside a three-dimensional box. The z component of the magnetic moment of each electron is $\pm\mu_B$. In the presence of a magnetic field B pointing in the z direction, each "up" state acquires an additional energy of $-\mu_B B$, while each "down" state acquires an additional energy of $+\mu_B B$.

(a) Explain why you would expect the magnetization of a degenerate electron gas to be substantially less than that of the electronic paramagnets studied in Chapters 3 and 6, for a given number of particles at a given field strength.

(b) Write down a formula for the density of states of this system in the presence of a magnetic field B, and interpret your formula graphically.

(c) The magnetization of this system is $\mu_B(N_\uparrow - N_\downarrow)$, where $N_\uparrow$ and $N_\downarrow$ are the numbers of electrons with up and down magnetic moments, respectively. Find a formula for the magnetization of this system at $T = 0$, in terms of N, μ_B, B, and the Fermi energy.

(d) Find the first temperature-dependent correction to your answer to part (c), in the limit $T \ll T_F$. You may assume that $\mu_B B \ll kT$; this implies that the presence of the magnetic field has negligible effect on the chemical potential μ. (To avoid confusing μ_B with μ, I suggest using an abbreviation such as δ for the quantity $\mu_B B$.)

7.4 Blackbody Radiation

As a next application of quantum statistics, I'd like to consider the electromagnetic radiation inside some "box" (like an oven or kiln) at a given temperature. First let me discuss what we would expect of such a system in classical (i.e., non-quantum) physics.

The Ultraviolet Catastrophe

In classical physics, we treat electromagnetic radiation as a continuous "field" that permeates all space. Inside a box, we can think of this field as a combination of various standing-wave patterns, as shown in Figure 7.18. Each standing-wave pattern behaves as a harmonic oscillator with frequency $f = c/\lambda$. Like a mechanical oscillator, each electromagnetic standing wave has two degrees of freedom,

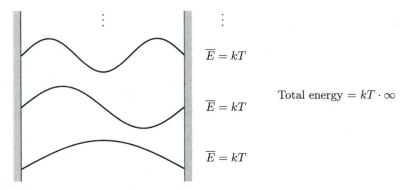

$\overline{E} = kT$

Total energy $= kT \cdot \infty$

$\overline{E} = kT$

$\overline{E} = kT$

Figure 7.18. We can analyze the electromagnetic field in a box as a superposition of standing-wave modes of various wavelengths. Each mode is a harmonic oscillator with some well-defined frequency. Classically, each oscillator should have an average energy of kT. Since the total number of modes is infinite, so is the total energy in the box.

with an average thermal energy of $2 \cdot \frac{1}{2}kT$. Since the total number of oscillators in the electromagnetic field is infinite, the total thermal energy should also be infinite. Experimentally, though, you're not blasted with an infinite amount of electromagnetic radiation every time you open the oven door to check the cookies. This disagreement between classical theory and experiment is called the **ultraviolet catastrophe** (because the infinite energy would come mostly from very short wavelengths).

The Planck Distribution

The solution to the ultraviolet catastrophe comes from quantum mechanics. (Historically, the ultraviolet catastrophe led to the *birth* of quantum mechanics.) In quantum mechanics, a harmonic oscillator can't have just any amount of energy; its allowed energy levels are

$$E_n = 0, \ hf, \ 2hf, \ \dots. \tag{7.69}$$

(As usual I'm measuring all energies relative to the ground-state energy. See Appendix A for more discussion of this point.) The partition function for a single oscillator is therefore

$$Z = 1 + e^{-\beta hf} + e^{-2\beta hf} + \cdots$$
$$= \frac{1}{1 - e^{-\beta hf}}, \tag{7.70}$$

and the average energy is

$$\overline{E} = -\frac{1}{Z}\frac{\partial Z}{\partial \beta} = \frac{hf}{e^{hf/kT} - 1}. \tag{7.71}$$

If we think of the energy as coming in "units" of hf, then the average *number* of units of energy in the oscillator is

$$\overline{n}_{\mathrm{Pl}} = \frac{1}{e^{hf/kT} - 1}. \tag{7.72}$$

This formula is called the **Planck distribution** (after Max Planck).

According to the Planck distribution, short-wavelength modes of the electromagnetic field, with $hf \gg kT$, are *exponentially* suppressed: They are "frozen out," and might as well not exist. Thus the total number of electromagnetic oscillators that effectively contribute to the energy inside the box is finite, and the ultraviolet catastrophe does not occur. Notice that this solution *requires* that the oscillator energies be quantized: It is the size of the energy units, compared to kT, that provides the exponential suppression factor.

Photons

"Units" of energy in the electromagnetic field can also be thought of as *particles*, called **photons**. They are bosons, so the number of them in any "mode" or wave pattern of the field ought to be given by the Bose-Einstein distribution:

$$\overline{n}_{\text{BE}} = \frac{1}{e^{(\epsilon - \mu)/kT} - 1}. \tag{7.73}$$

Here ϵ is the energy of each particle in the mode, that is, $\epsilon = hf$. Comparison with equation 7.72 therefore requires

$$\mu = 0 \qquad \text{for photons.} \tag{7.74}$$

But why should this be true? I'll give you two reasons, both based on the fact that photons can be created or destroyed in any quantity; their total number is not conserved.

First consider the Helmholtz free energy, which must attain the minimum possible value at equilibrium with T and V held fixed. In a system of photons, the number N of particles is not constrained, but rather takes whatever value will minimize F. If N then changes infinitesimally, F should be unchanged:

$$\left(\frac{\partial F}{\partial N} \right)_{T,V} = 0 \qquad \text{(at equilibrium).} \tag{7.75}$$

But this partial derivative is precisely equal to the chemical potential.

A second argument makes use of the condition for chemical equilibrium derived in Section 5.6. Consider a typical reaction in which a photon (γ) is created or absorbed by an electron:

$$e \longleftrightarrow e + \gamma. \tag{7.76}$$

As we saw in Section 5.6, the equilibrium condition for such a reaction is the same as the reaction equation, with the name of each species replaced by its chemical potential. In this case,

$$\mu_e = \mu_e + \mu_\gamma \qquad \text{(at equilibrium).} \tag{7.77}$$

In other words, the chemical potential for photons is zero.

By either argument, the chemical potential for a "gas" of photons inside a box at fixed temperature is zero, so the Bose-Einstein distribution reduces to the Planck distribution, as required.

Summing over Modes

The Planck distribution tells us how many photons are in any single "mode" (or "single-particle state") of the electromagnetic field. Next we might want to know the *total* number of photons inside the box, and also the total *energy* of all the photons. To compute either one, we have to sum over all possible states, just as we did for electrons. I'll compute the total energy, and let you compute the total number of photons in Problem 7.44.

Let's start in one dimension, with a "box" of length L. The allowed wavelengths and momenta are the same for photons as for any other particles:

$$\lambda = \frac{2L}{n}; \qquad p = \frac{hn}{2L}. \tag{7.78}$$

(Here n is a positive integer that labels which mode we're talking about, not to be confused with $\bar{n}_{\text{Pl}}$, the average number of photons in a given mode.) Photons, however, are ultrarelativistic particles, so their energies are given by

$$\epsilon = pc = \frac{hcn}{2L} \tag{7.79}$$

instead of $\epsilon = p^2/2m$. (You can also derive this result straight from the Einstein relation $\epsilon = hf$ between a photon's energy and its frequency. For light, $f = c/\lambda$, so $\epsilon = hc/\lambda = hcn/2L$.)

In three dimensions, momentum becomes a vector, with each component given by $h/2L$ times some integer. The energy is c times the *magnitude* of the momentum vector:

$$\epsilon = c\sqrt{p_x^2 + p_y^2 + p_z^2} = \frac{hc}{2L}\sqrt{n_x^2 + n_y^2 + n_z^2} = \frac{hcn}{2L}, \tag{7.80}$$

where in the last expression I'm using n for the magnitude of the $\vec{n}$ vector, as in Section 7.3.

Now the average energy in any particular mode is equal to ϵ times the occupancy of that mode, and the occupancy is given by the Planck distribution. To get the total energy in all modes, we sum over n_x, n_y, and n_z. We also need to slip in a factor of 2, since each wave shape can hold photons with two independent polarizations. So the total energy is

$$U = 2 \sum_{n_x} \sum_{n_y} \sum_{n_z} \epsilon\,\bar{n}_{\text{Pl}}(\epsilon) = \sum_{n_x, n_y, n_z} \frac{hcn}{L} \frac{1}{e^{hcn/2LkT} - 1}. \tag{7.81}$$

As in Section 7.3, we can convert the sums to integrals and carry out the integration in spherical coordinates (see Figure 7.11). This time, however, the upper limit on the integration over n is infinity:

$$U = \int_0^\infty dn \int_0^{\pi/2} d\theta \int_0^{\pi/2} d\phi \, n^2 \sin\theta \, \frac{hcn}{L} \frac{1}{e^{hcn/2LkT} - 1}. \tag{7.82}$$

Again the angular integrals give $\pi/2$, the surface area of an eighth of a unit sphere.

The Planck Spectrum

The integral over n looks a little nicer if we change variables to the photon energy, $\epsilon = hcn/2L$. We then get an overall factor of $L^3 = V$, so the total energy per unit volume is

$$\frac{U}{V} = \int_0^\infty \frac{8\pi \epsilon^3/(hc)^3}{e^{\epsilon/kT} - 1} \, d\epsilon. \tag{7.83}$$

Here the integrand has a nice interpretation: It is the energy density per unit photon energy, or the **spectrum** of the photons:

$$u(\epsilon) = \frac{8\pi}{(hc)^3} \frac{\epsilon^3}{e^{\epsilon/kT} - 1}. \tag{7.84}$$

This function, first derived by Planck, gives the relative intensity of the radiation as a function of photon energy (or as a function of frequency, if you change variables again to $f = \epsilon/h$). If you integrate $u(\epsilon)$ from ϵ_1 to ϵ_2, you get the energy per unit volume within that range of photon energies.

To actually evaluate the integral over ϵ, it's convenient to change variables again, to $x = \epsilon/kT$. Then equation 7.83 becomes

$$\frac{U}{V} = \frac{8\pi(kT)^4}{(hc)^3} \int_0^\infty \frac{x^3}{e^x - 1} \, dx. \tag{7.85}$$

The integrand is still proportional to the Planck spectrum; this function is plotted in Figure 7.19. The spectrum peaks at $x = 2.82$, or $\epsilon = 2.82kT$. Not surprisingly, higher temperatures tend to give higher photon energies. (This fact is called

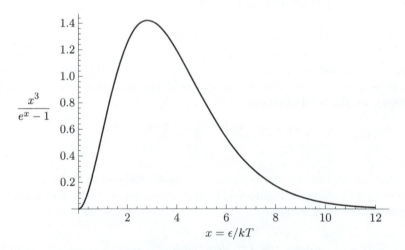

Figure 7.19. The Planck spectrum, plotted in terms of the dimensionless variable $x = \epsilon/kT = hf/kT$. The area under any portion of this graph, multiplied by $8\pi(kT)^4/(hc)^3$, equals the energy density of electromagnetic radiation within the corresponding frequency (or photon energy) range; see equation 7.85.

Wien's law.) You can measure the temperature inside an oven (or more likely, a kiln) by letting a bit of the radiation out and looking at its color. For instance, a typical clay-firing temperature of 1500 K gives a spectrum that peaks at $\epsilon = 0.36$ eV, in the near infrared. (Visible-light photons have higher energies, in the range of about 2–3 eV.)

Problem 7.37. Prove that the peak of the Planck spectrum is at $x = 2.82$.

Problem 7.38. It's not obvious from Figure 7.19 how the Planck spectrum changes as a function of temperature. To examine the temperature dependence, make a quantitative plot of the function $u(\epsilon)$ for $T = 3000$ K and $T = 6000$ K (both on the same graph). Label the horizontal axis in electron-volts.

Problem 7.39. Change variables in equation 7.83 to $\lambda = hc/\epsilon$, and thus derive a formula for the photon spectrum as a function of wavelength. Plot this spectrum, and find a numerical formula for the wavelength where the spectrum peaks, in terms of hc/kT. Explain why the peak does not occur at $hc/(2.82kT)$.

Problem 7.40. Starting from equation 7.83, derive a formula for the density of states of a photon gas (or any other gas of ultrarelativistic particles having two polarization states). Sketch this function.

Problem 7.41. Consider any two internal states, s_1 and s_2, of an atom. Let s_2 be the higher-energy state, so that $E(s_2) - E(s_1) = \epsilon$ for some positive constant ϵ. If the atom is currently in state s_2, then there is a certain probability per unit time for it to spontaneously decay down to state s_1, emitting a photon with energy ϵ. This probability per unit time is called the **Einstein A coefficient**:

$$A = \text{probability of spontaneous decay per unit time.}$$

On the other hand, if the atom is currently in state s_1 and we shine light on it with frequency $f = \epsilon/h$, then there is a chance that it will absorb a photon, jumping into state s_2. The probability for this to occur is proportional not only to the amount of time elapsed but also to the intensity of the light, or more precisely, the energy density of the light per unit frequency, $u(f)$. (This is the function which, when integrated over any frequency interval, gives the energy per unit volume within that frequency interval. For our atomic transition, all that matters is the value of $u(f)$ at $f = \epsilon/h$.) The probability of absorbing a photon, per unit time per unit intensity, is called the **Einstein B coefficient**:

$$B = \frac{\text{probability of absorption per unit time}}{u(f)}.$$

Finally, it is also possible for the atom to make a *stimulated* transition from s_2 down to s_1, again with a probability that is proportional to the intensity of light at frequency f. (Stimulated emission is the fundamental mechanism of the laser: Light Amplification by Stimulated Emission of Radiation.) Thus we define a third coefficient, B', that is analogous to B:

$$B' = \frac{\text{probability of stimulated emission per unit time}}{u(f)}.$$

As Einstein showed in 1917, knowing any one of these three coefficients is as good as knowing them all.

(a) Imagine a collection of many of these atoms, such that N_1 of them are in state s_1 and N_2 are in state s_2. Write down a formula for dN_1/dt in terms of A, B, B', N_1, N_2, and $u(f)$.

(b) Einstein's trick is to imagine that these atoms are bathed in *thermal* radiation, so that $u(f)$ is the Planck spectral function. At equilibrium, N_1 and N_2 should be constant in time, with their ratio given by a simple Boltzmann factor. Show, then, that the coefficients must be related by

$$B' = B \qquad \text{and} \qquad \frac{A}{B} = \frac{8\pi h f^3}{c^3}.$$

Total Energy

Enough about the spectrum—what about the total electromagnetic energy inside the box? Equation 7.85 is essentially the final answer, except for the integral over x, which is just some dimensionless number. From Figure 7.19 you can estimate that this number is about 6.5; a beautiful but very tricky calculation (see Appendix B) gives it exactly as $\pi^4/15$. Therefore the total energy density, summing over all frequencies, is

$$\frac{U}{V} = \frac{8\pi^5(kT)^4}{15(hc)^3}. \tag{7.86}$$

The most important feature of this result is its dependence on the *fourth* power of the temperature. If you double the temperature of your oven, the amount of electromagnetic energy inside increases by a factor of $2^4 = 16$.

Numerically, the total electromagnetic energy inside a typical oven is quite small. At cookie-baking temperature, 375°F or about 460 K, the energy per unit volume comes out to 3.5×10^{-5} J/m^3. This is *tiny* compared to the thermal energy of the air inside the oven.

Formula 7.86 may look complicated, but you could have guessed the answer, aside from the numerical coefficient, by dimensional analysis. The average energy per photon must be something of order kT, so the total energy must be proportional to NkT, where N is the total number of photons. Since N is extensive, it must be proportional to the volume V of the container; thus the total energy must be of the form

$$U = (\text{constant}) \cdot \frac{V\,kT}{\ell^3}, \tag{7.87}$$

where ℓ is something with units of length. (If you want, you can pretend that each photon occupies a volume of ℓ^3.) But the only relevant length in the problem is the typical de Broglie wavelength of the photons, $\lambda = h/p = hc/E \propto hc/kT$. Plugging this in for ℓ yields equation 7.86, aside from the factor of $8\pi^5/15$.

Problem 7.42. Consider the electromagnetic radiation inside a kiln, with a volume of 1 m^3 and a temperature of 1500 K.

(a) What is the total energy of this radiation?

(b) Sketch the spectrum of the radiation as a function of photon energy.

(c) What fraction of all the energy is in the *visible* portion of the spectrum, with wavelengths between 400 nm and 700 nm?

Problem 7.43. At the surface of the sun, the temperature is approximately 5800 K.

    (a) How much energy is contained in the electromagnetic radiation filling a cubic meter of space at the sun's surface?

    (b) Sketch the spectrum of this radiation as a function of photon energy. Mark the region of the spectrum that corresponds to visible wavelengths, between 400 nm and 700 nm.

    (c) What fraction of the energy is in the visible portion of the spectrum? (Hint: Do the integral numerically.)

Entropy of a Photon Gas

Besides the total energy of a photon gas, we might want to know a number of other quantities, for instance, the total number of photons present or the total entropy. These two quantities turn out to be equal, up to a constant factor. Let me now compute the entropy.

The easiest way to compute the entropy is from the heat capacity. For a box of thermal photons with volume V,

$$C_V = \left(\frac{\partial U}{\partial T}\right)_V = 4aT^3, \tag{7.88}$$

where a is an abbreviation for $8\pi^5 k^4 V/15(hc)^3$. This expression is good all the way down to absolute zero, so we can integrate it to find the absolute entropy. Introducing the symbol T' for the integration variable,

$$S(T) = \int_0^T \frac{C_V(T')}{T'}\, dT' = 4a \int_0^T (T')^2\, dT' = \frac{4}{3}aT^3 = \frac{32\pi^5}{45} V \left(\frac{kT}{hc}\right)^3 k. \tag{7.89}$$

The total *number* of photons is given by the same formula, with a different numerical coefficient, and without the final k (see Problem 7.44).

The Cosmic Background Radiation

The grandest example of a photon gas is the radiation that fills the entire observable universe, with an almost perfect thermal spectrum at a temperature of 2.73 K. Interpreting this temperature is a bit tricky, however: There is no longer any mechanism to keep the photons in thermal equilibrium with each other or with anything else; the radiation is instead thought to be left over from a time when the universe was filled with ionized gas that interacted strongly with electromagnetic radiation. At that time, the temperature was more like 3000 K; since then the universe has expanded a thousandfold in all directions, and the photon wavelengths have been stretched out accordingly (Doppler-shifted, if you care to think of it this way), preserving the shape of the spectrum but shifting the effective temperature down to 2.73 K.

The photons making up the cosmic background radiation have rather low energies: The spectrum peaks at $\epsilon = 2.82kT = 6.6 \times 10^{-4}$ eV. This corresponds to wavelengths of about a millimeter, in the far infrared. These wavelengths don't penetrate our atmosphere, but the long-wavelength tail of the spectrum, in the microwave region of a few centimeters, can be detected without much difficulty. It was discovered accidentally by radio astronomers in 1965. Figure 7.20 shows a more recent set of measurements over a wide range of wavelengths, made from above earth's atmosphere by the *Cosmic Background Explorer* satellite.

According to formula 7.86, the total energy in the cosmic background radiation is only 0.26 MeV/m^3. This is to be contrasted with the average energy density of ordinary matter, which on cosmic scales is of the order of a proton per cubic meter or 1000 MeV/m^3. (Ironically, the density of the exotic background radiation is known to three significant figures, while the average density of ordinary matter is uncertain by nearly a factor of 10.) On the other hand, the *entropy* of the background radiation is much greater than that of ordinary matter: According to equation 7.89, every cubic meter of space contains a photon entropy of $(2.89 \times 10^9)k$, nearly three billion "units" of entropy. The entropy of ordinary matter is not easy to calculate precisely, but if we pretend that this matter is an ordinary ideal gas we can estimate that its entropy is Nk times some small number, in other words, only a few k per cubic meter.

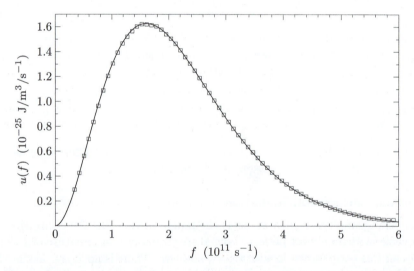

Figure 7.20. Spectrum of the cosmic background radiation, as measured by the *Cosmic Background Explorer* satellite. Plotted vertically is the energy density per unit frequency, in SI units. Note that a frequency of 3×10^{11} s^{-1} corresponds to a wavelength of $\lambda = c/f = 1.0$ mm. Each square represents a measured data point. The point-by-point uncertainties are too small to show up on this scale; the size of the squares instead represents a liberal estimate of the uncertainty due to systematic effects. The solid curve is the theoretical Planck spectrum, with the temperature adjusted to 2.735 K to give the best fit. From J. C. Mather et al., *Astrophysical Journal Letters* **354**, L37 (1990); adapted courtesy of NASA/GSFC and the COBE Science Working Group. Subsequent measurements from this experiment and others now give a best-fit temperature of 2.728 ± 0.002 K.

Problem 7.44. Number of photons in a photon gas.

(a) Show that the number of photons in equilibrium in a box of volume V at temperature T is

$$N = 8\pi V \left(\frac{kT}{hc}\right)^3 \int_0^\infty \frac{x^2}{e^x - 1}\, dx.$$

The integral cannot be done analytically; either look it up in a table or evaluate it numerically.

(b) How does this result compare to the formula derived in the text for the entropy of a photon gas? (What is the entropy per photon, in terms of k?)

(c) Calculate the number of photons per cubic meter at the following temperatures: 300 K; 1500 K (a typical kiln); 2.73 K (the cosmic background radiation).

Problem 7.45. Use the formula $P = -(\partial U / \partial V)_{S,N}$ to show that the pressure of a photon gas is $1/3$ times the energy density (U/V). Compute the pressure exerted by the radiation inside a kiln at 1500 K, and compare to the ordinary gas pressure exerted by the air. Then compute the pressure of the radiation at the center of the sun, where the temperature is 15 million K. Compare to the gas pressure of the ionized hydrogen, whose density is approximately 10^5 kg/m^3.

Problem 7.46. Sometimes it is useful to know the free energy of a photon gas.

(a) Calculate the (Helmholtz) free energy directly from the definition $F = U - TS$. (Express the answer in terms of T and V.)

(b) Check the formula $S = -(\partial F / \partial T)_V$ for this system.

(c) Differentiate F with respect to V to obtain the pressure of a photon gas. Check that your result agrees with that of the previous problem.

(d) A more interesting way to calculate F is to apply the formula $F = -kT \ln Z$ separately to each mode (that is, each effective oscillator), then sum over all modes. Carry out this calculation, to obtain

$$F = 8\pi V \frac{(kT)^4}{(hc)^3} \int_0^\infty x^2 \ln(1 - e^{-x})\, dx.$$

Integrate by parts, and check that your answer agrees with part (a).

Problem 7.47. In the text I claimed that the universe was filled with ionized gas until its temperature cooled to about 3000 K. To see why, assume that the universe contains only photons and hydrogen atoms, with a constant ratio of 10^9 photons per hydrogen atom. Calculate and plot the fraction of atoms that were ionized as a function of temperature, for temperatures between 0 and 6000 K. How does the result change if the ratio of photons to atoms is 10^8, or 10^{10}? (Hint: Write everything in terms of dimensionless variables such as $t = kT/I$, where I is the ionization energy of hydrogen.)

Problem 7.48. In addition to the cosmic background radiation of photons, the universe is thought to be permeated with a background radiation of neutrinos (ν) and antineutrinos ($\bar{\nu}$), currently at an effective temperature of 1.95 K. There are three species of neutrinos, each of which has an antiparticle, with only one allowed polarization state for each particle or antiparticle. For parts (a) through (c) below, assume that all three species are exactly massless.

(a) It is reasonable to assume that for each species, the concentration of neutrinos equals the concentration of antineutrinos, so that their chemical potentials are equal: $\mu_\nu = \mu_{\bar\nu}$. Furthermore, neutrinos and antineutrinos can be produced and annihilated in pairs by the reaction

$$\nu + \bar\nu \leftrightarrow 2\gamma$$

(where γ is a photon). Assuming that this reaction is at equilibrium (as it would have been in the very early universe), prove that $\mu = 0$ for both the neutrinos and the antineutrinos.

(b) If neutrinos are massless, they must be highly relativistic. They are also fermions: They obey the exclusion principle. Use these facts to derive a formula for the total energy density (energy per unit volume) of the neutrino-antineutrino background radiation. (Hint: There are very few differences between this "neutrino gas" and a photon gas. Antiparticles still have positive energy, so to include the antineutrinos all you need is a factor of 2. To account for the three species, just multiply by 3.) To evaluate the final integral, first change to a dimensionless variable and then use a computer or look it up in a table or consult Appendix B.

(c) Derive a formula for the *number* of neutrinos per unit volume in the neutrino background radiation. Evaluate your result numerically for the present neutrino temperature of 1.95 K.

(d) It is possible that neutrinos have very small, but nonzero, masses. This wouldn't have affected the production of neutrinos in the early universe, when mc^2 would have been negligible compared to typical thermal energies. But today, the total mass of all the background neutrinos could be significant. Suppose, then, that just one of the three species of neutrinos (and the corresponding antineutrino) has a nonzero mass m. What would mc^2 have to be (in eV), in order for the total mass of neutrinos in the universe to be comparable to the total mass of ordinary matter?

Problem 7.49. For a brief time in the early universe, the temperature was hot enough to produce large numbers of electron-positron pairs. These pairs then constituted a third type of "background radiation," in addition to the photons and neutrinos (see Figure 7.21). Like neutrinos, electrons and positrons are fermions. Unlike neutrinos, electrons and positrons are known to be massive (each with the same mass), and each has two independent polarization states. During the time period of interest the densities of electrons and positrons were approximately equal, so it is a good approximation to set the chemical potentials equal to zero as in

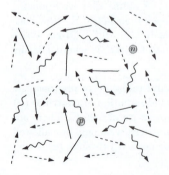

Figure 7.21. When the temperature was greater than the electron mass times c^2/k, the universe was filled with three types of radiation: electrons and positrons (solid arrows); neutrinos (dashed); and photons (wavy). Bathed in this radiation were a few protons and neutrons, roughly one for every billion radiation particles.

the previous problem. Recall from special relativity that the energy of a massive particle is $\epsilon = \sqrt{(pc)^2 + (mc^2)^2}$.

(a) Show that the energy density of electrons and positrons at temperature T is given by

$$\frac{U}{V} = \frac{16\pi(kT)^4}{(hc)^3} u(T),$$

where

$$u(T) = \int_0^\infty \frac{x^2\sqrt{x^2 + (mc^2/kT)^2}}{e^{\sqrt{x^2+(mc^2/kT)^2}} + 1}\, dx.$$

(b) Show that $u(T)$ goes to zero when $kT \ll mc^2$, and explain why this is a reasonable result.

(c) Evaluate $u(T)$ in the limit $kT \gg mc^2$, and compare to the result of the previous problem for the neutrino radiation.

(d) Use a computer to calculate and plot $u(T)$ at intermediate temperatures.

(e) Use the method of Problem 7.46, part (d), to show that the free energy density of the electron-positron radiation is

$$\frac{F}{V} = -\frac{16\pi(kT)^4}{(hc)^3} f(T),$$

where

$$f(T) = \int_0^\infty x^2 \ln\left(1 + e^{-\sqrt{x^2+(mc^2/kT)^2}}\right) dx.$$

Evaluate $f(T)$ in both limits, and use a computer to calculate and plot $f(T)$ at intermediate temperatures.

(f) Write the entropy of the electron-positron radiation in terms of the functions $u(T)$ and $f(T)$. Evaluate the entropy explicitly in the high-T limit.

Problem 7.50. The results of the previous problem can be used to explain why the current temperature of the cosmic neutrino background (Problem 7.48) is 1.95 K rather than 2.73 K. Originally the temperatures of the photons and the neutrinos would have been equal, but as the universe expanded and cooled, the interactions of neutrinos with other particles soon became negligibly weak. Shortly thereafter, the temperature dropped to the point where kT/c^2 was no longer much greater than the electron mass. As the electrons and positrons disappeared during the next few minutes, they "heated" the photon radiation but not the neutrino radiation.

(a) Imagine that the universe has some finite total volume V, but that V is increasing with time. Write down a formula for the total entropy of the electrons, positrons, and photons as a function of V and T, using the auxilliary functions $u(T)$ and $f(T)$ introduced in the previous problem. Argue that this total entropy would have been conserved in the early universe, assuming that no other species of particles interacted with these.

(b) The entropy of the neutrino radiation would have been separately conserved during this time period, because the neutrinos were unable to interact with anything. Use this fact to show that the neutrino temperature T_ν and the photon temperature T are related by

$$\left(\frac{T}{T_\nu}\right)^3 \left[\frac{2\pi^4}{45} + u(T) + f(T)\right] = \text{constant}$$

as the universe expands and cools. Evaluate the constant by assuming that $T = T_\nu$ when the temperatures are very high.

(c) Calculate the ratio T/T_ν in the limit of low temperature, to confirm that the present neutrino temperature should be 1.95 K.

(d) Use a computer to plot the ratio T/T_ν as a function of T, for kT/mc^2 ranging from 0 to 3.*

Photons Escaping through a Hole

So far in this section I have analyzed the gas of photons *inside* an oven or any other box in thermal equilibrium. Eventually, though, we'd like to understand the photons *emitted* by a hot object. To begin, let's ask what happens if you start with a photon gas in a box, then poke a hole in the box to let some photons out (see Figure 7.22).

All photons travel at the same speed (in vacuum), regardless of their wavelengths. So low-energy photons will escape through the hole with the same probability as high-energy photons, and thus the spectrum of the photons coming out will look the same as the spectrum of the photons inside. What's harder to figure out is the total *amount* of radiation that escapes; the calculation doesn't involve much physics, but the geometry is rather tricky.

The photons that escape now, during a time interval dt, were once pointed at the hole from somewhere within a hemispherical shell, as shown in Figure 7.23. The radius R of the shell depends on how long ago we're looking, while the thickness of the shell is $c\,dt$. I'll use spherical coordinates to label various points on the shell, as shown. The angle θ ranges from 0, at the left end of the shell, to $\pi/2$, at the extreme edges on the right. There's also an azimuthal angle ϕ, not shown, which ranges from 0 to 2π as you go from the top edge of the shell into the page, down

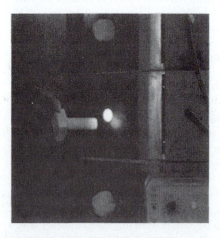

Figure 7.22. When you open a hole in a container filled with radiation (here a kiln), the spectrum of the light that escapes is the same as the spectrum of the light inside. The total amount of energy that escapes is proportional to the size of the hole and to the amount of time that passes.

*Now that you've finished this problem, you'll find it relatively easy to work out the dynamics of the early universe, to determine *when* all this happened. The basic idea is to assume that the universe is expanding at "escape velocity." Everything you need to know is in Weinberg (1977).

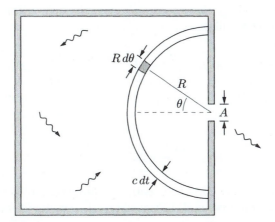

Figure 7.23. The photons that escape now were once somewhere within a hemispherical shell inside the box. From a given point in this shell, the probability of escape depends on the distance from the hole and the angle θ.

to the bottom, out of the page, and back to the top.

Now consider the shaded chunk of the shell shown Figure 7.23. Its volume is

$$\text{volume of chunk} = (R\,d\theta) \times (R\sin\theta\,d\phi) \times (c\,dt). \tag{7.90}$$

(The depth of the chunk, perpendicular to the page, is $R\sin\theta\,d\phi$, since $R\sin\theta$ is the radius of a ring of constant θ swept out as ϕ ranges from 0 to 2π.) The energy density of the photons within this chunk is given by equation 7.86:

$$\frac{U}{V} = \frac{8\pi^5}{15}\frac{(kT)^4}{(hc)^3}. \tag{7.91}$$

In what follows I'll simply call this quantity U/V; the total energy in the chunk is thus

$$\text{energy in chunk} = \frac{U}{V}\,c\,dt\,R^2\sin\theta\,d\theta\,d\phi. \tag{7.92}$$

But not all the energy in this chunk of space will escape through the hole, because most of the photons are pointed in the wrong direction. The probability of a photon being pointed in the *right* direction is equal to the apparent area of the hole, as viewed from the chunk, divided by the total area of an imaginary sphere of radius R centered on the chunk:

$$\text{probability of escape} = \frac{A\cos\theta}{4\pi R^2}. \tag{7.93}$$

Here A is the area of the hole, and $A\cos\theta$ is its foreshortened area, as seen from the chunk. The amount of energy that escapes from this chunk is therefore

$$\text{energy escaping from chunk} = \frac{A\cos\theta}{4\pi}\frac{U}{V}\,c\,dt\,\sin\theta\,d\theta\,d\phi. \tag{7.94}$$

To find the *total* energy that escapes through the hole in the time interval dt, we just integrate over θ and ϕ:

$$\text{total energy escaping} = \int_0^{2\pi} d\phi \int_0^{\pi/2} d\theta \, \frac{A\cos\theta}{4\pi} \frac{U}{V} c\, dt \, \sin\theta$$

$$= 2\pi \frac{A}{4\pi} \frac{U}{V} c\, dt \int_0^{\pi/2} \cos\theta \, \sin\theta \, d\theta \qquad (7.95)$$

$$= \frac{A}{4} \frac{U}{V} c\, dt.$$

The amount of energy that escapes is naturally proportional to the area A of the hole, and also to the duration dt of the time interval. If we divide by these quantities we get the *power* emitted per unit area:

$$\text{power per unit area} = \frac{c}{4} \frac{U}{V}. \qquad (7.96)$$

Aside from the factor of $1/4$, you could have guessed this result using dimensional analysis: To turn energy/volume into power/area, you have to multiply by something with units of distance/time, and the only relevant speed in the problem is the speed of light.

Plugging in formula 7.91 for the energy density inside the box, we obtain the more explicit result

$$\text{power per unit area} = \frac{2\pi^5}{15} \frac{(kT)^4}{h^3 c^2} = \sigma T^4, \qquad (7.97)$$

where σ is known as the **Stefan-Boltzmann constant**,

$$\sigma = \frac{2\pi^5 k^4}{15 h^3 c^2} = 5.67 \times 10^{-8} \frac{\text{W}}{\text{m}^2 \, \text{K}^4}. \qquad (7.98)$$

(This number isn't hard to memorize: Just think "5–6–7–8," and don't forget the minus sign.) The dependence of the power radiated on the fourth power of the temperature is known as **Stefan's law**, and was discovered empirically in 1879.

Radiation from Other Objects

Although I derived Stefan's law for photons emitted from a hole in a box, it also applies to photons emitted by any nonreflecting ("black") surface at temperature T. Such radiation is therefore called **blackbody radiation**. The proof that a black object emits photons exactly as does a hole in a box is amazingly simple.

Suppose you have a hole in a box, on one hand, and a black object, on the other hand, both at the same temperature, facing each other as in Figure 7.24. Each object emits photons, some of which are absorbed by the other. If the objects are the same size, each will absorb the same fraction of the other's radiation. Now suppose that the blackbody does *not* emit the same amount of power as the hole; perhaps

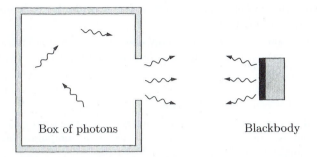

Figure 7.24. A thought experiment to demonstrate that a perfectly black surface emits radiation identical to that emitted by a hole in a box of thermal photons.

it emits somewhat less. Then more energy will flow from the hole to the blackbody than from the blackbody to the hole, and the blackbody will gradually get hotter. Oops! This process would violate the second law of thermodynamics. And if the blackbody emits *more* radiation than the hole, then the blackbody gradually cools off while the box with the hole gets hotter; again, this can't happen.

So the total power emitted by the blackbody, per unit area at any given temperature, must be the same as that emitted by the hole. But we can say more. Imagine inserting a filter, which allows only a certain range of wavelengths to pass through, between the hole and the blackbody. Again, if one object emits more radiation at these wavelengths than the other, its temperature will decrease while the other's temperature increases, in violation of the second law. Thus the entire spectrum of radiation emitted by the blackbody must be the same as for the hole.

If an object is *not* black, so that it reflects some photons instead of absorbing them, things get a bit more complicated. Let's say that out of every three photons (at some given wavelength) that hit the object, it reflects one back and absorbs the other two. Now, in order to remain in thermal equilibrium with the hole, it only needs to emit two photons, which join the reflected photon on its way back. More generally, if e is the fraction of photons absorbed (at some given wavelength), then e is also the fraction emitted, in comparison to a perfect blackbody. This number e is called the **emissivity** of the material. It equals 1 for a perfect blackbody, and equals 0 for a perfectly reflective surface. Thus, a good reflector is a poor emitter, and vice versa. Generally the emissivity depends upon the wavelength of the light, so the spectrum of radiation emitted will differ from a perfect blackbody spectrum. If we use a weighted average of e over all relevant wavelengths, then the total power radiated by an object can be written

$$\text{power} = \sigma e A T^4, \tag{7.99}$$

where A is the object's surface area.

Problem 7.51. The tungsten filament of an incandescent light bulb has a temperature of approximately 3000 K. The emissivity of tungsten is approximately 1/3, and you may assume that it is independent of wavelength.

(a) If the bulb gives off a total of 100 watts, what is the surface area of its filament in square millimeters?

(b) At what value of the photon energy does the peak in the bulb's spectrum occur? What is the wavelength corresponding to this photon energy?

(c) Sketch (or use a computer to plot) the spectrum of light given off by the filament. Indicate the region on the graph that corresponds to visible wavelengths, between 400 and 700 nm.

(d) Calculate the fraction of the bulb's energy that comes out as visible light. (Do the integral numerically on a calculator or computer.) Check your result qualitatively from the graph of part (c).

(e) To increase the efficiency of an incandescent bulb, would you want to raise or lower the temperature? (Some incandescent bulbs *do* attain slightly higher efficiency by using a different temperature.)

(f) Estimate the maximum possible efficiency (i.e., fraction of energy in the visible spectrum) of an incandescent bulb, and the corresponding filament temperature. Neglect the fact that tungsten melts at 3695 K.

Problem 7.52.

(a) Estimate (roughly) the total power radiated by your body, neglecting any energy that is returned by your clothes and environment. (Whatever the color of your skin, its emissivity at infrared wavelengths is quite close to 1; almost any nonmetal is a near-perfect blackbody at these wavelengths.)

(b) Compare the total energy radiated by your body in one day (expressed in kilocalories) to the energy in the food you eat. Why is there such a large discrepancy?

(c) The sun has a mass of 2×10^{30} kg and radiates energy at a rate of 3.9×10^{26} watts. Which puts out more power *per units mass*—the sun or your body?

Problem 7.53. A black hole is a blackbody if ever there was one, so it should emit blackbody radiation, called **Hawking radiation**. A black hole of mass M has a total energy of Mc^2, a surface area of $16\pi G^2 M^2/c^4$, and a temperature of $hc^3/16\pi^2 kGM$ (as shown in Problem 3.7).

(a) Estimate the typical wavelength of the Hawking radiation emitted by a one-solar-mass (2×10^{30} kg) black hole. Compare your answer to the size of the black hole.

(b) Calculate the total power radiated by a one-solar-mass black hole.

(c) Imagine a black hole in empty space, where it emits radiation but absorbs nothing. As it loses energy, its mass must decrease; one could say it "evaporates." Derive a differential equation for the mass as a function of time, and solve this equation to obtain an expression for the lifetime of a black hole in terms of its initial mass.

(d) Calculate the lifetime of a one-solar-mass black hole, and compare to the estimated age of the known universe (10^{10} years).

(e) Suppose that a black hole that was created early in the history of the universe finishes evaporating today. What was its initial mass? In what part of the electromagnetic spectrum would most of its radiation have been emitted?

The Sun and the Earth

From the amount of solar radiation received by the earth (1370 W/m^2, known as the **solar constant**) and the earth's distance from the sun (150 million kilometers), it's pretty easy to calculate the sun's total energy output or **luminosity**: 3.9×10^{26} watts. The sun's radius is a little over 100 times the earth's: 7.0×10^8 m; so its surface area is 6.1×10^{18} m^2. From this information, assuming an emissivity of 1 (which is not terribly accurate but good enough for our purposes), we can calculate the sun's surface temperature:

$$T = \left(\frac{\text{luminosity}}{\sigma A} \right)^{1/4} = 5800 \text{ K}. \tag{7.100}$$

Knowing the temperature, we can predict that the spectrum of sunlight should peak at a photon energy of

$$\epsilon = 2.82 \, kT = 1.41 \text{ eV}, \tag{7.101}$$

which corresponds to a wavelength of 880 nm, in the near infrared. This is a testable prediction, and it agrees with experiment: The sun's spectrum is approximately given by the Planck formula, with a peak at this energy. Since the peak is so close to the red end of the visible spectrum, much of the sun's energy is emitted as visible light. (If you've learned elsewhere that the sun's spectrum peaks in the middle of the visible spectrum at about 500 nm, and you're worried about the discrepancy, go back and work Problem 7.39.)

A tiny fraction of the sun's radiation is absorbed by the earth, warming the earth's surface to a temperature suitable for life. But the earth doesn't just keep getting hotter and hotter; it also *emits* radiation into space, at the same rate, on average. This balance between absorption and emission gives us a way to estimate the earth's equilibrium surface temperature.

As a first crude estimate, let's pretend that the earth is a perfect blackbody at all wavelengths. Then the power absorbed is the solar constant times the earth's cross-sectional area as viewed from the sun, πR^2. The power emitted, meanwhile, is given by Stefan's law, with A being the full surface area of the earth, $4\pi R^2$, and T being the effective average surface temperature. Setting the power absorbed equal to the power emitted gives

$$(\text{solar constant}) \cdot \pi R^2 = 4\pi R^2 \sigma T^4$$

$$\Rightarrow T = \left(\frac{1370 \text{ W/m}^2}{4 \cdot 5.67 \times 10^{-8} \text{ W/m}^2 \cdot \text{K}^4} \right)^{1/4} = 279 \text{ K}. \tag{7.102}$$

This is extremely close to the measured average temperature of 288 K (15°C).

However, the earth is *not* a perfect blackbody. About 30% of the sunlight striking the earth is reflected directly back into space, mostly by clouds. Taking reflection into account brings the earth's predicted average temperature down to a frigid 255 K.

Since a poor absorber is also a poor emitter, you might think we could bring the earth's predicted temperature back up by taking the imperfect emissivity into account on the right-hand side of equation 7.102. Unfortunately, this doesn't work. There's no particular reason why the earth's emissivity should be the same for the infrared light emitted as for the visible light absorbed, and in fact, the earth's surface (like almost any nonmetal) is a very efficient emitter at infrared wavelengths. But there's another mechanism that saves us: Water vapor and carbon dioxide in earth's atmosphere make the atmosphere mostly opaque at wavelengths above a few microns, so if you look at the earth from space with an eye sensitive to infrared light, what you see is mostly the atmosphere, not the surface. The equilibrium temperature of 255 K applies (roughly) to the atmosphere, while the surface below is heated both by the incoming sunlight and by the atmospheric "blanket." If we model the atmosphere as a single layer that is transparent to visible light but opaque to infrared, we get the situation shown in Figure 7.25. Equilibrium requires that the energy of the incident sunlight (minus what is reflected) be equal to the energy emitted upward by the atmosphere, which in turn is equal to the energy radiated downward by the atmosphere. Therefore the earth's surface receives twice as much energy (in this simplified model) as it would from sunlight alone. According to equation 7.102, this mechanism raises the surface temperature by a factor of $2^{1/4}$, to 303 K. This is a bit high, but then, the atmosphere isn't just a single perfectly opaque layer. By the way, this mechanism is called the **greenhouse effect**, even though most greenhouses depend primarily on a different mechanism (namely, limiting convective cooling).

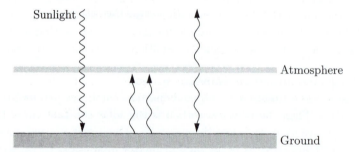

Figure 7.25. Earth's atmosphere is mostly transparent to incoming sunlight, but opaque to the infrared light radiated upward by earth's surface. If we model the atmosphere as a single layer, then equilibrium requires that earth's surface receive as much energy from the atmosphere as from the sun.

Problem 7.54. The sun is the only star whose size we can easily measure directly; astronomers therefore estimate the sizes of other stars using Stefan's law.

(a) The spectrum of Sirius A, plotted as a function of energy, peaks at a photon energy of 2.4 eV, while Sirius A is approximately 24 times as luminous as the sun. How does the radius of Sirius A compare to the sun's radius?

(b) Sirius B, the companion of Sirius A (see Figure 7.12), is only 3% as luminous as the sun. Its spectrum, plotted as a function of energy, peaks at about 7 eV. How does its radius compare to that of the sun?

(c) The spectrum of the star Betelgeuse, plotted as a function of energy, peaks at a photon energy of 0.8 eV, while Betelgeuse is approximately 10,000 times as luminous as the sun. How does the radius of Betelgeuse compare to the sun's radius? Why is Betelgeuse called a "red supergiant"?

Problem 7.55. Suppose that the concentration of infrared-absorbing gases in earth's atmosphere were to double, effectively creating a second "blanket" to warm the surface. Estimate the equilibrium surface temperature of the earth that would result from this catastrophe. (Hint: First show that the lower atmospheric blanket is warmer than the upper one by a factor of $2^{1/4}$. The surface is warmer than the lower blanket by a smaller factor.)

Problem 7.56. The planet Venus is different from the earth in several respects. First, it is only 70% as far from the sun. Second, its thick clouds reflect 77% of all incident sunlight. Finally, its atmosphere is much more opaque to infrared light.

(a) Calculate the solar constant at the location of Venus, and estimate what the average surface temperature of Venus would be if it had no atmosphere and did not reflect any sunlight.

(b) Estimate the surface temperature again, taking the reflectivity of the clouds into account.

(c) The opaqueness of Venus's atmosphere at infrared wavelengths is roughly 70 times that of earth's atmosphere. You can therefore model the atmosphere of Venus as 70 successive "blankets" of the type considered in the text, with each blanket at a different equilibrium temperature. Use this model to estimate the surface temperature of Venus. (Hint: The temperature of the top layer is what you found in part (b). The next layer down is warmer by a factor of $2^{1/4}$. The *next* layer down is warmer by a smaller factor. Keep working your way down until you see the pattern.)

7.5 Debye Theory of Solids

In Section 2.2 I introduced the **Einstein model** of a solid crystal, in which each atom is treated as an independent three-dimensional harmonic oscillator. In Problem 3.25, you used this model to derive a prediction for the heat capacity,

$$C_V = 3Nk \frac{(\epsilon/kT)^2 e^{\epsilon/kT}}{(e^{\epsilon/kT} - 1)^2} \qquad \text{(Einstein model)}, \qquad (7.103)$$

where N is the number of *atoms* and $\epsilon = hf$ is the universal size of the units of energy for the identical oscillators. When $kT \gg \epsilon$, the heat capacity approaches a constant value, $3Nk$, in agreement with the equipartition theorem. Below $kT \approx \epsilon$, the heat capacity falls off, approaching zero as the temperature goes to zero. This prediction agrees with experiment to a first approximation, but not in detail. In particular, equation 7.103 predicts that the heat capacity goes to zero *exponentially* in the limit $T \to 0$, whereas experiments show that the true low-temperature behavior is cubic: $C_V \propto T^3$.

The problem with the Einstein model is that the atoms in a crystal do *not* vibrate independently of each other. If you wiggle one atom, its neighbors will also start to wiggle, in a complicated way that depends on the frequency of oscillation.

There are low-frequency modes of oscillation in which large groups of atoms are all moving together, and also high-frequency modes in which atoms are moving opposite to their neighbors. The units of energy come in different sizes, proportional to the frequencies of the modes of vibration. Even at very low temperatures, when the high-frequency modes are frozen out, a few low-frequency modes are still active. This is the reason why the heat capacity goes to zero less dramatically than the Einstein model predicts.

In many ways, the modes of oscillation of a solid crystal are similar to the modes of oscillation of the electromagnetic field in vacuum. This similarity suggests that we try to adapt our recent treatment of electromagnetic radiation to the mechanical oscillations of the crystal. Mechanical oscillations are also called sound waves, and behave very much like light waves. There are a few differences, however:

- Sound waves travel much slower than light waves, at a speed that depends on the stiffness and density of the material. I'll call this speed c_s, and treat it as a constant, neglecting the fact that it can depend on wavelength and direction.

- Whereas light waves must be transversely polarized, sound waves can also be longitudinally polarized. (In seismology, transversely polarized waves are called shear waves, or S-waves, while longitudinally polarized waves are called pressure waves, or P-waves.) So instead of two polarizations we have three. For simplicity, I'll pretend that all three polarizations have the same speed.

- Whereas light waves can have arbitrarily short wavelengths, sound waves in solids cannot have wavelengths shorter than twice the atomic spacing.

The first two differences are easy to take into account. The third will require some thought.

Aside from these three differences, sound waves behave almost identically to light waves. Each mode of oscillation has a set of equally spaced energy levels, with the unit of energy equal to

$$\epsilon = hf = \frac{hc_s}{\lambda} = \frac{hc_s n}{2L}. \tag{7.104}$$

In the last expression, L is the length of the crystal and $n = |\vec{n}|$ is the magnitude of the vector in n-space specifying the shape of the wave. When this mode is in equilibrium at temperature T, the number of units of energy it contains, on average, is given by the Planck distribution:

$$\overline{n}_{\text{Pl}} = \frac{1}{e^{\epsilon/kT} - 1}. \tag{7.105}$$

(This $\overline{n}$ is not to be confused with the n in the previous equation.) As with electromagnetic waves, we can think of these units of energy as particles obeying Bose-Einstein statistics with $\mu = 0$. This time the "particles" are called **phonons**.

To calculate the total thermal energy of the crystal, we add up the energies of all allowed modes:

$$U = 3 \sum_{n_x} \sum_{n_y} \sum_{n_z} \epsilon\, \overline{n}_{\text{Pl}}(\epsilon). \tag{7.106}$$

The factor of 3 counts the three polarization states for each $\vec{n}$. The next step will be to convert the sum to an integral. But first we'd better worry about what values of $\vec{n}$ are being summed over.

If these were electromagnetic oscillations, there would be an infinite number of allowed modes and each sum would go to infinity. But in a crystal, the atomic spacing puts a strict lower limit on the wavelength. Consider a lattice of atoms in just one dimension (see Figure 7.26). Each mode of oscillation has its own distinct shape, with the number of "bumps" equal to n. Because each bump must contain at least one atom, n cannot exceed the number of atoms in a row. If the three-dimensional crystal is a perfect cube, then the number of atoms along any direction is $\sqrt[3]{N}$, so each sum in equation 7.106 should go from 1 to $\sqrt[3]{N}$. In other words, we're summing over a *cube* in n-space. If the crystal itself is not a perfect cube, then neither is the corresponding volume of n-space. Still, however, the sum will run over a region in n-space whose total volume is N.

Now comes the tricky approximation. Summing (or integrating) over a cube or some other complicated region of n-space is no fun, because the function we're summing depends on n_x, n_y, and n_z in a very complicated way (an exponential of a square root). On the other hand, the function depends on the *magnitude* of $\vec{n}$ in a simpler way, and it doesn't depend on the *angle* in n-space at all. So Peter Debye got the clever idea to pretend that the relevant region of n-space is a sphere, or rather, an eighth of a sphere. To preserve the total number of degrees of freedom, he chose a sphere whose total volume is N. You can easily show that the radius of the sphere has to be

$$n_{\max} = \left(\frac{6N}{\pi}\right)^{1/3}. \tag{7.107}$$

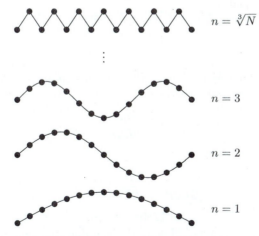

$$n = \sqrt[3]{N}$$

$$\vdots$$

$$n = 3$$

$$n = 2$$

$$n = 1$$

Figure 7.26. Modes of oscillation of a row of atoms in a crystal. If the crystal is a cube, then the number of atoms along any row is $\sqrt[3]{N}$. This is also the total number of modes along this direction, because each "bump" in the wave form must contain at least one atom.

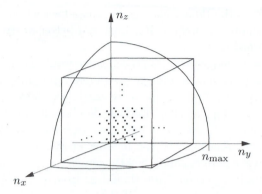

Figure 7.27. The sum in equation 7.106 is technically over a cube in n-space whose width is $\sqrt[3]{N}$. As an approximation, we instead sum over an eighth-sphere with the same total volume.

Figure 7.27 shows the cube in n-space, and the sphere that approximates it.

Remarkably, Debye's approximation is exact in *both* the high-temperature and low-temperature limits. At high temperature, all that matters is the total number of modes, that is, the total number of degrees of freedom; this number is preserved by choosing the sphere to have the correct volume. At low temperature, modes with large $\vec{n}$ are frozen out anyway, so we can count them however we like. At intermediate temperatures, we'll get results that are not exact, but they'll still be surprisingly good.

When we make Debye's approximation, and convert the sums to integrals in spherical coordinates, equation 7.106 becomes

$$U = 3 \int_0^{n_{\max}} dn \int_0^{\pi/2} d\theta \int_0^{\pi/2} d\phi \; n^2 \sin\theta \; \frac{\epsilon}{e^{\epsilon/kT} - 1}. \tag{7.108}$$

The angular integrals give $\pi/2$ (yet again), leaving us with

$$U = \frac{3\pi}{2} \int_0^{n_{\max}} \frac{hc_s}{2L} \frac{n^3}{e^{hc_s n/2LkT} - 1} \, dn. \tag{7.109}$$

This integral cannot be done analytically, but it's at least a little cleaner if we change to the dimensionless variable

$$x = \frac{hc_s n}{2LkT}. \tag{7.110}$$

The upper limit on the integral will then be

$$x_{\max} = \frac{hc_s n_{\max}}{2LkT} = \frac{hc_s}{2kT} \left(\frac{6N}{\pi V} \right)^{1/3} \equiv \frac{T_{\mathrm{D}}}{T}, \tag{7.111}$$

where the last equality defines the **Debye temperature**, T_{D}—essentially an abbreviation for all the constants. Making the variable change and collecting all the constants is now straightforward. When the smoke clears, we obtain

$$U = \frac{9NkT^4}{T_{\mathrm{D}}^3} \int_0^{T_{\mathrm{D}}/T} \frac{x^3}{e^x - 1} \, dx. \tag{7.112}$$

At this point you can do the integral on a computer if you like, for any desired temperature. Without a computer, though, we can still check the low-temperature and high-temperature limits.

When $T \gg T_{\mathrm{D}}$, the upper limit of the integral is much less than 1, so x is always very small and we can approximate $e^x \approx 1 + x$ in the denominator. The 1 cancels, leaving the x to cancel one power of x in the numerator. The integral then gives simply $\frac{1}{3}(T_{\mathrm{D}}/T)^3$, leading to the final result

$$U = 3NkT \qquad \text{when } T \gg T_{\mathrm{D}}, \qquad (7.113)$$

in agreement with the equipartition theorem (and the Einstein model). The heat capacity in this limit is just $C_V = 3Nk$.

When $T \ll T_{\mathrm{D}}$, the upper limit on the integral is so large that by the time we get to it, the integrand is dead (due to the e^x in the denominator). So we might as well replace the upper limit by infinity—the extra modes we're adding don't contribute anyway. In this approximation, the integral is the same as the one we did for the photon gas (equation 7.85), and evaluates to $\pi^4/15$. So the total energy is

$$U = \frac{3\pi^4}{5} \frac{NkT^4}{T_{\mathrm{D}}^3} \qquad \text{when } T \ll T_{\mathrm{D}}. \qquad (7.114)$$

To get the heat capacity, differentiate with respect to T:

$$C_V = \frac{12\pi^4}{5} \left(\frac{T}{T_{\mathrm{D}}}\right)^3 Nk \qquad \text{when } T \ll T_{\mathrm{D}}. \qquad (7.115)$$

The prediction $C_V \propto T^3$ agrees beautifully with low-temperature experiments on almost any solid material. For metals, though, there is also a linear contribution to the heat capacity from the conduction electrons, as described in Section 7.3. The total heat capacity at low temperature is therefore

$$C = \gamma T + \frac{12\pi^4 Nk}{5T_{\mathrm{D}}^3} T^3 \qquad (\text{metal, } T \ll T_{\mathrm{D}}), \qquad (7.116)$$

where $\gamma = \pi^2 Nk^2/2\epsilon_{\mathrm{F}}$ in the free electron model. Figure 7.28 shows plots of C/T

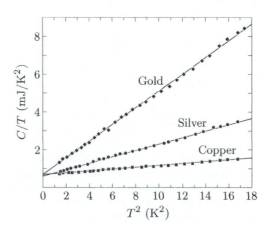

Figure 7.28. Low-temperature measurements of the heat capacities (per mole) of copper, silver, and gold. Adapted with permission from William S. Corak et al., *Physical Review* **98**, 1699 (1955).

vs. T^2 for three familiar metals. The linearity of the data confirms the Debye theory of lattice vibrations, while the intercepts give us the experimental values of γ.

At intermediate temperatures, you have to do a numerical integral to get the total thermal energy in the crystal. If what you really want is the heat capacity, it's best to differentiate equation 7.109 analytically, then change variables to x. The result is

$$C_V = 9Nk\left(\frac{T}{T_D}\right)^3 \int_0^{T_D/T} \frac{x^4\,e^x}{(e^x - 1)^2}\,dx. \tag{7.117}$$

A computer-generated plot of this function is shown in Figure 7.29. For comparison, the Einstein model prediction, equation 7.103, is also plotted, with the constant ϵ chosen to make the curves agree at relatively high temperatures. As you can see, the two curves still differ significantly at low temperatures. Figure 1.14 shows further comparisons of experimental data to the prediction of the Debye model.

The Debye temperature of any particular substance can be predicted from the speed of sound in that substance, using equation 7.111. Usually, however, one obtains a better fit to the data by choosing T_D so that the measured heat capacity best fits the theoretical prediction. Typical values of T_D range from 88 K for lead (which is soft and dense) to 1860 K for diamond (which is stiff and light). Since the heat capacity reaches 95% of its maximum value at $T = T_D$, the Debye temperature gives you a rough idea of when you can get away with just using the equipartition theorem. When you can't, Debye's formula usually gives a good, but not great, estimate of the heat capacity over the full range of temperatures. To do better, we'd have to do a lot more work, taking into account the fact that the speed of a phonon depends on its wavelength, polarization, and direction of travel with respect to the crystal axes. That kind of analysis belongs in a book on solid state physics.

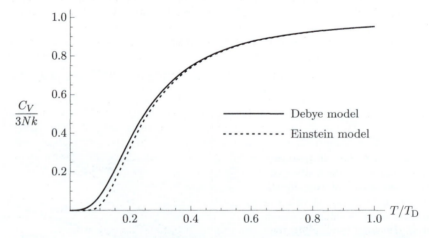

Figure 7.29. The Debye prediction for the heat capacity of a solid, with the prediction of the Einstein model plotted for comparison. The constant ϵ in the Einstein model has been chosen to obtain the best agreement with the Debye model at high temperatures. Note that the Einstein curve is much flatter than the Debye curve at low temperatures.

Problem 7.57. Fill in the steps to derive equations 7.112 and 7.117.

Problem 7.58. The speed of sound in copper is 3560 m/s. Use this value to calculate its theoretical Debye temperature. Then determine the experimental Debye temperature from Figure 7.28, and compare.

Problem 7.59. Explain in some detail why the three graphs in Figure 7.28 all intercept the vertical axis in about the same place, whereas their slopes differ considerably.

Problem 7.60. Sketch the heat capacity of copper as a function of temperature from 0 to 5 K, showing the contributions of lattice vibrations and conduction electrons separately. At what temperature are these two contributions equal?

Problem 7.61. The heat capacity of liquid ^{4}He below 0.6 K is proportional to T^3, with the measured value $C_V/Nk = (T/4.67 \text{ K})^3$. This behavior suggests that the dominant excitations at low temperature are long-wavelength phonons. The only important difference between phonons in a liquid and phonons in a solid is that a liquid cannot transmit transversely polarized waves—sound waves must be longitudinal. The speed of sound in liquid ^{4}He is 238 m/s, and the density is 0.145 g/cm^3. From these numbers, calculate the phonon contribution to the heat capacity of ^{4}He in the low-temperature limit, and compare to the measured value.

Problem 7.62. Evaluate the integrand in equation 7.112 as a power series in x, keeping terms through x^4. Then carry out the integral to find a more accurate expression for the energy in the high-temperature limit. Differentiate this expression to obtain the heat capacity, and use the result to estimate the percent deviation of C_V from $3Nk$ at $T = T_D$ and $T = 2T_D$.

Problem 7.63. Consider a two-dimensional solid, such as a stretched drumhead or a layer of mica or graphite. Find an expression (in terms of an integral) for the thermal energy of a square chunk of this material of area $A = L^2$, and evaluate the result approximately for very low and very high temperatures. Also find an expression for the heat capacity, and use a computer or a calculator to plot the heat capacity as a function of temperature. Assume that the material can only vibrate perpendicular to its own plane, i.e., that there is only one "polarization."

Problem 7.64. A **ferromagnet** is a material (like iron) that magnetizes spontaneously, even in the absence of an externally applied magnetic field. This happens because each elementary dipole has a strong tendency to align parallel to its neighbors. At $T = 0$ the magnetization of a ferromagnet has the maximum possible value, with all dipoles perfectly lined up; if there are N atoms, the total magnetization is typically $\sim 2\mu_B N$, where μ_B is the Bohr magneton. At somewhat higher temperatures, the excitations take the form of **spin waves**, which can be visualized classically as shown in Figure 7.30. Like sound waves, spin waves are quantized: Each wave mode can have only integer multiples of a basic energy unit. In analogy with phonons, we think of the energy units as particles, called **magnons**. Each magnon reduces the total spin of the system by one unit of $h/2\pi$, and therefore reduces the magnetization by $\sim 2\mu_B$. However, whereas the frequency of a sound wave is inversely proportional to its wavelength, the frequency of a spin wave is proportional to the *square* of $1/\lambda$ (in the limit of long wavelengths). Therefore, since $\epsilon = hf$ and $p = h/\lambda$ for any "particle," the energy of a magnon is proportional

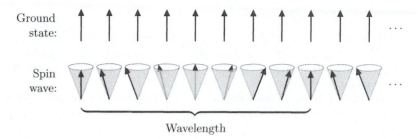

Ground state: / Spin wave: / Wavelength

Figure 7.30. In the ground state of a ferromagnet, all the elementary dipoles point in the same direction. The lowest-energy excitations above the ground state are **spin waves**, in which the dipoles precess in a conical motion. A long-wavelength spin wave carries very little energy, because the difference in direction between neighboring dipoles is very small.

to the square of its momentum. In analogy with the energy-momentum relation for an ordinary nonrelativistic particle, we can write $\epsilon = p^2/2m^*$, where m^* is a constant related to the spin-spin interaction energy and the atomic spacing. For iron, m^* turns out to equal 1.24×10^{-29} kg, about 14 times the mass of an electron. Another difference between magnons and phonons is that each magnon (or spin wave mode) has only one possible polarization.

(a) Show that at low temperatures, the number of magnons per unit volume in a three-dimensional ferromagnet is given by

$$\frac{N_m}{V} = 2\pi \left(\frac{2m^* kT}{h^2} \right)^{3/2} \int_0^\infty \frac{\sqrt{x}}{e^x - 1} \, dx.$$

Evaluate the integral numerically.

(b) Use the result of part (a) to find an expression for the fractional reduction in magnetization, $(M(0) - M(T))/M(0)$. Write your answer in the form $(T/T_0)^{3/2}$, and estimate the constant T_0 for iron.

(c) Calculate the heat capacity due to magnetic excitations in a ferromagnet at low temperature. You should find $C_V/Nk = (T/T_1)^{3/2}$, where T_1 differs from T_0 only by a numerical constant. Estimate T_1 for iron, and compare the magnon and phonon contributions to the heat capacity. (The Debye temperature of iron is 470 K.)

(d) Consider a *two*-dimensional array of magnetic dipoles at low temperature. Assume that each elementary dipole can still point in any (three-dimensional) direction, so spin waves are still possible. Show that the integral for the total number of magnons diverges in this case. (This result is an indication that there can be no spontaneous magnetization in such a two-dimensional system. However, in Section 8.2 we will consider a different two-dimensional model in which magnetization *does* occur.)

7.6 Bose-Einstein Condensation

The previous two sections treated bosons (photons and phonons) that can be created in arbitrary numbers—whose total number is determined by the condition of thermal equilibrium. But what about more "ordinary" bosons, such as atoms with integer spin, whose number is fixed from the outset?

I've saved this case for last because it is more difficult. In order to apply the Bose-Einstein distribution we'll have to determine the chemical potential, which (rather than being fixed at zero) is now a nontrivial function of the density and temperature. Determining μ will require some careful analysis, but is worth the trouble: We'll find that it behaves in a most peculiar way, indicating that a gas of bosons will abruptly "condense" into the ground state as the temperature goes below a certain critical value.

It's simplest to first consider the limit $T \to 0$. At zero temperature, all the atoms will be in the lowest-energy available state, and since arbitrarily many bosons are allowed in any given state, this means that *every* atom will be in the ground state. (Here again, when I say simply "state" I mean a single-particle state.) For atoms confined to a box of volume $V = L^3$, the energy of the ground state is

$$\epsilon_0 = \frac{h^2}{8mL^2}(1^2 + 1^2 + 1^2) = \frac{3h^2}{8mL^2}, \tag{7.118}$$

which works out to a *very* small energy provided that L is macroscopic. At any temperature, the average number of atoms in this state, which I'll call N_0, is given by the Bose-Einstein distribution:

$$N_0 = \frac{1}{e^{(\epsilon_0 - \mu)/kT} - 1}. \tag{7.119}$$

When T is sufficiently low, N_0 will be quite large. In this case, the denominator of this expression must be very small, which implies that the exponential is very close to 1, which implies that the exponent, $(\epsilon_0 - \mu)/kT$, is very small. We can therefore expand the exponential in a Taylor series and keep only the first two terms, to obtain

$$N_0 = \frac{1}{1 + (\epsilon_0 - \mu)/kT - 1} = \frac{kT}{\epsilon_0 - \mu} \qquad \text{(when } N_0 \gg 1\text{)}. \tag{7.120}$$

The chemical potential μ, therefore, must be equal to ϵ_0 at $T = 0$, and just a *tiny* bit less than ϵ_0 when T is nonzero but still sufficiently small that nearly all of the atoms are in the ground state. The remaining question is this: How low must the temperature be, in order for N_0 to be large?

The general condition that determines μ is that the sum of the Bose-Einstein distribution over *all* states must add up to the total number of atoms, N:

$$N = \sum_{\text{all } s} \frac{1}{e^{(\epsilon_s - \mu)/kT} - 1}. \tag{7.121}$$

In principle, we could keep guessing values of μ until this sum works out correctly (and repeat the process for each value of T). In practice, it's usually easier to convert the sum to an integral:

$$N = \int_0^\infty g(\epsilon) \frac{1}{e^{(\epsilon-\mu)/kT} - 1} \, d\epsilon. \tag{7.122}$$

This approximation should be valid when $kT \gg \epsilon_0$, so that the number of terms that contribute significantly to the sum is large. The function $g(\epsilon)$ is the **density of states**: the number of single-particle states per unit energy. For spin-zero bosons confined in a box of volume V, this function is the same as what we used for electrons in Section 7.3 (equation 7.51) but divided by 2 because now there is only one spin orientation:

$$g(\epsilon) = \frac{2}{\sqrt{\pi}} \left(\frac{2\pi m}{h^2}\right)^{3/2} V \sqrt{\epsilon}. \tag{7.123}$$

Figure 7.31 shows graphs of the density of states, the Bose-Einstein distribution (drawn for μ slightly less than zero), and the product of the two, which is the distribution of particles as a function of energy.

Unfortunately, the integral 7.122 cannot be performed analytically. Therefore we must guess values of μ until we find one that works, doing the integral numerically each time. The most interesting (and easiest) guess is $\mu = 0$, which should work (to a good approximation) at temperatures that are low enough for N_0 to be large. Plugging in $\mu = 0$ and changing variables to $x = \epsilon/kT$ gives

$$\begin{aligned} N &= \frac{2}{\sqrt{\pi}} \left(\frac{2\pi m}{h^2}\right)^{3/2} V \int_0^\infty \frac{\sqrt{\epsilon} \, d\epsilon}{e^{\epsilon/kT} - 1} \\ &= \frac{2}{\sqrt{\pi}} \left(\frac{2\pi m kT}{h^2}\right)^{3/2} V \int_0^\infty \frac{\sqrt{x} \, dx}{e^x - 1}. \end{aligned} \tag{7.124}$$

The integral over x is equal to 2.315; combining this number with the factor of $2/\sqrt{\pi}$ yields the formula

$$N = 2.612 \left(\frac{2\pi m kT}{h^2}\right)^{3/2} V. \tag{7.125}$$

This result is obviously wrong: Everything on the right-hand side is independent

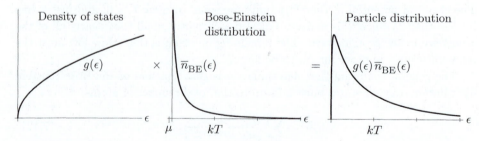

Figure 7.31. The distribution of bosons as a function of energy is the product of two functions, the density of states and the Bose-Einstein distribution.

of temperature except T, so it says that the number of atoms depends on the temperature, which is absurd. In fact, there can be only one particular temperature for which equation 7.125 is correct; I'll call this temperature T_c:

$$N = 2.612\left(\frac{2\pi mkT_c}{h^2}\right)^{3/2}V, \quad \text{or} \quad kT_c = 0.527\left(\frac{h^2}{2\pi m}\right)\left(\frac{N}{V}\right)^{2/3}. \qquad (7.126)$$

But what's wrong with equation 7.125 when $T \neq T_c$? At temperatures *higher* than T_c, the chemical potential must be significantly less than zero; from equation 7.122 you can see that a negative value of μ will yield a result for N that is smaller than the right-hand side of equation 7.125, as desired. At temperatures *lower* than T_c, on the other hand, the solution to the paradox is more subtle; in this case, replacing the discrete sum 7.121 with the integral 7.122 is invalid.

Look carefully at the integrand in equation 7.124. As ϵ goes to zero, the density of states (proportional to $\sqrt{\epsilon}$) goes to zero while the Bose-Einstein distribution blows up (in proportion to $1/\epsilon$). Although the product is an integrable function, it is not at all clear that this infinite spike at $\epsilon = 0$ correctly represents the sum 7.121 over the actual discretely spaced states. In fact, we have already seen in equation 7.120 that the number of atoms in the ground state can be enormous when $\mu \approx 0$, and this enormous number is not included in our integral. On the other hand, the integral *should* correctly represent the number of particles in the vast majority of the states, away from the spike, where $\epsilon \gg \epsilon_0$. If we imagine cutting off the integral at a lower limit that is somewhat greater than ϵ_0 but much less than kT, we'll still obtain *approximately* the same answer,

$$N_{\text{excited}} = 2.612\left(\frac{2\pi mkT}{h^2}\right)^{3/2}V \qquad (\text{when } T < T_c). \qquad (7.127)$$

This is then the number of atoms in excited states, *not* including the ground state. (Whether this expression correctly accounts for the few *lowest* excited states, just above the ground state in energy, is not completely clear. If we assume that the difference between N and the preceding expression for N_{excited} is sufficiently large, then it follows that μ must be much closer to the ground-state energy than to the energy of the first excited state, and therefore that no excited state contains anywhere near as many atoms as the ground state. However, there will be a narrow range of temperatures, just below T_c, where this condition is not met. When the total number of atoms is not particularly large, this range of temperatures might not even be so narrow. These issues are explored in Problem 7.66.)

So the bottom line is this: At temperatures higher than T_c, the chemical potential is negative and essentially all of the atoms are in excited states. At temperatures lower than T_c, the chemical potential is very close to zero and the number of atoms in excited states is given by equation 7.127; this formula can be rewritten more simply as

$$N_{\text{excited}} = \left(\frac{T}{T_c}\right)^{3/2}N \qquad (T < T_c). \qquad (7.128)$$

The rest of the atoms must be in the ground state, so

$$N_0 = N - N_{\text{excited}} = \left[1 - \left(\frac{T}{T_c} \right)^{3/2} \right] N \qquad (T < T_c). \tag{7.129}$$

Figure 7.32 shows a graph of N_0 and N_{excited} as functions of temperature; Figure 7.33 shows the temperature dependence of the chemical potential.

The abrupt accumulation of atoms in the ground state at temperatures below T_c is called **Bose-Einstein condensation**. The transition temperature T_c is called the **condensation temperature**, while the ground-state atoms themselves are called the **condensate**. Notice from equation 7.126 that the condensation temperature is (aside from the factor of 2.612) precisely the temperature at which the quantum volume $(v_Q = (h^2/2\pi mkT)^{3/2})$ equals the average volume per particle (V/N). In other words, if we imagine the atoms being in wavefunctions that are as localized in space as possible (as in Figure 7.4), then condensation begins to occur

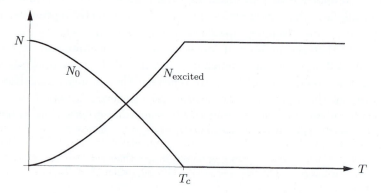

Figure 7.32. Number of atoms in the ground state (N_0) and in excited states, for an ideal Bose gas in a three-dimensional box. Below T_c the number of atoms in excited states is proportional to $T^{3/2}$.

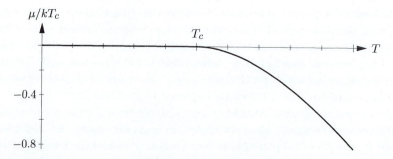

Figure 7.33. Chemical potential of an ideal Bose gas in a three-dimensional box. Below the condensation temperature, μ differs from zero by an amount that is too small to show on this scale. Above the condensation temperature μ becomes negative; the values plotted here were calculated numerically as described in Problem 7.69.

just as the wavefunctions begin to overlap significantly. (The condensate atoms themselves have wavefunctions that occupy the entire container, which I won't try to draw.)

Numerically, the condensation temperature turns out to be very small in all realistic experimental situations. However, it's not as low as we might have guessed. If you put a *single* particle into a box of volume V, it's reasonably likely to be found in the ground state only when kT is of order ϵ_0 or smaller (so that the excited states, which have energies of $2\epsilon_0$ and higher, are significantly less probable). However, if you put a *large* number of identical bosons into the same box, you can get most of them into the ground state at temperatures only somewhat less than T_c, which is much higher: From equations 7.118 and 7.126 we see that kT_c is greater than ϵ_0 by a factor of order $N^{2/3}$. The hierarchy of energy scales—$(\epsilon_0 - \mu) \ll \epsilon_0 \ll kT_c$—is depicted schematically in Figure 7.34.

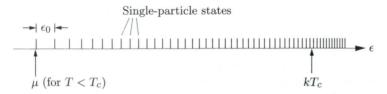

Figure 7.34. Schematic representation of the energy scales involved in Bose-Einstein condensation. The short vertical lines mark the energies of various single-particle states. (Aside from growing closer together (on average) with increasing energy, the locations of these lines are not quantitatively accurate.) The condensation temperature (times k) is many times larger than the spacing between the lowest energy levels, while the chemical potential, when $T < T_c$, is only a tiny amount below the ground-state energy.

Real-World Examples

Bose-Einstein condensation of a gas of weakly interacting atoms was first achieved in 1995, using rubidium-87.[*] In this experiment, roughly 10^4 atoms were confined (using the laser cooling and trapping technique described in Section 4.4) in a volume of order 10^{-15} m^3. A large fraction of the atoms were observed to condense into the ground state at a temperature of about 10^{-7} K, a hundred times greater than the temperature at which a *single* isolated atom would have a good chance of being in the ground state. Figure 7.35 shows the velocity distribution of the atoms in this experiment, at temperatures above, just below, and far below the condensation temperature. As of 1999, Bose-Einstein condensation has also been achieved with dilute gases of atomic sodium, lithium, and hydrogen.

---

[*]For a beautiful description of this experiment see Carl E. Wieman, "The Richtmyer Memorial Lecture: Bose-Einstein Condensation in an Ultracold Gas," *American Journal of Physics* **64**, 847–855 (1996).

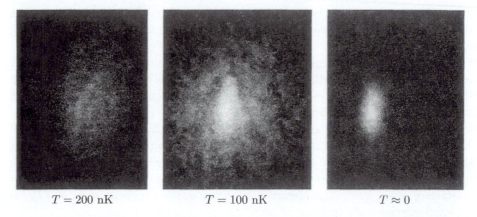

$$T = 200 \text{ nK} \qquad\qquad T = 100 \text{ nK} \qquad\qquad T \approx 0$$

Figure 7.35. Evidence for Bose-Einstein condensation of rubidium-87 atoms. These images were made by turning off the magnetic field that confined the atoms, letting the gas expand for a moment, and then shining light on the expanded cloud to map its distribution. Thus, the positions of the atoms in these images give a measure of their *velocities* just before the field was turned off. Above the condensation temperature (left), the velocity distribution is broad and isotropic, in accord with the Maxwell-Boltzmann distribution. Below the condensation temperature (center), a substantial fraction of the atoms fall into a small, elongated region in velocity space. These atoms make up the condensate; the elongation occurs because the trap is narrower in the vertical direction, causing the ground-state wavefunction to be narrower in position space and thus wider in velocity space. At the lowest temperatures achieved (right), essentially all of the atoms are in the ground-state wavefunction. From Carl E. Wieman, *American Journal of Physics* **64**, 854 (1996).

Bose-Einstein condensation also occurs in systems where particle interactions are significant, so that the quantitative treatment of this section is not very accurate. The most famous example is liquid helium-4, which forms a **superfluid** phase, with essentially zero viscosity, at temperatures below 2.17 K (see Figure 5.13). More precisely, the liquid below this temperature is a mixture of normal and superfluid components, with the superfluid becoming more predominant as the temperature decreases. This behavior suggests that the superfluid component is a Bose-Einstein condensate; indeed, a naive calculation, ignoring interatomic forces, predicts a condensation temperature only slightly greater than the observed value (see Problem 7.68). Unfortunately, the superfluid property itself cannot be understood without accounting for interactions between the helium atoms.

If the superfluid component of helium-4 is a Bose-Einstein condensate, then you would think that helium-3, which is a fermion, would have no such phase. And indeed, it has no superfluid transition anywhere near 2 K. Below 3 *milli*kelvin, however, ^{3}He turns out to have not one but *two* distinct superfluid phases.* How

---

*These phases were discovered in the early 1970s. To achieve such low temperatures the experimenters used a helium dilution refrigerator (see Section 4.4) in combination with the cooling technique described in Problem 5.34.

is this possible for a system of fermions? It turns out that the "particles" that condense are actually *pairs* of ^{3}He atoms, held together by the interaction of their nuclear magnetic moments with the surrounding atoms.[*] A pair of fermions has integer spin and is therefore a boson. An analogous phenomenon occurs in a superconductor, where pairs of electrons are held together through interactions with the vibrating lattice of ions. At low temperature these pairs "condense" into a superconducting state, yet another example of Bose-Einstein condensation.[†]

Why Does it Happen?

Now that I've shown you that Bose-Einstein condensation *does* happen, let me return to the question of *why* it happens. The derivation above was based entirely on the Bose-Einstein distribution function—a powerful tool, but not terribly intuitive. It's not hard, though, to gain some understanding of this phenomenon using more elementary methods.

Suppose that, instead of a collection of identical bosons, we have a collection of N *distinguishable* particles all confined inside a box. (Perhaps they're all painted different colors or something.) Then, if the particles don't interact with each other, we can treat each one of them as a separate system using Boltzmann statistics. At temperature T, a given particle has a decent chance of occupying any single-particle state whose energy is of order kT, and the number of such states will be quite large under any realistic conditions. (This number is essentially equal to the single-particle partition function, Z_1.) The probability of the particle being in the ground state is therefore very small, namely $1/Z_1$. Since this conclusion applies separately to each one of the N distinguishable particles, only a tiny fraction of the particles will be found in the ground state. There is no Bose-Einstein condensation.

It's useful to analyze this same situation from a different perspective, treating the entire system all at once, rather than one particle at a time. From this viewpoint, each *system* state has its own probability and its own Boltzmann factor. The system state with all the particles in the ground state has a Boltzmann factor of 1 (taking the ground-state energy to be zero for simplicity), while a system state with total energy U has a Boltzmann factor of $e^{-U/kT}$. According to the conclusion of the previous paragraph, the dominant system states are those for which nearly all of the particles are in excited states with energies of order kT; the total system energy is therefore $U \sim NkT$, so the Boltzmann factor of a typical system state is something like $e^{-NkT/kT} = e^{-N}$. This is a *very* small number! How can it be that the system prefers these states, rather than condensing into the ground state with its much larger Boltzmann factor?

The answer is that while any *particular* system state with energy of order NkT is highly improbable, the *number* of such states is so huge that taken together they

---

[*]For an overview of the physics of both isotopes of liquid helium, see Wilks and Betts (1987).

[†]For review articles on Bose-Einstein condensation in a variety of systems, see A. Griffin, D. W. Snoke, and S. Stringari, eds., *Bose-Einstein Condensation* (Cambridge University Press, Cambridge, 1995).

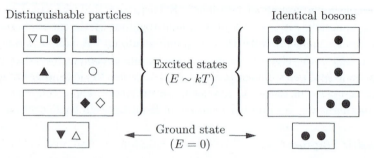

Distinguishable particles   Identical bosons

Excited states
$(E \sim kT)$

Ground state
$(E = 0)$

Figure 7.36. When most particles are in excited states, the Boltzmann factor for the entire system is always very small (of order e^{-N}). For distinguishable particles, the number of arrangements among these states is so large that system states of this type are still very probable. For identical bosons, however, the number of arrangements is much smaller.

are quite probable after all (see Figure 7.36). The number of ways of arranging N distinguishable particles among Z_1 single-particle states is Z_1^N, which overwhelms the Boltzmann factor e^{-N} provided that $Z_1 \gg 1$.

Now let's return to the case of identical bosons. Here again, if essentially all the particles are in single-particle states with energies of order kT, then the system state has a Boltzmann factor of order e^{-N}. But now, the *number* of such system states is much smaller. This number is essentially the number of ways of arranging N indistinguishable particles among Z_1 single-particle states, which is mathematically the same as the number of ways of arranging N units of energy among Z_1 oscillators in an Einstein solid:

$$\begin{pmatrix} \text{number of} \\ \text{system states} \end{pmatrix} \sim \begin{pmatrix} N+Z_1-1 \\ N \end{pmatrix} \sim \begin{cases} (eZ_1/N)^N & \text{when } Z_1 \gg N; \\ (eN/Z_1)^{Z_1} & \text{when } Z_1 \ll N. \end{cases} \quad (7.130)$$

When the number of available single-particle states is much larger than the number of bosons, the combinatoric factor is again large enough to overwhelm the Boltzmann factor e^{-N}, so system states with essentially all the bosons in excited states will again predominate. On the other hand, when the number of available single-particle states is much smaller than the number of bosons, the combinatoric factor is not large enough to compensate for the Boltzmann factor, so these system states, even all taken together, will be exponentially improbable. (This last conclusion is not quite clear from looking at the formulas, but here is a simple numerical example: When $N = 100$ and $Z_1 = 25$, a system state with all the bosons in excited states has a Boltzmann factor of order $e^{-100} = 4 \times 10^{-44}$, while the number of such system states is only $\binom{124}{100} = 3 \times 10^{25}$.) In general, the combinatoric factor will be sufficiently large to get about one boson, on average, into each available excited state. Any remaining bosons condense into the ground state, because of the way the Boltzmann factor favors system states with lower energy.

So the explanation of Bose-Einstein condensation lies in the combinatorics of counting arrangements of identical particles: Since the number of distinct ways of arranging identical particles among the excited states is relatively small, the ground state becomes much more favored than if the particles were distinguishable. You may still be wondering, though, how we *know* that bosons of a given species are truly identical and must therefore be counted in this way. Or alternatively, how do we *know* that the fundamental assumption, which gives all distinct states (of the system plus its environment) the same statistical weight, applies to systems of identical bosons? These questions have good theoretical answers, but the answers require an understanding of quantum mechanics that is beyond the scope of this book. Even then, the answers are not completely airtight—there is still the possibility that *some* undiscovered type of interaction may be able to distinguish supposedly identical bosons from each other, causing a Bose-Einstein condensate to spontaneously evaporate. So far, the experimental fact is that such interactions do not seem to exist. Let us therefore invoke Occam's Razor and conclude, if only tentatively, that bosons of a given species are truly indistinguishable; as David Griffiths has said,* even God cannot tell them apart.

Problem 7.65. Evaluate the integral in equation 7.124 numerically, to confirm the value quoted in the text.

Problem 7.66. Consider a collection of 10,000 atoms of rubidium-87, confined inside a box of volume $(10^{-5} \text{ m})^3$.

(a) Calculate ϵ_0, the energy of the ground state. (Express your answer in both joules and electron-volts.)

(b) Calculate the condensation temperature, and compare kT_c to ϵ_0.

(c) Suppose that $T = 0.9T_c$. How many atoms are in the ground state? How close is the chemical potential to the ground-state energy? How many atoms are in each of the (threefold-degenerate) first excited states?

(d) Repeat parts (b) and (c) for the case of 10^6 atoms, confined to the same volume. Discuss the conditions under which the number of atoms in the ground state will be much greater than the number in the first excited state.

Problem 7.67. In the first achievement of Bose-Einstein condensation with atomic hydrogen,[†] a gas of approximately 2×10^{10} atoms was trapped and cooled until its peak density was 1.8×10^{14} atoms/cm³. Calculate the condensation temperature for this system, and compare to the measured value of 50 μK.

Problem 7.68. Calculate the condensation temperature for liquid helium-4, pretending that the liquid is a gas of noninteracting atoms. Compare to the observed temperature of the superfluid transition, 2.17 K. (The density of liquid helium-4 is 0.145 g/cm³.)

Problem 7.69. If you have a computer system that can do numerical integrals, it's not particularly difficult to evaluate μ for $T > T_c$.

(a) As usual when solving a problem on a computer, it's best to start by putting everything in terms of dimensionless variables. So define $t = T/T_c$,

---

Introduction to Quantum Mechanics (Prentice-Hall, Englewood Cliffs, NJ, 1995), page 179.

[†]Dale G. Fried et al., *Physical Review Letters* **81**, 3811 (1998).

$c = \mu/kT_c$, and $x = \epsilon/kT_c$. Express the integral that defines μ, equation 7.122, in terms of these variables. You should obtain the equation

$$2.315 = \int_0^{\infty} \frac{\sqrt{x}\,dx}{e^{(x-c)/t} - 1}.$$

(b) According to Figure 7.33, the correct value of c when $T = 2T_c$ is approximately -0.8. Plug in these values and check that the equation above is approximately satisfied.

(c) Now vary μ, holding T fixed, to find the precise value of μ for $T = 2T_c$. Repeat for values of T/T_c ranging from 1.2 up to 3.0, in increments of 0.2. Plot a graph of μ as a function of temperature.

Problem 7.70. Figure 7.37 shows the heat capacity of a Bose gas as a function of temperature. In this problem you will calculate the shape of this unusual graph.

(a) Write down an expression for the total energy of a gas of N bosons confined to a volume V, in terms of an integral (analogous to equation 7.122).

(b) For $T < T_c$ you can set $\mu = 0$. Evaluate the integral numerically in this case, then differentiate the result with respect to T to obtain the heat capacity. Compare to Figure 7.37.

(c) Explain why the heat capacity must approach $\frac{3}{2}Nk$ in the high-T limit.

(d) For $T > T_c$ you can evaluate the integral using the values of μ calculated in Problem 7.69. Do this to obtain the energy as a function of temperature, then numerically differentiate the result to obtain the heat capacity. Plot the heat capacity, and check that your graph agrees with Figure 7.37.

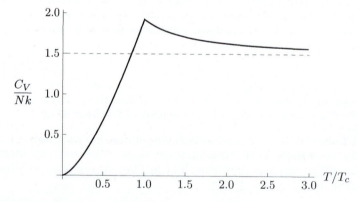

Figure 7.37. Heat capacity of an ideal Bose gas in a three-dimensional box.

Problem 7.71. Starting from the formula for C_V derived in Problem 7.70(b), calculate the entropy, Helmholtz free energy, and pressure of a Bose gas for $T < T_c$. Notice that the pressure is independent of volume; how can this be the case?

Problem 7.72. For a gas of particles confined inside a *two*-dimensional box, the density of states is constant, independent of ϵ (see Problem 7.28). Investigate the behavior of a gas of noninteracting bosons in a two-dimensional box. You should find that the chemical potential remains significantly less than zero as long as T is significantly greater than zero, and hence that there is no abrupt condensation of particles into the ground state. Explain how you know that this is the case, and describe what *does* happen to this system as the temperature decreases. What property must $g(\epsilon)$ have in order for there to be an abrupt Bose-Einstein condensation?

Problem 7.73. Consider a gas of N identical spin-0 bosons confined by an isotropic three-dimensional harmonic oscillator potential. (In the rubidium experiment discussed above, the confining potential was actually harmonic, though not isotropic.) The energy levels in this potential are $\epsilon = nhf$, where n is any nonnegative integer and f is the classical oscillation frequency. The degeneracy of level n is $(n + 1)(n + 2)/2$.

- **(a)** Find a formula for the density of states, $g(\epsilon)$, for an atom confined by this potential. (You may assume $n \gg 1$.)

- **(b)** Find a formula for the condensation temperature of this system, in terms of the oscillation frequency f.

- **(c)** This potential effectively confines particles inside a volume of roughly the cube of the oscillation amplitude. The oscillation amplitude, in turn, can be estimated by setting the particle's total energy (of order kT) equal to the potential energy of the "spring." Making these associations, and neglecting all factors of 2 and π and so on, show that your answer to part (b) is roughly equivalent to the formula derived in the text for the condensation temperature of bosons confined inside a box with rigid walls.

Problem 7.74. Consider a Bose gas confined in an isotropic harmonic trap, as in the previous problem. For this system, because the energy level structure is much simpler than that of a three-dimensional box, it is feasible to carry out the sum in equation 7.121 *numerically*, without approximating it as an integral.*

- **(a)** Write equation 7.121 for this system as a sum over energy levels, taking degeneracy into account. Replace T and μ with the dimensionless variables $t = kT/hf$ and $c = \mu/hf$.

- **(b)** Program a computer to calculate this sum for any given values of t and c. Show that, for $N = 2000$, equation 7.121 is satisfied at $t = 15$ provided that $c = -10.534$. (Hint: You'll need to include approximately the first 200 energy levels in the sum.)

- **(c)** For the same parameters as in part (b), plot the number of particles in each energy level as a function of energy.

- **(d)** Now reduce t to 14, and adjust the value of c until the sum again equals 2000. Plot the number of particles as a function of energy.

- **(e)** Repeat part (d) for $t = 13, 12, 11$, and 10. You should find that the required value of c increases toward zero but never quite reaches it. Discuss the results in some detail.

---

*This problem is based on an article by Martin Ligare, *American Journal of Physics* **66**, 185–190 (1998).

Problem 7.75. Consider a gas of noninteracting spin-0 bosons at *high* temperatures, when $T \gg T_c$. (Note that "high" in this sense can still mean below 1 K.)

(a) Show that, in this limit, the Bose-Einstein distribution function can be written approximately as

$$\bar{n}_{\mathrm{BE}} = e^{-(\epsilon-\mu)/kT}\left[1 + e^{-(\epsilon-\mu)/kT} + \cdots\right].$$

(b) Keeping only the terms shown above, plug this result into equation 7.122 to derive the first quantum correction to the chemical potential for a gas of bosons.

(c) Use the properties of the grand free energy (Problems 5.23 and 7.7) to show that the pressure of any system is given by $P = (kT/V)\ln \mathcal{Z}$, where $\mathcal{Z}$ is the grand partition function. Argue that, for a gas of noninteracting particles, $\ln \mathcal{Z}$ can be computed as the sum over all modes (or single-particle states) of $\ln \mathcal{Z}_i$, where $\mathcal{Z}_i$ is the grand partition function for the ith mode.

(d) Continuing with the result of part (c), write the sum over modes as an integral over energy, using the density of states. Evaluate this integral explicitly for a gas of noninteracting bosons in the high-temperature limit, using the result of part (b) for the chemical potential and expanding the logarithm as appropriate. When the smoke clears, you should find

$$P = \frac{NkT}{V}\left(1 - \frac{Nv_Q}{4\sqrt{2}V}\right),$$

again neglecting higher-order terms. Thus, quantum statistics results in a *lowering* of the pressure of a boson gas, as one might expect.

(e) Write the result of part (d) in the form of the virial expansion introduced in Problem 1.17, and read off the second virial coefficient, $B(T)$. Plot the predicted $B(T)$ for a hypothetical gas of noninteracting helium-4 atoms.

(f) Repeat this entire problem for a gas of spin-1/2 fermions. (Very few modifications are necessary.) Discuss the results, and plot the predicted virial coefficient for a hypothetical gas of noninteracting helium-3 atoms.

Ten percent or more of a complete stellar inventory consists of white dwarfs, just sitting there, radiating away the thermal (kinetic) energy of their carbon and oxygen nuclei from underneath very thin skins of hydrogen and helium. They will continue this uneventful course until the universe recontracts, their baryons decay, or they collapse to black holes by barrier penetration. (Likely time scales for these three outcomes are 10^{14}, 10^{33}, and $10^{10^{76}}$ —years for the first two and for the third one it doesn't matter.)

—Virginia Trimble, *SLAC Beam Line*
21, 3 (fall, 1991).

8 Systems of Interacting Particles

An **ideal** system, in statistical mechanics, is one in which the particles (be they molecules, electrons, photons, phonons, or magnetic dipoles) do not exert significant forces on each other. All of the systems considered in the previous two chapters were "ideal" in this sense. But the world would be a boring place if everything in it were ideal. Gases would never condense into liquids, and no material would magnetize spontaneously, for example. So it's about time we considered some nonideal systems.

Predicting the behavior of a nonideal system, consisting of many mutually interacting particles, is not easy. You can't just break the system down into lots of independent subsystems (particles or modes), treat these subsystems one at a time, and then sum over subsystems as we did in the previous two chapters. Instead you have to treat the whole system all at once. Usually this means that you can't calculate thermodynamic quantities exactly—you have to resort to approximation. Applying suitable approximation schemes to various systems of interacting particles has become a major component of modern statistical mechanics. Moreover, analogous approximation schemes are widely used in other research fields, especially in the application of quantum mechanics to multiparticle systems.

In this chapter I will introduce just two examples of interacting systems: a gas of weakly interacting molecules, and an array of magnetic dipoles that tend to align parallel to their neighbors. For each of these systems there is an approximation method (diagrammatic perturbation theory and Monte Carlo simulation, respectively) that not only solves the problem at hand, but has also proved useful in tackling a much wider variety of problems in theoretical physics.*

---

*The two sections of this chapter are independent of each other; feel free to read them in either order. Also, aside from a few problems, nothing in this chapter depends on Chapter 7.

8.1 Weakly Interacting Gases

In Section 5.3 we made a first attempt at understanding nonideal gases, using the van der Waals equation. That equation is very successful qualitatively, even predicting the condensation of a dense gas into a liquid. But it is not very accurate quantitatively, and its connection to fundamental molecular interactions is tenuous at best. So, can we do better? Specifically, can we predict the behavior of a nonideal gas from first principles, using the powerful tools of statistical mechanics?

The answer is yes, but it's not easy. At least at the level of this book, a fundamental calculation of the properties of a nonideal gas is feasible only in the limit of low density, when the interactions between molecules are still relatively weak. In this section I'll carry out such a calculation, ultimately deriving a correction to the ideal gas law that is valid in the low-density limit. This approach won't help us understand the liquid-gas phase transformation, but at least the results will be quantitatively accurate within their limited range of validity. In short, we're trading generality for accuracy and rigor.

The Partition Function

As always, we begin by writing down the partition function. Taking the viewpoint of Section 2.5 and Problem 6.51, let us characterize the "state" of a molecule by its position and momentum vectors. Then the partition function for a *single* molecule is

$$Z_1 = \frac{1}{h^3} \int d^3r\, d^3p\, e^{-\beta E}, \tag{8.1}$$

where the single integral sign actually represents six integrals, three over the position components (denoted d^3r) and three over the momentum components (denoted d^3p). The region of integration includes all momentum vectors, but only those position vectors that lie within a box of volume V. The factor of $1/h^3$ is needed to give us a unitless number that counts the independent wavefunctions. For simplicity I've omitted any sum over internal states (such as rotational states) of the molecule.

For a single molecule with no internal degrees of freedom, equation 8.1 is equivalent to what I wrote in Section 6.7 for an ideal gas (as shown in Problem 6.51). For a gas of N identical molecules, the corresponding expression is easy to write down but rather frightening to look at:

$$Z = \frac{1}{N!}\frac{1}{h^{3N}} \int d^3r_1 \cdots d^3r_N\, d^3p_1 \cdots d^3p_N\, e^{-\beta U}. \tag{8.2}$$

Now there are $6N$ integrals, over the position and momentum components of all N molecules. There are also N factors of $1/h^3$, and a prefactor of $1/N!$ to account for the indistinguishability of identical molecules. The Boltzmann factor contains the total energy U of the entire system.

If this were an *ideal* gas, then U would just be a sum of kinetic energy terms,

$$U_{\text{kin}} = \frac{|\vec{p}_1|^2}{2m} + \frac{|\vec{p}_2|^2}{2m} + \cdots + \frac{|\vec{p}_N|^2}{2m}. \tag{8.3}$$

For a *non*ideal gas, though, there is also potential energy, due to the interactions between molecules. Denoting the entire potential energy as U_{pot}, the partition function can be written as

$$Z = \frac{1}{N!} \frac{1}{h^{3N}} \int d^3 r_1 \cdots d^3 r_N \, d^3 p_1 \cdots d^3 p_N \, e^{-\beta |\vec{p}_1|^2 / 2m} \cdots e^{-\beta |\vec{p}_N|^2 / 2m} \, e^{-\beta U_{\text{pot}}}.$$

(8.4)

Now the good news is, the $3N$ momentum integrals are easy to evaluate. Because the potential energy depends only on the positions of the molecules, not on their momenta, each momentum $\vec{p}_i$ appears only in the kinetic energy Boltzmann factor $e^{-\beta |\vec{p}_i|^2 / 2m}$, and the integral over this momentum can be evaluated exactly as for an ideal gas, yielding the same result:

$$\int d^3 p_i \, e^{-\beta |\vec{p}_i|^2 / 2m} = \left(\sqrt{2\pi m k T} \right)^3.$$

(8.5)

Assembling N of these factors gives us

$$Z = \frac{1}{N!} \left(\frac{\sqrt{2\pi m k T}}{h} \right)^{3N} \int d^3 r_1 \cdots d^3 r_N \, e^{-\beta U_{\text{pot}}}$$

$$= Z_{\text{ideal}} \cdot \frac{1}{V^N} \int d^3 r_1 \cdots d^3 r_N \, e^{-\beta U_{\text{pot}}},$$

(8.6)

where Z_{ideal} is the partition function of an ideal gas, equation 6.85. Thus, our task is reduced to evaluating the rest of this expression,

$$Z_c = \frac{1}{V^N} \int d^3 r_1 \cdots d^3 r_N \, e^{-\beta U_{\text{pot}}},$$

(8.7)

called the **configuration integral** (because it involves an integral over all configurations, or positions, of the molecules).

The Cluster Expansion

In order to write the configuration integral more explicitly, let me assume that the potential energy of the gas can be written as a sum of potential energies due to interactions between *pairs* of molecules:

$$U_{\text{pot}} = u_{12} + u_{13} + \cdots + u_{1N} + u_{23} + \cdots + u_{N-1,N}$$

$$= \sum_{\text{pairs}} u_{ij}.$$

(8.8)

Each term u_{ij} represents the potential energy due to the interaction of molecule i with molecule j, and I'll assume that it depends only on the distance between these two molecules, $|\vec{r}_i - \vec{r}_j|$. This is a significant simplification. For one thing, I'm neglecting any possible dependence of the potential energy on the orientation of a molecule. For another, I'm neglecting the fact that when two molecules are close together they distort each other, thus altering the interaction of either of them with a third molecule. Still, this "simplification" doesn't make the configuration integral look any prettier; we now have

$$Z_c = \frac{1}{V^N} \int d^3r_1 \cdots d^3r_N \prod_{\text{pairs}} e^{-\beta u_{ij}}, \tag{8.9}$$

where the $\prod$ symbol denotes a product over all distinct pairs i, j.

Eventually we'll need to assume an explicit formula for the function u_{ij}. For now, though, all we need to know is that it goes to zero as the distance between molecules i and j becomes large. Especially in a low-density gas, practically all pairs of molecules will be far enough apart that $u_{ij} \ll kT$, and therefore the Boltzmann factor $e^{-\beta u_{ij}}$ is extremely close to 1. With this in mind, the next step is to isolate the *deviation* of each Boltzmann factor from 1, by writing

$$e^{-\beta u_{ij}} = 1 + f_{ij}, \tag{8.10}$$

which defines a new quantity f_{ij}, called the **Mayer f-function**. The product of all these Boltzmann factors is then

$$\prod_{\text{pairs}} e^{-\beta u_{ij}} = \prod_{\text{pairs}} (1 + f_{ij})$$
$$= (1 + f_{12})(1 + f_{13}) \cdots (1 + f_{1N})(1 + f_{23}) \cdots (1 + f_{N-1,N}). \tag{8.11}$$

If we imagine multiplying out all these factors, the first term will just be 1. Then there will be a bunch of terms with just one f-function, then a bunch with two f-functions, and so on:

$$\prod_{\text{pairs}} e^{-\beta u_{ij}} = 1 + \sum_{\text{pairs}} f_{ij} + \sum_{\substack{\text{distinct} \\ \text{pairs}}} f_{ij} f_{kl} + \cdots. \tag{8.12}$$

Plugging this expansion back into the configuration integral yields

$$Z_c = \frac{1}{V^N} \int d^3r_1 \cdots d^3r_N \left(1 + \sum_{\text{pairs}} f_{ij} + \sum_{\substack{\text{distinct} \\ \text{pairs}}} f_{ij} f_{kl} + \cdots \right). \tag{8.13}$$

Our hope is that the terms in this series will become less important as they contain more f-functions, so we can get away with evaluating only the first term or two.

The very first term in equation 8.13, with no f-functions, is easy to evaluate: Each d^3r integral yields a factor of the volume of the box, so

$$\frac{1}{V^N} \int d^3r_1 \cdots d^3r_N \, (1) = 1. \tag{8.14}$$

In each of the terms with one f-function, all but two of the integrals yield trivial factors of V; for instance,

$$\frac{1}{V^N} \int d^3r_1 \cdots d^3r_N \, f_{12} = \frac{1}{V^N} V^{N-2} \int d^3r_1 \, d^3r_2 \, f_{12}$$
$$= \frac{1}{V^2} \int d^3r_1 \, d^3r_2 \, f_{12}, \tag{8.15}$$

because f_{12} depends only on $\vec{r}_1$ and $\vec{r}_2$. Actually, since it doesn't matter which molecules we call 1 and 2, every one of the terms with one f-function is exactly equal to this one, and the sum of all of them is equal to this expression times the number of distinct pairs, $N(N-1)/2$:

$$\frac{1}{V^N} \int d^3r_1 \cdots d^3r_N \left(\sum_{\text{pairs}} f_{ij} \right) = \frac{1}{2} \frac{N(N-1)}{V^2} \int d^3r_1 \, d^3r_2 \, f_{12}. \tag{8.16}$$

Before going on, I'd like to introduce a pictorial abbreviation for this expression, which will also give us a physical interpretation of it. The picture is simply a pair of dots, representing molecules 1 and 2, connected by a line, representing the interaction between these molecules:

$$\begin{matrix}\bullet \\ | \\ \bullet\end{matrix} = \frac{1}{2} \frac{N(N-1)}{V^2} \int d^3r_1 \, d^3r_2 \, f_{12}. \tag{8.17}$$

The rules for translating the picture into the formula are as follows:

1. Number the dots starting with 1, and for each dot i, write down the expression $(1/V) \int d^3r_i$. Multiply by N for the first dot, $N-1$ for the second dot, $N-2$ for the third dot, and so on.

2. For a line connecting dots i and j, write down a factor f_{ij}.

3. Divide by the **symmetry factor** of the diagram, which is the number of ways of numbering the dots without changing the corresponding product of f-functions. (Equivalently, the symmetry factor is the number of permutations of dots that leave the diagram unchanged.)

For the simple diagram in equation 8.17, these rules give precisely the expression written; the symmetry factor is 2, because $f_{12} = f_{21}$. Physically, this diagram represents a configuration in which only two molecules are interacting with each other.

Now consider the terms in the configuration integral (8.13) with two f-functions. Each of these terms involves two pairs of molecules, and these pairs could have one molecule in common, or none. In a term in which the pairs share a molecule, all but three of the integrals give trivial factors of V; the number of such terms is $N(N-1)(N-2)/2$, so the sum of these terms is equal to the diagram

$$\begin{matrix}\bullet \quad \bullet \\ \diagdown \diagup \\ \bullet\end{matrix} = \frac{1}{2} \frac{N(N-1)(N-2)}{V^3} \int d^3r_1 \, d^3r_2 \, d^3r_3 \, f_{12} f_{23}. \tag{8.18}$$

This diagram represents a configuration in which one molecule simultaneously interacts with two others. In a term in which the pairs do not share a molecule there are four left-over integrals; the number of such terms is $N(N-1)(N-2)(N-3)/8$, so the sum of these terms is

$$\left(\begin{matrix}\bullet \\ | \\ \bullet\end{matrix} \; \begin{matrix}\bullet \\ | \\ \bullet\end{matrix} \right) = \frac{1}{8} \frac{N(N-1)(N-2)(N-3)}{V^4} \int d^3r_1 \, d^3r_2 \, d^3r_3 \, d^3r_4 \, f_{12} f_{34}. \tag{8.19}$$

This diagram represents two simultaneous interactions between pairs of molecules. For either diagram, the rules given above yield precisely the correct expression.

By now you can probably guess that the entire configuration integral can be written as a sum of diagrams:

$$Z_c = 1 + \text{⬤} + \text{⟨⟩} + (\text{⬤⬤}) + \triangle + \text{⊤} + \text{⊓}$$
$$+ (\text{⬤⟨⟩}) + (\text{⬤⬤⬤}) + \cdots . \tag{8.20}$$

Every possible diagram occurs exactly once in this sum, with the constraints that every dot must be connected to at least one other dot, and that no pair of dots can be connected more than once. I won't try to prove that the combinatoric factors work out exactly right in all cases, but they do. This representation of the configuration integral is an example of a **diagrammatic perturbation series**: The first term, 1, represents the "ideal" case of a gas of noninteracting molecules; the remaining terms, represented by diagrams, depict the interactions that "perturb" the system away from the ideal limit. We expect that the simpler diagrams will be more important than the more complicated ones, at least for a low-density gas in which simultaneous interactions of large numbers of molecules should be rare. Although you would never want to *calculate* the more complicated diagrams, they still give a way to visualize interactions involving arbitrary numbers of molecules.

But even for a low-density gas, we can't get away with keeping only the first couple of terms in the diagrammatic expansion of Z_c. We'll soon see that even the simplest two-dot diagram evaluates to a number much larger than 1, so the subseries

$$1 + \text{⬤} + (\text{⬤⬤}) + (\text{⬤⬤⬤}) + \cdots \tag{8.21}$$

does not converge until after many terms, when the symmetry factors grow to be quite large. Physically, this is because simultaneous interactions of many isolated pairs are very common when N is large. Fortunately, though, this sum can be simplified. At least as a first approximation, we can set $N = N - 1 = N - 2 = \cdots$. Then, because a disconnected diagram containing n identical subdiagrams gets a symmetry factor of $n!$ in the denominator, this series is simply

$$1 + \text{⬤} + \frac{1}{2}\left(\text{⬤}\right)^2 + \frac{1}{3!}\left(\text{⬤}\right)^3 + \cdots = \exp\left(\text{⬤}\right). \tag{8.22}$$

In other words, the disconnected diagrams gather themselves into a simple exponential function of the basic connected diagram. But that's not all. At the next level of approximation, when we keep terms that are smaller by a factor of N, it turns out that the series 8.22 even includes the connected diagram 8.18 (see Problem 8.7). Similar cancellations occur among the more complicated diagrams, with the end result being that Z_c can be written as the exponential of the sum of only those diagrams that are connected, and that would remain connected if any single dot were to be removed:

$$Z_c = \exp\left(\text{⬤} + \triangle + \square + \boxtimes_{\text{half}} + \boxtimes + \cdots\right). \tag{8.23}$$

This formula isn't quite exact, but the terms that are omitted go to zero in the thermodynamic limit, $N \to \infty$ with N/V fixed. Similarly, at this stage it is valid to set $N = N - 1 = N - 2 = \cdots$ in all subsequent calculations. Unfortunately, the general proof of this formula is beyond the scope of this book.

Each diagram in equation 8.23 is called a **cluster**, because it represents a cluster of simultaneously interacting molecules. The formula itself is called the **cluster expansion** for the configuration integral. The cluster expansion is a well-behaved series: For a low-density gas, a cluster diagram with more dots is always smaller than one with fewer dots.

Now let's put the pieces back together. Recall from equation 8.6 that the whole partition function for the gas is equal to the configuration integral times the partition function for an ideal gas:

$$Z = Z_{\text{ideal}} \cdot Z_c. \tag{8.24}$$

In order to compute the pressure, we really want to know the Helmholtz free energy,

$$F = -kT \ln Z = -kT \ln Z_{\text{ideal}} - kT \ln Z_c. \tag{8.25}$$

We computed Z_{ideal} in Section 6.7; plugging in that result and the cluster expansion for Z_c, we obtain

$$F = -NkT \ln\left(\frac{V}{N v_Q}\right) - kT \left(\vdots + \triangle + \square + \cdots \right). \tag{8.26}$$

The pressure is therefore

$$P = -\left(\frac{\partial F}{\partial V}\right)_{N,T} = \frac{NkT}{V} + kT \frac{\partial}{\partial V}\left(\vdots + \triangle + \square + \cdots \right). \tag{8.27}$$

Thus, if we can evaluate some of the cluster diagrams explicitly, we can improve upon the ideal gas law.

Problem 8.1. For each of the diagrams shown in equation 8.20, write down the corresponding formula in terms of f-functions, and explain why the symmetry factor gives the correct overall coefficient.

Problem 8.2. Draw all the diagrams, connected or disconnected, representing terms in the configuration integral with four factors of f_{ij}. You should find 11 diagrams in total, of which five are connected.

Problem 8.3. Keeping only the first two diagrams in equation 8.23, and approximating $N \approx N - 1 \approx N - 2 \approx \cdots$, expand the exponential in a power series through the third power. Multiply each term out, and show that all the numerical coefficients give precisely the correct symmetry factors for the disconnected diagrams.

Problem 8.4. Draw all the connected diagrams containing four dots. There are six diagrams in total; be careful to avoid drawing two diagrams that look superficially different but are actually the same. Which of the diagrams would remain connected if any single dot were removed?

The Second Virial Coefficient

Let's now consider just the simplest, two-dot diagram:

$$\text{[diagram]} = \frac{1}{2}\frac{N^2}{V^2}\int d^3r_1\, d^3r_2\, f_{12}. \tag{8.28}$$

Because the f-function depends only on the distance between the two molecules, let me define $\vec{r} \equiv \vec{r}_2 - \vec{r}_1$ and change variables in the second integral from $\vec{r}_2$ to $\vec{r}$:

$$\text{[diagram]} = \frac{1}{2}\frac{N^2}{V^2}\int d^3r_1\left(\int d^3r\, f(r)\right), \tag{8.29}$$

where

$$f(r) = e^{-\beta u(r)} - 1 \tag{8.30}$$

and $u(r)$ is the potential energy due to the interaction of any pair of molecules, as a function of the distance between their centers. To evaluate the integral over $\vec{r}$ we'll have to do a bit of work, but we can say one thing about it already: The result will be some intensive quantity that is independent of $\vec{r}_1$ and of V. This is because $f(r)$ goes to zero when r is only a few times larger than the size of a molecule, and the chance of r_1 being within this distance of the wall of the box is negligible. So whatever the value of this integral, the remaining integral over $\vec{r}_1$ will simply give a factor of V:

$$\text{[diagram]} = \frac{1}{2}\frac{N^2}{V}\int d^3r\, f(r). \tag{8.31}$$

Having all the V's written explicitly, we can plug into equation 8.27 for the pressure:

$$\begin{aligned}
P &= \frac{NkT}{V} + kT\frac{\partial}{\partial V}\left(\frac{1}{2}\frac{N^2}{V}\int d^3r\, f_{12}(r)\right) + \cdots \\
&= \frac{NkT}{V} - kT\cdot\frac{1}{2}\frac{N^2}{V^2}\int d^3r\, f(r) + \cdots \\
&= \frac{NkT}{V}\left(1 - \frac{1}{2}\frac{N}{V}\int d^3r\, f(r) + \cdots\right).
\end{aligned} \tag{8.32}$$

It is conventional to write this series in the form of the **virial expansion**, introduced in Problem 1.17:

$$P = \frac{NkT}{V}\left(1 + \frac{B(T)}{(V/N)} + \frac{C(T)}{(V/N)^2} + \cdots\right). \tag{8.33}$$

We are now in a position to compute the **second virial coefficient**, $B(T)$:

$$B(T) = -\frac{1}{2}\int d^3r\, f(r). \tag{8.34}$$

To evaluate this last triple integral I'll use spherical coordinates, where the measure of the integral is

$$d^3r = (dr)(r\, d\theta)(r\sin\theta\, d\phi) \tag{8.35}$$

(see Figure 7.11). The integrand $f(r)$ is independent of the angles θ and ϕ, so the angular integrals give simply 4π, the area of a unit sphere. Even more explicitly, then,

$$B(T) = -2\pi \int_0^\infty r^2 f(r)\, dr = -2\pi \int_0^\infty r^2 (e^{-\beta u(r)} - 1)\, dr. \qquad (8.36)$$

This is as far as we can go without an explicit formula for the intermolecular potential energy, $u(r)$.

To model the intermolecular potential energy realistically, we want a function that is weakly attractive at large distances and strongly repulsive at short distances (as discussed in Section 5.3). For molecules with no permanent electric dipole moment, the long-distance force arises from a spontaneously fluctuating dipole moment in one molecule, which induces a dipole moment in the other and then attracts it; one can show that this force varies as $1/r^7$, so the corresponding potential energy varies as $1/r^6$. The exact formula used to model the repulsive part of the potential turns out not to be critical; for mathematical convenience a term proportional to $1/r^{12}$ is most often used. The sum of these attractive and repulsive terms gives what is called the **Lennard-Jones 6-12 potential**; with appropriately named constants it can be written

$$u(r) = u_0 \left[\left(\frac{r_0}{r} \right)^{12} - 2 \left(\frac{r_0}{r} \right)^6 \right]. \qquad (8.37)$$

Figure 8.1 shows a plot of this function, and of the corresponding Mayer f-function for three different temperatures. The parameter r_0 represents the distance between the molecular centers when the energy is a minimum—very roughly, the diameter of a molecule. The parameter u_0 is the maximum depth of the potential well.

If you plug the Lennard-Jones potential function into equation 8.36 for the second virial coefficient and integrate numerically for various temperatures, you

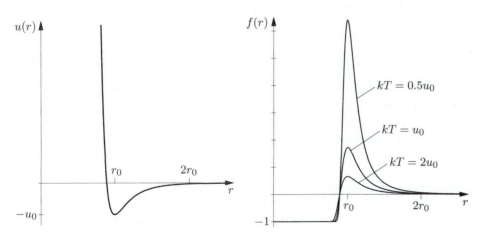

Figure 8.1. Left: The Lennard-Jones intermolecular potential function, with a strong repulsive region at small distances and a weak attractive region at somewhat larger distances. Right: The corresponding Mayer f-function, for three different temperatures.

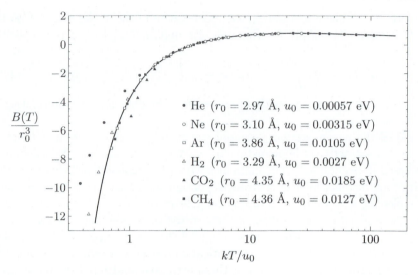

Figure 8.2. Measurements of the second virial coefficients of selected gases, compared to the prediction of equation 8.36 with $u(r)$ given by the Lennard-Jones function. Note that the horizontal axis is logarithmic. The constants r_0 and u_0 have been chosen separately for each gas to give the best fit. For carbon dioxide, the poor fit is due to the asymmetric shape of the molecules. For hydrogen and helium, the discrepancies at low temperatures are due to quantum effects. Data from J. H. Dymond and E. B. Smith, *The Virial Coefficients of Pure Gases and Mixtures: A Critical Compilation* (Oxford University Press, Oxford, 1980).

obtain the solid curve shown in Figure 8.2. At low temperatures the integral of the f-function is dominated by its large upward spike at r_0, that is, by the attractive potential well. A positive average f leads to a negative virial coefficient, indicating that the pressure is lower than that of an ideal gas. At high temperatures, however, the negative potential well shows up less dramatically in f, so the integral is dominated by the negative portion of f that comes from the repulsive short-distance interaction; then the virial coefficient is positive and the pressure is greater than that of an ideal gas. At *very* high temperatures, though, this effect is lessened somewhat by the ability of high-energy molecules to partially penetrate into the region of repulsion.

Figure 8.2 also shows experimental values of $B(T)$ for several gases, plotted after choosing r_0 and u_0 for each gas to obtain the best fit to the theoretical curve. For most simple gases, the shape of $B(T)$ predicted by the Lennard-Jones potential agrees very well with experiments. (For molecules with strongly asymmetric shapes and/or permanent dipole moments, other potential functions would be more appropriate, while for the light gases hydrogen and helium, quantum-mechanical effects become important at low temperatures.*) This agreement tells us that the

---

*As shown in Problem 7.75, the contribution of quantum *statistics* to $B(T)$ should be *negative* for a gas of bosons like hydrogen or helium. However, there is another quantum effect not considered in that problem. The de Broglie wave of one molecule cannot pen-

Lennard-Jones potential is a reasonably accurate model of intermolecular interactions, while the values of r_0 and u_0 that are used to fit the data give us quantitative information about the sizes and polarizabilities of the molecules. Here, as always, statistical mechanics works in both directions: From our theoretical understanding of microscopic physics, it gives us predictions for the bulk behavior of large numbers of molecules; and from measurements of the properties of bulk matter, it lets us infer a great deal about the molecules themselves.

In principle, we could now go on to compute the third and higher virial coefficients of a low-density gas using the cluster expansion. In practice, though, we would encounter two major problems. The first is that the remaining diagrams in equation 8.27 are very difficult to evaluate explicitly. But worse, the second problem is that when clusters of three or more molecules interact, it is often *not* valid to write the potential energy as a sum of pair-wise interactions as I did in equation 8.8. Both of these problems can be overcome,* but a proper calculation of the third virial coefficient is far beyond the scope of this book.

> **Problem 8.5.** By changing variables as in the text, express the diagram in equation 8.18 in terms of the same integral as in equation 8.31. Do the same for the last two diagrams in the first line of equation 8.20. Which diagrams cannot be written in terms of this basic integral?

> **Problem 8.6.** You can *estimate* the size of any diagram by realizing that $f(r)$ is of order 1 out to a distance of about the diameter of a molecule, and $f \approx 0$ beyond that. Hence, a three-dimensional integral of a product of f's will generally give a result that is of the order of the volume of a molecule. Estimate the sizes of all the diagrams shown explicitly in equation 8.20, and explain why it was necessary to rewrite the series in exponential form.

> **Problem 8.7.** Show that, if you don't make too many approximations, the exponential series in equation 8.22 includes the three-dot diagram in equation 8.18. There will be some leftover terms; show that these vanish in the thermodynamic limit.

> **Problem 8.8.** Show that the nth virial coefficient depends on the diagrams in equation 8.23 that have n dots. Write the third virial coefficient, $C(T)$, in terms of an integral of f-functions. Why it would be difficult to carry out this integral?

---

etrate the physical volume of another, so when the average de Broglie wavelength (ℓ_Q) becomes larger than the physical diameter (r_0), there is more repulsion than there would be classically. The effect of quantum statistics would become dominant only at still lower temperatures, when ℓ_Q is comparable to the average distance between molecules. Both hydrogen and helium have the inconvenient habit of liquefying before such low temperatures are reached. For a thorough discussion of virial coefficients, including quantum effects, see Joseph O. Hirschfelder, Charles F. Curtiss, and R. Byron Bird, *Molecular Theory of Gases and Liquids* (Wiley, New York, 1954).

*For a discussion of the computation of the third virial coefficient and comparisons between theory and experiment, see Reichl (1998).

Problem 8.9. Show that the Lennard-Jones potential reaches its minimum value at $r = r_0$, and that its value at this minimum is $-u_0$. At what value of r does the potential equal zero?

Problem 8.10. Use a computer to calculate and plot the second virial coefficient for a gas of molecules interacting via the Lennard-Jones potential, for values of kT/u_0 ranging from 1 to 7. On the same graph, plot the data for nitrogen given in Problem 1.17, choosing the parameters r_0 and u_0 so as to obtain a good fit.

Problem 8.11. Consider a gas of "hard spheres," which do not interact at all unless their separation distance is less than r_0, in which case their interaction energy is infinite. Sketch the Mayer f-function for this gas, and compute the second virial coefficient. Discuss the result briefly.

Problem 8.12. Consider a gas of molecules whose interaction energy $u(r)$ is infinite for $r < r_0$ and negative for $r > r_0$, with a minimum value of $-u_0$. Suppose further that $kT \gg u_0$, so you can approximate the Boltzmann factor for $r > r_0$ using $e^x \approx 1 + x$. Show that under these conditions the second virial coefficient has the form $B(T) = b - (a/kT)$, the same as what you found for a van der Waals gas in Problem 1.17. Write the van der Waals constants a and b in terms of r_0 and $u(r)$, and discuss the results briefly.

Problem 8.13. Use the cluster expansion to write the total energy of a monatomic nonideal gas in terms of a sum of diagrams. Keeping only the first diagram, show that the energy is approximately

$$U \approx \frac{3}{2}NkT + \frac{N^2}{V} \cdot 2\pi \int_0^\infty r^2\, u(r)\, e^{-\beta u(r)}\, dr.$$

Use a computer to evaluate this integral numerically, as a function of T, for the Lennard-Jones potential. Plot the temperature-dependent part of the correction term, and explain the shape of the graph physically. Discuss the correction to the heat capacity at constant volume, and compute this correction numerically for argon at room temperature and atmospheric pressure.

Problem 8.14. In this section I've formulated the cluster expansion for a gas with a fixed number of particles, using the "canonical" formalism of Chapter 6. A somewhat cleaner approach, however, is to use the "grand canonical" formalism introduced in Section 7.1, in which we allow the system to exchange particles with a much larger reservoir.

(a) Write down a formula for the grand partition function ($\mathcal{Z}$) of a weakly interacting gas in thermal and diffusive equilibrium with a reservoir at fixed T and μ. Express $\mathcal{Z}$ as a sum over all possible particle numbers N, with each term involving the ordinary partition function $Z(N)$.

(b) Use equations 8.6 and 8.20 to express $Z(N)$ as a sum of diagrams, then carry out the sum over N, diagram by diagram. Express the result as a sum of similar diagrams, but with a new rule 1 that associates the expression $(\lambda/v_Q) \int d^3 r_i$ with each dot, where $\lambda = e^{\beta\mu}$. Now, with the awkward factors of $N(N-1)\cdots$ taken care of, you should find that the sum of all diagrams organizes itself into exponential form, resulting in the formula

$$\mathcal{Z} = \exp\left(\frac{\lambda V}{v_Q} + \mathord{\text{\scriptsize ▮}} + \mathord{\triangle} + \mathord{\triangle} + \mathord{\square} + \cdots\right).$$

Note that the exponent contains all connected diagrams, including those that can be disconnected by removal of a single line.

(c) Using the properties of the grand partition function (see Problem 7.7), find diagrammatic expressions for the average number of particles and the pressure of this gas.

(d) Keeping only the first diagram in each sum, express $\overline{N}(\mu)$ and $P(\mu)$ in terms of an integral of the Mayer f-function. Eliminate μ to obtain the same result for the pressure (and the second virial coefficient) as derived in the text.

(e) Repeat part (d) keeping the three-dot diagrams as well, to obtain an expression for the third virial coefficient in terms of an integral of f-functions. You should find that the Λ-shaped diagram cancels, leaving only the triangle diagram to contribute to $C(T)$.

8.2   The Ising Model of a Ferromagnet

In an ideal *para*magnet, each microscopic magnetic dipole responds only to the external magnetic field (if any); the dipoles have no inherent tendency to point parallel (or antiparallel) to their immediate neighbors. In the real world, however, atomic dipoles *are* influenced by their neighbors: There is always some preference for neighboring dipoles to align either parallel or antiparallel. In some materials this preference is due to ordinary magnetic forces between the dipoles. In the more dramatic examples (such as iron), however, the alignment of neighboring dipoles is due to complicated quantum-mechanical effects involving the Pauli exclusion principle. Either way, there is a contribution to the energy that is greater or less, depending on the relative alignment of neighboring dipoles.

When neighboring dipoles align parallel to each other, even in the absence of an external field, we call the material a **ferromagnet** (in honor of iron, the most familiar example). When neighboring dipoles align antiparallel, we call the material an **antiferromagnet** (examples include Cr, NiO, and FeO). In this section I'll discuss ferromagnets, although most of the same ideas can also be applied to antiferromagnets.

The long-range order of a ferromagnet manifests itself as a net nonzero magnetization. Raising the temperature, however, causes random fluctuations that decrease the overall magnetization. For every ferromagnet there is a certain critical temperature, called the **Curie temperature**, at which the net magnetization becomes zero (when there is no external field). Above the Curie temperature a ferromagnet becomes a paramagnet. The Curie temperature of iron is 1043 K, considerably higher than that of most other ferromagnets.

Even below the Curie temperature, you may not notice that a piece of iron is magnetized. This is because a large chunk of iron ordinarily divides itself into **domains** that are microscopic in size but still contain billions of atomic dipoles. Within each domain the material is magnetized, but the magnetic field created by all the dipoles in one domain gives neighboring domains a tendency to magnetize in the opposite direction. (Put two ordinary bar magnets side by side and you'll see why.) Because there are so many domains, with about as many pointing one way as another, the material as a whole has no net magnetization. However, if you heat a chunk of iron in the presence of an external magnetic field, this field can overcome the interaction between domains and cause essentially *all* the dipoles to line up parallel. Remove the external field after the material has cooled to room temperature and the ferromagnetic interaction prevents any significant realigning. You then have a "permanent" magnet.

In this section I'd like to model the behavior of a ferromagnet, or rather, of a single domain within a ferromagnet. I'll account for the tendency of neighboring dipoles to align parallel to each other, but I'll neglect any long-range magnetic interactions between dipoles. To simplify the problem further, I'll assume that the material has a preferred axis of magnetization, and that each atomic dipole can only point parallel or antiparallel to this axis.[*] This simplified model of a magnet is called the **Ising model**, after Ernst Ising, who studied it in the 1920s.[†] Figure 8.3 shows one possible state of a two-dimensional Ising model on a 10×10 square lattice.

Notation: Let N be the total number of atomic dipoles, and let s_i be the current state of the ith dipole, with the convention that $s_i = 1$ when this dipole is pointing up, and $s_i = -1$ when this dipole is pointing down. The energy due to the interaction of a pair of neighboring dipoles will be $-\epsilon$ when they are parallel and $+\epsilon$ when they are antiparallel. Either way, we can write this energy as $-\epsilon s_i s_j$,

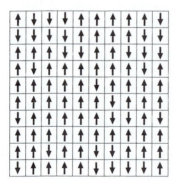

Figure 8.3. One of the many possible states of a two-dimensional Ising model on a 10×10 square lattice.

---

[*]I should point out that in many respects this model is *not* an accurate representation of a real ferromagnet. Even if there really is a preferred axis of magnetization, and even if the elementary dipoles each have only two possible orientations along this direction, quantum mechanics is more subtle than this naive model. Because we do not measure the orientation of each individual dipole, it is only the *sum* of their magnetic moments that is quantized—not the moment of each individual particle. At low temperatures, for instance, the relevant states of a real ferromagnet are long-wavelength "magnons" (described in Problem 7.64), in which all the dipoles are nearly parallel and a unit of opposite alignment is spread over many dipoles. The Ising model therefore does not yield accurate predictions for the low-temperature behavior of a ferromagnet. Fortunately, it turns out to be much more accurate near the Curie temperature.

[†]For a good historical overview of the Ising model see Stephen G. Brush, "History of the Lenz-Ising Model," *Reviews of Modern Physics* **39**, 883–893 (1967).

assuming that dipoles i and j are neighbors. Then the total energy of the system from *all* the nearest-neighbor interactions is

$$U = -\epsilon \sum_{\substack{\text{neighboring} \\ \text{pairs } i,j}} s_i s_j. \tag{8.38}$$

To predict the thermal behavior of this system, we should try to calculate the partition function,

$$Z = \sum_{\{s_i\}} e^{-\beta U}, \tag{8.39}$$

where the sum is over all possible sets of dipole alignments. For N dipoles, each with two possible alignments, the number of terms in this sum is 2^N, usually a *very* large number. Adding up all the terms by brute force is not going to be practical.

Problem 8.15. For a two-dimensional Ising model on a square lattice, each dipole (except on the edges) has four "neighbors"—above, below, left, and right. (Diagonal neighbors are normally not included.) What is the total energy (in terms of ϵ) for the particular state of the 4×4 square lattice shown in Figure 8.4?

Figure 8.4. One particular state of an Ising model on a 4×4 square lattice (Problem 8.15).

Problem 8.16. Consider an Ising model of 100 elementary dipoles. Suppose you wish to calculate the partition function for this system, using a computer that can compute one billion terms of the partition function per second. How long must you wait for the answer?

Problem 8.17. Consider an Ising model of just two elementary dipoles, whose mutual interaction energy is $\pm\epsilon$. Enumerate the states of this system and write down their Boltzmann factors. Calculate the partition function. Find the probabilities of finding the dipoles parallel and antiparallel, and plot these probabilities as a function of kT/ϵ. Also calculate and plot the average energy of the system. At what temperatures are you more likely to find both dipoles pointing up than to find one up and one down?

Exact Solution in One Dimension

So far I haven't specified how our atomic dipoles are to be arranged in space, or how many nearest neighbors each of them has. To simulate a real ferromagnet, I should arrange them in three dimensions on a crystal lattice. But I'll start with a much simpler arrangement, with the dipoles strung out along a one-dimensional line (see Figure 8.5). Then each has only two nearest neighbors, and we can actually carry out the partition sum exactly.

For a one-dimensional Ising model (with no external magnetic field), the energy is

$$U = -\epsilon\big(s_1 s_2 + s_2 s_3 + s_3 s_4 + \cdots s_{N-1} s_N\big), \tag{8.40}$$

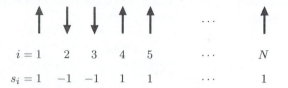

$$i = 1 \quad 2 \quad 3 \quad 4 \quad 5 \qquad \cdots \qquad N$$

$$s_i = 1 \quad -1 \quad -1 \quad 1 \quad 1 \qquad \cdots \qquad 1$$

Figure 8.5. A one-dimensional Ising model with N elementary dipoles.

and the partition function can be written

$$Z = \sum_{s_1} \sum_{s_2} \cdots \sum_{s_N} e^{\beta \epsilon s_1 s_2} e^{\beta \epsilon s_2 s_3} \cdots e^{\beta \epsilon s_{N-1} s_N}, \tag{8.41}$$

where each sum runs over the values -1 and 1. Notice that the final sum, over s_N, is

$$\sum_{s_N} e^{\beta \epsilon s_{N-1} s_N} = e^{\beta \epsilon} + e^{-\beta \epsilon} = 2 \cosh \beta \epsilon, \tag{8.42}$$

regardless of whether s_{N-1} is $+1$ or -1. With this sum done, the sum over s_{N-1} can now be evaluated in the same way, then the sum over s_{N-2}, and so on down to s_2, yielding $N - 1$ factors of $2 \cosh \beta \epsilon$. The remaining sum over s_1 gives another factor of 2, so the partition function is

$$Z = 2^N (\cosh \beta \epsilon)^{N-1} \approx (2 \cosh \beta \epsilon)^N, \tag{8.43}$$

where the last approximation is valid when N is large.

So we've got the partition function. Now what? Well, let's find the average energy as a function of temperature. By a straightforward calculation you can show that

$$\overline{U} = -\frac{\partial}{\partial \beta} \ln Z = -N \epsilon \tanh \beta \epsilon, \tag{8.44}$$

which goes to $-N\epsilon$ as $T \to 0$ and to 0 as $T \to \infty$. Therefore the dipoles must be randomly aligned at high temperature (so that half the neighboring pairs are parallel and half are antiparallel), but lined up parallel to each other at $T = 0$ (achieving the minimum possible energy).

If you're getting a sense of *déjà vu*, don't be surprised. Yes indeed, both Z and $\overline{U}$ for this system are exactly the same as for a two-state paramagnet, if you replace the magnetic interaction energy μB with the neighbor-neighbor interaction energy ϵ. Here, however, the dipoles like to line up with each other, instead of with an external field.

Notice that, while this system *does* become more ordered (less random) as its temperature decreases, the order sets in gradually. The behavior of $\overline{U}$ as a function of T is perfectly smooth, with no abrupt transition at a nonzero critical temperature. Apparently, the one-dimensional Ising model does *not* behave like a real three-dimensional ferromagnet in this crucial respect. Its tendency to magnetize is not great enough, because each dipole has only two nearest neighbors.

So our next step should be to consider Ising models in higher dimensions. Unfortunately, though, such models are *much* harder to solve. The two-dimensional Ising model on a square lattice was first solved in the 1940s by Lars Onsager. Onsager evaluated the exact partition function as $N \to \infty$ in closed form, and found that this model *does* have a critical temperature, just like a real ferromagnet. Because Onsager's solution is extremely difficult mathematically, I will not attempt to present it in this book. In any case, nobody has ever found an exact solution to the *three*-dimensional Ising model. The most fruitful approach from here, therefore, is to give up on exact solutions and rely instead on approximations.

Problem 8.18. Starting from the partition function, calculate the average energy of the one-dimensional Ising model, to verify equation 8.44. Sketch the average energy as a function of temperature.

The Mean Field Approximation

Next I'd like to present a very crude approximation, which can be used to "solve" the Ising model in any dimensionality. This approximation won't be very accurate, but it does give some qualitative insight into what's happening and why the dimensionality matters.

Let's concentrate on just a single dipole, somewhere in the middle of the lattice. I'll label this dipole i, so its alignment is s_i which can be -1 or 1. Let n be the number of nearest neighbors that this dipole has:

$$n = \begin{cases} 2 & \text{in one dimension;} \\ 4 & \text{in two dimensions (square lattice);} \\ 6 & \text{in three dimensions (simple cubic lattice);} \\ 8 & \text{in three dimensions (body-centered cubic lattice);} \\ 12 & \text{in three dimensions (face-centered cubic lattice).} \end{cases} \quad (8.45)$$

Imagine that the alignments of these neighboring dipoles are temporarily frozen, but that our dipole i is free to point up or down. If it points up, then the interaction energy between this dipole and its neighbors is

$$E_\uparrow = -\epsilon \sum_{\text{neighbors}} s_{\text{neighbor}} = -\epsilon\, n\, \overline{s}, \quad (8.46)$$

where $\overline{s}$ is the *average* alignment of the neighbors (see Figure 8.6). Similarly, if

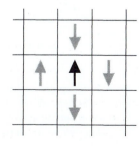

Figure 8.6. The four neighbors of this particular dipole have an average s value of $(+1-3)/4 = -1/2$. If the central dipole points up, the energy due to its interactions with its neighbors is $+2\epsilon$, while if it points down, the energy is -2ϵ.

dipole i points down, then the interaction energy is

$$E_\downarrow = +\epsilon\, n\,\bar{s}. \tag{8.47}$$

The partition function for just this dipole is therefore

$$Z_i = e^{\beta\epsilon n\bar{s}} + e^{-\beta\epsilon n\bar{s}} = 2\cosh(\beta\epsilon n\bar{s}), \tag{8.48}$$

and the average expected value of its spin alignment is

$$\bar{s}_i = \frac{1}{Z_i}\left[(1)e^{\beta\epsilon n\bar{s}} + (-1)e^{-\beta\epsilon n\bar{s}}\right] = \frac{2\sinh(\beta\epsilon n\bar{s})}{2\cosh(\beta\epsilon n\bar{s})} = \tanh(\beta\epsilon n\bar{s}). \tag{8.49}$$

Now look at both sides of this equation. On the left is $\bar{s}_i$, the thermal average value of the alignment of any typical dipole (except those on the edge of the lattice, which we'll neglect). On the right is $\bar{s}$, the average of the actual instantaneous alignments of this dipole's n neighbors. The idea of the **mean field approximation** is to assume (or pretend) that these two quantities are the same: $\bar{s}_i = \bar{s}$. In other words, we assume that at every moment, the alignments of all the dipoles are such that every neighborhood is "typical"—there are no fluctuations that cause the magnetization in any neighborhood to be more or less than the expected thermal average. (This approximation is similar to the one I used to derive the van der Waals equation in Section 5.3. There it was the density, rather than the spin alignment, whose average value was not allowed to vary from place to place within the system.)

In the mean field approximation, then, we have the relation

$$\bar{s} = \tanh(\beta\epsilon n\bar{s}), \tag{8.50}$$

where $\bar{s}$ is now the average dipole alignment over the entire system. This is a transcendental equation, so we can't just solve for $\bar{s}$ in terms of $\beta\epsilon n$. The best approach is to plot both sides of the equation and look for a graphical solution (see Figure 8.7). Notice that the larger the value of $\beta\epsilon n$, the steeper the slope of the hyperbolic tangent function near $\bar{s} = 0$. This means that our equation can have either one solution or three, depending on the value of $\beta\epsilon n$.

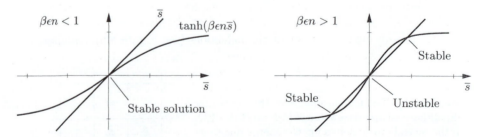

Figure 8.7. Graphical solution of equation 8.50. The slope of the tanh function at the origin is $\beta\epsilon n$. When this quantity is less than 1, there is only one solution, at $\bar{s} = 0$; when this quantity is greater than 1, the $\bar{s} = 0$ solution is unstable but there are also two nontrivial stable solutions.

When $\beta \epsilon n < 1$, that is, when $kT > n\epsilon$, the only solution is at $\bar{s} = 0$; there is no net magnetization. If a thermal fluctuation were to momentarily increase the value of $\bar{s}$, then the hyperbolic tangent function, which dictates what $\bar{s}$ *should* be, would be less than the current value of $\bar{s}$, so $\bar{s}$ would tend to decrease back to zero. The solution $\bar{s} = 0$ is stable.

When $\beta \epsilon n > 1$, that is, when $kT < n\epsilon$, we still have a solution at $\bar{s} = 0$ and we also have two more solutions, at positive and negative values of $\bar{s}$. But the solution at $\bar{s} = 0$ is unstable: A small positive fluctuation of $\bar{s}$ would cause the hyperbolic tangent function to exceed the current value of $\bar{s}$, driving $\bar{s}$ to even higher values. The stable solutions are the other two, which are symmetrically located because the system has no inherent tendency toward positive or negative magnetization. Thus, the system will acquire a net nonzero magnetization, which is equally likely to be positive or negative. When a system has a built-in symmetry such as this, yet must choose one state or another at low temperatures, we say that the symmetry is **spontaneously broken**.

The critical temperature T_c below which the system becomes magnetized is

$$kT_c = n\epsilon, \tag{8.51}$$

proportional to both the neighbor-neighbor interaction energy and to the number of neighbors. This result is no surprise: The more neighbors each dipole has, the greater the tendency of the whole system to magnetize. Notice, though, that even a *one*-dimensional Ising model should magnetize below a temperature of $2\epsilon/k$, according to this analysis. Yet we already saw from the exact solution that there is no abrupt transition in the behavior of a one-dimensional Ising model; it magnetizes only as the temperature goes to zero. Apparently, the mean field approximation is no good at all in one dimension.* Fortunately, the accuracy improves as the dimensionality increases.

Problem 8.19. The critical temperature of iron is 1043 K. Use this value to make a rough estimate of the dipole-dipole interaction energy ϵ, in electron-volts.

Problem 8.20. Use a computer to plot $\bar{s}$ as a function of kT/ϵ, as predicted by mean field theory, for a two-dimensional Ising model (with a square lattice).

Problem 8.21. At $T = 0$, equation 8.50 says that $\bar{s} = 1$. Work out the first temperature-dependent correction to this value, in the limit $\beta \epsilon n \gg 1$. Compare to the low-temperature behavior of a real ferromagnet, treated in Problem 7.64.

Problem 8.22. Consider an Ising model in the presence of an external magnetic field B, which gives each dipole an additional energy of $-\mu_B B$ if it points up and $+\mu_B B$ if it points down (where μ_B is the dipole's magnetic moment). Analyze this system using the mean field approximation to find the analogue of equation 8.50. Study the solutions of the equation graphically, and discuss the magnetization of this system as a function of both the external field strength and the temperature. Sketch the region in the T-B plane for which the equation has three solutions.

---

*There do exist more complicated versions of the mean field approximation that lack this fatal flaw, predicting correctly that the one-dimensional Ising model magnetizes only at $T = 0$. See, for example, Pathria (1996).

Problem 8.23. The Ising model can be used to simulate other systems besides ferromagnets; examples include antiferromagnets, binary alloys, and even fluids. The Ising model of a fluid is called a **lattice gas**. We imagine that space is divided into a lattice of sites, each of which can be either occupied by a gas molecule or unoccupied. The system has no kinetic energy, and the only potential energy comes from interactions of molecules on adjacent sites. Specifically, there is a contribution of $-u_0$ to the energy for each pair of neighboring sites that are both occupied.

(a) Write down a formula for the *grand* partition function for this system, as a function of u_0, T, and μ.

(b) Rearrange your formula to show that it is identical, up to a multiplicative factor that does not depend on the state of the system, to the *ordinary* partition function for an Ising ferromagnet in the presence of an external magnetic field B, provided that you make the replacements $u_0 \to 4\epsilon$ and $\mu \to 2\mu_B B - 8\epsilon$. (Note that μ is the chemical potential of the gas while μ_B is the magnetic moment of a dipole in the magnet.)

(c) Discuss the implications. Which states of the magnet correspond to low-density states of the lattice gas? Which states of the magnet correspond to high-density states in which the gas has condensed into a liquid? What shape does this model predict for the liquid-gas phase boundary in the P-T plane?

Problem 8.24. In this problem you will use the mean field approximation to analyze the behavior of the Ising model near the critical point.

(a) Prove that, when $x \ll 1$, $\tanh x \approx x - \frac{1}{3}x^3$.

(b) Use the result of part (a) to find an expression for the magnetization of the Ising model, in the mean field approximation, when T is very close to the critical temperature. You should find $M \propto (T_c - T)^\beta$, where β (not to be confused with $1/kT$) is a **critical exponent**, analogous to the β defined for a fluid in Problem 5.55. Onsager's exact solution shows that $\beta = 1/8$ in two dimensions, while experiments and more sophisticated approximations show that $\beta \approx 1/3$ in three dimensions. The mean field approximation, however, predicts a larger value.

(c) The magnetic susceptibility χ is defined as $\chi \equiv (\partial M/\partial B)_T$. The behavior of this quantity near the critical point is conventionally written as $\chi \propto (T - T_c)^{-\gamma}$, where γ is another critical exponent. Find the value of γ in the mean field approximation, and show that it does not depend on whether T is slightly above or slightly below T_c. (The exact value of γ in two dimensions turns out to be 7/4, while in three dimensions $\gamma \approx 1.24$.)

Monte Carlo Simulation

Consider a medium-sized, two-dimensional Ising model on a square lattice, with 100 or so elementary dipoles (as shown in Figure 8.3). Although even the fastest computer could never compute the probabilities of *all* the possible states of this system, maybe it isn't necessary to consider all of them—perhaps a random sampling of only a million or so states would be enough. This is the idea of **Monte Carlo summation** (or integration), a technique named after the famous European gambling center. The procedure is to generate a random sampling of as many states as possible, compute the Boltzmann factors for these states, and then use this random sample to compute the average energy, magnetization, and other thermodynamic quantities.

Unfortunately, the procedure just outlined does not work well for the Ising model. Even if we consider as many as one billion states, this is only a *tiny* fraction—about one in 10^{21}—of all the states for a modest 10×10 lattice. And at low temperatures, when the system wants to magnetize, the *important* states (with nearly all of the dipoles pointing in the same direction) constitute such a small fraction of the total that we are likely to miss them entirely. Sampling the states purely at random just isn't efficient enough; for this reason it's sometimes called the *naive* Monte Carlo method.

A better idea is to use the Boltzmann factors themselves as a guide during the random generation of a subset of states to sample. A specific algorithm that does this is as follows: Start with any state whatsoever. Then choose a dipole at random and consider the possibility of flipping it. Compute the energy difference, ΔU, that would result from the flip. If $\Delta U \le 0$, so the system's energy would decrease or remain unchanged, go ahead and flip this dipole to generate the next system state. If $\Delta U > 0$, so the system's energy would increase, decide at random whether to flip the dipole, with the probability of the flip being $e^{-\Delta U/kT}$. If the dipole does *not* get flipped, then the new system state will be the same as the previous one. Either way, continue by choosing another dipole at random and repeat the process, over and over again, until every dipole has had many chances to be flipped. This algorithm is called the **Metropolis algorithm**, after Nicholas Metropolis, the first author of a 1953 article* that presented a calculation of this type. This technique is also called Monte Carlo summation with **importance sampling**.

The Metropolis algorithm generates a subset of system states in which low-energy states occur more frequently than high-energy states. To see in more detail why the algorithm works, consider just two states, 1 and 2, which differ only by the flipping of a single dipole. Let U_1 and U_2 be the energies of these states, and let us number the states so that $U_1 \le U_2$. If the system is initially in state 2, then the probability of making a transition to state 1 is $1/N$, simply the probability that the correct dipole will be chosen at random among all the others. If the system is initially in state 1, then the probability of making a transition to state 2 is $(1/N)e^{-(U_2-U_1)/kT}$, according to the Metropolis algorithm. The ratio of these two transition probabilities is therefore

$$\frac{\mathcal{P}(1 \to 2)}{\mathcal{P}(2 \to 1)} = \frac{(1/N)e^{-(U_2-U_1)/kT}}{(1/N)} = \frac{e^{-U_2/kT}}{e^{-U_1/kT}}, \qquad (8.52)$$

simply the ratio of the Boltzmann factors of the two states. If these were the *only*

---

*N. Metropolis, A. W. Rosenbluth, M. N. Rosenbluth, A. H. Teller, and E. Teller, "Equation of State Calculations for Fast Computing Machines," *Journal of Chemical Physics* **21**, 1087–1092 (1953). In this article the authors use their algorithm to calculate the pressure of a two-dimensional gas of 224 hard disks. This rather modest calculation required several days of computing time on what was then a state-of-the-art computer.

two states available to the system, then the frequencies with which they occur would be in exactly this ratio, as Boltzmann statistics demands.*

Next consider two other states, 3 and 4, that differ from 1 and 2 by the flipping of some other dipole. The system can now go between 1 and 2 through the indirect process $1 \leftrightarrow 3 \leftrightarrow 4 \leftrightarrow 2$, whose forward and backward rates have the ratio

$$\frac{\mathcal{P}(1 \to 3 \to 4 \to 2)}{\mathcal{P}(2 \to 4 \to 3 \to 1)} = \frac{e^{-U_3/kT}}{e^{-U_1/kT}} \frac{e^{-U_4/kT}}{e^{-U_3/kT}} \frac{e^{-U_2/kT}}{e^{-U_4/kT}} = \frac{e^{-U_2/kT}}{e^{-U_1/kT}}, \qquad (8.53)$$

again as demanded by Boltzmann statistics. The same conclusion applies to transitions involving any number of steps, and to transitions between states that differ by the flipping of more than one dipole. Thus, the Metropolis algorithm does indeed generate states with the correct Boltzmann probabilities.

Strictly speaking, though, this conclusion applies only after the algorithm has been running infinitely long, so that every state has been generated many times. *We* want to run the algorithm for a relatively short time, so that most states are never generated at all! Under these circumstances we have no guarantee that the subset of states actually generated will accurately represent the full collection of all system states. In fact, it's hard to even define what is *meant* by an "accurate" representation. In the case of the Ising model, our main concerns are that the randomly generated states give an accurate picture of the expected energy and magnetization of the system. The most noticeable exception in practice will be that at low temperatures, the Metropolis algorithm will rapidly push the system into a "metastable" state in which nearly all of the dipoles are parallel to their neighbors. Although such a state *is* quite probable according to Boltzmann statistics, it may take a very long time for the algorithm to generate other probable states that differ significantly, such as a state in which every dipole is flipped. (In this way the Metropolis algorithm is analogous to what happens in the real world, where a large system never has time to explore all possible microstates, and the relaxation time for achieving true thermodynamic equilibrium can sometimes be very long.)

With this limitation in mind, let's now go on and implement the Metropolis algorithm. The algorithm can be programmed in almost any traditional computer language, and in many nontraditional languages as well. Rather than singling out one particular language, let me instead present the algorithm in "pseudocode," which you can translate into the language of your choice. A pseudocode program for a basic two-dimensional Ising simulation is shown in Figure 8.8. This program produces only graphical output, showing the lattice as an array of colored squares— one color for dipoles pointing up, another color for dipoles pointing down. Each time a dipole is flipped the color of a square changes, so you can see exactly what sequence of states is being generated.

The program uses a two-dimensional array called s(i,j) to store the values of the spin orientations; the indices i and j each go from 1 to the value of size, which can be changed to simulate lattices of different sizes. The temperature T,

---

*When the transition rates between two states have the correct ratio, we say that the transitions are in **detailed balance**.

```
program ising                              Monte Carlo simulation of a 2D Ising
                                           model using the Metropolis algorithm

size = 10                                  Width of square lattice
T = 2.5                                    Temperature in units of ε/k
initialize
for iteration = 1 to 100*size^2 do         Main iteration loop
  i = int(rand*size+1)                     Choose a random row number
  j = int(rand*size+1)                     and a random column number
  deltaU(i,j,Ediff)                        Compute ΔU of hypothetical flip
  if Ediff <= 0 then                       If flipping reduces the energy ...
    s(i,j) = -s(i,j)                        then flip it!
    colorsquare(i,j)
  else
    if rand < exp(-Ediff/T) then           otherwise the Boltzmann factor
      s(i,j) = -s(i,j)                      gives the probability of flipping
      colorsquare(i,j)
    end if
  end if
next iteration                             Now go back and start over ...
end program

subroutine deltaU(i,j,Ediff)               Compute ΔU of flipping a dipole
                                           (note periodic boundary conditions)
  if i = 1 then top = s(size,j) else top = s(i-1,j)
  if i = size then bottom = s(1,j) else bottom = s(i+1,j)
  if j = 1 then left = s(i,size) else left = s(i,j-1)
  if j = size then right = s(i,1) else right = s(i,j+1)
  Ediff = 2*s(i,j)*(top+bottom+left+right)
end subroutine

subroutine initialize                      Initialize to a random array
  for i = 1 to size
    for j = 1 to size
      if rand < .5 then s(i,j) = 1 else s(i,j) = -1
      colorsquare(i,j)
    next j
  next i
end subroutine

subroutine colorsquare(i,j)                Color a square according to s value
                                           (implementation depends on system)
```

Figure 8.8. A pseudocode program to simulate a two-dimensional Ising model, using the Metropolis algorithm.

measured in units of ϵ/k, can also be changed for different runs. After setting these two constants, the program calls the subroutine `initialize` to assign the initial value of each `s` randomly.*

The heart of the program is the "main iteration loop," which executes the Metropolis algorithm 100 times per dipole so that each dipole will have many chances to be flipped. The value 100 can be changed as appropriate. (Note that `*` represents multiplication, while `^` represents exponentiation.) Within the loop, we first choose a dipole at random; the function `rand` is assumed to return a random real number between 0 and 1, while `int()` returns the largest integer less than or equal to its argument. The subroutine `deltaU`, defined later in the program, computes the energy change upon hypothetically flipping the chosen dipole; this energy change (in units of ϵ) is returned as `Ediff`. If `Ediff` is negative or zero, we flip the dipole, while if `Ediff` is positive, we use it to compute a Boltzmann factor and compare this to a random number to decide whether to flip the dipole. If the dipole gets flipped, we call the subroutine `colorsquare` to change the color of the corresponding square on the screen.

The subroutine `deltaU` requires further explanation. There is always a problem, when a simulation uses a relatively small lattice, in dealing with "edge effects." In the Ising model, dipoles on the edge of the lattice are less constrained to align with their neighbors than are dipoles elsewhere. If we're modeling a very small system whose size is the same as that of our simulated lattice, then we should treat the edges as edges, with fewer neighbors per dipole. But if we're really interested in the behavior of much larger systems, then we should try to minimize edge effects. One way to do this is to make the lattice "wrap around," treating the right edge as if it were immediately left of the left edge and the bottom edge as if it were immediately above the top edge. Physically this would be like putting the array of dipoles on the surface of a torus. Another interpretation of this wrapping is to imagine that the lattice is flat and infinite in all directions, but that its state is always perfectly periodic, so that moving up, down, left, or right by a certain amount (the value of `size`) always takes you to an equivalent place where the dipoles have exactly the same alignments at all times. Based on this latter interpretation, we say that we are using **periodic boundary conditions**. Back to the subroutine `deltaU`, notice that it correctly identifies all four nearest neighbors, whether or not the chosen dipole is on an edge. The change in energy upon flipping is then twice the product of `s(i,j)` with `s` of the neighbor, summed over the four neighbors.

To convert my pseudocode into a real program that runs on a real computer, you first need to pick a computer system and a programming language. The syntax for arithmetic operations, variable assignments, if-then constructions, and for-next loops will vary from language to language, but almost any common programming language should provide easy ways to do these things. Some languages require that

---

*In principle, the initial state can by anything. In practice, the choice of initial state can be important if you don't want to wait forever for the system to equilibrate to a "typical" state. A random initial state works well at high temperatures; a completely magnetized initial state would work better at low temperatures.

variables be declared and given a type (such as integer or real) at the beginning of the program. Variables that are accessed both in the main program and in subroutines may require special treatment. The least standardized element of all is the handling of graphics; the contents of the subroutine `colorsquare` will vary wildly from system to system. Nevertheless, I hope that you will have little trouble implementing this program on your favorite computer and getting it to run.

Running the `ising` program is great fun: You get to watch the squares constantly changing colors as the system tries to find states with relatively large Boltzmann factors. It is tempting, in fact, to imagine that you are watching a simulation of what really happens in a magnet, as the dipoles change their alignments back and forth with the passage of time. Because of this similarity, a Monte Carlo program using importance sampling is usually called a Monte Carlo **simulation**. But please remember that we have made no attempt to simulate the real time-dependent behavior of a magnet. Instead we have implemented a "pseudodynamics," which flips only one dipole at a time and otherwise ignores the true time-dependent dynamics of the system. The only realistic property of our pseudodynamics is that it generates states with probabilities proportional to their Boltzmann factors, just as the real dynamics of a magnet presumably does.

Figure 8.9 shows some graphical output from the `ising` program for a 20×20 lattice. The first image shows a random initial state generated by the program, while the remaining images each show the final state at the end of a run of 40,000 iterations (100 per dipole), for various temperatures. Although these snapshots are no substitute for watching the program in action, they do show what a typical state at each temperature looks like. At T = 10 the final state is still almost random, with only a slight tendency for dipoles to align with their neighbors. At successively lower temperatures the dipoles tend to form larger and larger clusters* of positive and negative magnetization until, at T = 2.5, the clusters are about as large as the lattice itself. At T = 2 a single cluster has taken over the whole lattice, and we would say that the system is "magnetized." Small clusters of dipoles will still occasionally flip, but they don't last long; we would have to wait a very long time for the whole lattice to flip to a (just as probable) state of opposite magnetization. The T = 1.5 run happens to have settled into the opposite magnetization, and at this temperature fluctuations of individual dipoles are becoming uncommon. At T = 1 we might expect the system to magnetize completely and stay that way, and indeed, sometimes it does. About half the time, however, it instead becomes stuck in a metastable state with two domains, one positive and the other negative, as shown in the figure.

Based on these results, we can conclude that this system has a critical temperature somewhere between 2.0 and 2.5, in units of ϵ/k. Recall that the mean field approximation predicts a critical temperature of $4\epsilon/k$—not bad qualitatively, though off by nearly a factor of 2. But a 20×20 lattice is really quite small; what

---

*I'm making no attempt here to precisely define a "cluster"—just look at the pictures and use your intuition. A careful definition of the "size" of a cluster is given in Problem 8.29.

Figure 8.9. Graphical output from eight runs of the `ising` program, at successively lower temperatures. Each black square represents an "up" dipole and each white square represents a "down" dipole. The variable `T` is the temperature in units of ϵ/k.

happens in larger, more realistic simulations?

The answer isn't hard to guess. As long as the temperature is sufficiently high, so that the size of a typical cluster is much smaller than the size of the lattice, the behavior of the system is pretty much independent of the lattice size. But a larger lattice allows for larger clusters, so near the critical temperature we should use as large a lattice as possible. With sufficiently long runs with large lattices one can show that the size of the largest clusters approaches infinity at a temperature of $2.27\epsilon/k$ (see Figure 8.10). This, then, is the *true* critical temperature in the thermodynamic limit. And indeed, this result agrees with Onsager's exact solution

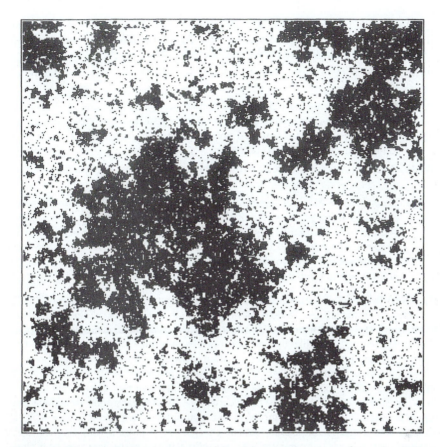

Figure 8.10. A typical state generated by the `ising` program after a few billion iterations on a 400×400 lattice at T = 2.27 (the critical temperature). Notice that there are clusters of all possible sizes, from individual dipoles up to the size of the lattice itself.

of the two-dimensional Ising model.

Similar simulations have been performed for the *three*-dimensional Ising model, although this requires much more computer time and the results are harder to display. For a simple cubic lattice one finds a critical temperature of approximately $4.5\epsilon/k$, again somewhat less than the prediction of the mean field approximation. The Monte Carlo method can also be applied to more complicated models of ferromagnets and to a huge variety of other systems including fluids, alloys, interfaces, nuclei, and subnuclear particles.

Problem 8.25. In Problem 8.15 you manually computed the energy of a particular state of a 4×4 square lattice. Repeat that computation, but this time apply periodic boundary conditions.

Problem 8.26. Implement the `ising` program on your favorite computer, using your favorite programming language. Run it for various lattice sizes and temperatures and observe the results. In particular:

(a) Run the program with a 20×20 lattice at T $= 10, 5, 4, 3,$ and 2.5, for at least 100 iterations per dipole per run. At each temperature make a rough estimate of the size of the largest clusters.

(b) Repeat part (a) for a 40×40 lattice. Are the cluster sizes any different? Explain.

(c) Run the program with a 20×20 lattice at T $= 2, 1.5,$ and 1. Estimate the average magnetization (as a percentage of total saturation) at each of these temperatures. Disregard runs in which the system gets stuck in a metastable state with two domains.

(d) Run the program with a 10×10 lattice at T $= 2.5$. Watch it run for 100,000 iterations or so. Describe and explain the behavior.

(e) Use successively larger lattices to estimate the typical cluster size at temperatures from 2.5 down to 2.27 (the critical temperature). The closer you are to the critical temperature, the larger a lattice you'll need and the longer the program will have to run. Quit when you realize that there are better ways to spend your time. Is it plausible that the cluster size goes to infinity as the temperature approaches the critical temperature?

Problem 8.27. Modify the `ising` program to compute the average energy of the system over all iterations. To do this, first add code to the `initialize` subroutine to compute the initial energy of the lattice; then, whenever a dipole is flipped, change the energy variable by the appropriate amount. When computing the average energy, be sure to average over *all* iterations, not just those iterations in which a dipole is actually flipped (why?). Run the program for a 5×5 lattice for T values from 4 down to 1 in reasonably small intervals, then plot the average energy as a function of T. Also plot the heat capacity. Use at least 1000 iterations per dipole for each run, preferably more. If your computer is fast enough, repeat for a 10×10 lattice and for a 20×20 lattice. Discuss the results. (Hint: Rather than starting over at each temperature with a random initial state, you can save time by starting with the final state generated at the previous, nearby temperature. For the larger lattices you may wish to save time by considering only a smaller temperature interval, perhaps from 3 down to 1.5.)

Problem 8.28. Modify the `ising` program to compute the total magnetization (that is, the sum of all the s values) for each iteration, and to tally how often each possible magnetization value occurs during a run, plotting the results as a histogram. Run the program for a 5×5 lattice at a variety of temperatures, and discuss the results. Sketch a graph of the most likely magnetization value as a function of temperature. If your computer is fast enough, repeat for a 10×10 lattice.

Problem 8.29. To quantify the clustering of alignments within an Ising magnet, we define a quantity called the **correlation function**, $c(r)$. Take any two dipoles i and j, separated by a distance r, and compute the product of their states: $s_i s_j$. This product is 1 if the dipoles are parallel and -1 if the dipoles are antiparallel. Now average this quantity over *all* pairs that are separated by a fixed distance r, to obtain a measure of the tendency of dipoles to be "correlated" over this distance. Finally, to remove the effect of any overall magnetization of the system, subtract off the square of the average s. Written as an equation, then, the correlation function is

$$c(r) = \overline{s_i s_j} - \overline{s_i}^2,$$

where it is understood that the first term averages over all pairs at the fixed distance r. Technically, the averages should also be taken over all possible states of the system, but don't do this yet.

(a) Add a routine to the `ising` program to compute the correlation function for the current state of the lattice, averaging over all pairs separated either vertically or horizontally (but not diagonally) by r units of distance, where r varies from 1 to half the lattice size. Have the program execute this routine periodically and plot the results as a bar graph.

(b) Run this program at a variety of temperatures, above, below, and near the critical point. Use a lattice size of at least 20, preferably larger (especially near the critical point). Describe the behavior of the correlation function at each temperature.

(c) Now add code to compute the *average* correlation function over the duration of a run. (However, it's best to let the system "equilibrate" to a typical state before you begin accumulating averages.) The **correlation length** is defined as the distance over which the correlation function decreases by a factor of e. Estimate the correlation length at each temperature, and plot a graph of the correlation length vs. T.

Problem 8.30. Modifiy the `ising` program to simulate a *one*-dimensional Ising model.

(a) For a lattice size of 100, observe the sequence of states generated at various temperatures and discuss the results. According to the exact solution (for an infinite lattice), we expect this system to magnetize only as the temperature goes to zero; is the behavior of your program consistent with this prediction? How does the typical cluster size depend on temperature?

(b) Modify your program to compute the average energy as in Problem 8.27. Plot the energy and heat capacity vs. temperature and compare to the exact result for an infinite lattice.

(c) Modify your program to compute the magnetization as in Problem 8.28. Determine the most likely magnetization for various temperatures, and discuss your results.

Problem 8.31. Modify the `ising` program to simulate a *three*-dimensional Ising model with a simple cubic lattice. In whatever way you can, try to show that this system has a critical point at around T = 4.5.

Problem 8.32. Imagine taking a two-dimensional Ising lattice and dividing the sites into 3×3 "blocks," as shown in Figure 8.11. In a **block spin transformation**, we replace the nine dipoles in each block with a single dipole, whose state is determined by "majority rule": If more than half of the original dipoles point up, then the new dipole points up, while if more than half of the original dipoles point down, then the new dipole points down. By applying this transformation to the entire lattice, we reduce it to a new lattice whose width is 1/3 the original width. This transformation is one version of a **renormalization group transformation**, a powerful technique for studying the behavior of systems near their critical points.*

---

*For more about the renormalization group and its applications, see Kenneth G. Wilson, "Problems in Physics with Many Scales of Length," *Scientific American* **241**, 158–179 (August, 1979).

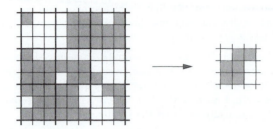

Figure 8.11. In a **block spin transformation**, we replace each block of nine dipoles with a single dipole whose orientation is determined by "majority rule."

(a) Add a routine to the `ising` program to apply a block spin transformation to the current state of the lattice, drawing the transformed lattice alongside the original. (Leave the original lattice unchanged.) Have the program execute this routine periodically, so you can observe the evolution of both lattices.

(b) Run your modified program with a 90 × 90 original lattice, at a variety of temperatures. After the system has equilibrated to a "typical" state at each temperature, compare the transformed lattice to a typical 30 × 30 piece of the original lattice. In general you should find that the transformed lattice resembles an original lattice at a *different* temperature. Let us call this temperature the "transformed temperature." When is the transformed temperature greater than the original temperature, and when is it less?

(c) Imagine starting with a very large lattice and applying many block spin transformations in succession, each time taking the system to a new effective temperature. Argue that, no matter what the original temperature, this procedure will eventually take you to one of three **fixed points**: zero, infinity, or the critical temperature. For what initial temperatures will you end up at each fixed point? [Comment: Think about the implications of the fact that the critical temperature is a fixed point of the block spin transformation. If averaging over the small-scale state of the system leaves the dynamics unchanged, then many aspects of the behavior of this system must be independent of any specific microscopic details. This implies that many different physical systems (magnets, fluids, and so on) should have essentially the same critical behavior. More specifically, the different systems will have the same "critical exponents," such as those defined in Problems 5.55 and 8.24. There are, however, two parameters that can still affect the critical behavior. One is the dimensionality of the space that the system is in (3 for most real-world systems); the other is the dimensionality of the "vector" that defines the magnetization (or the analogous "order parameter") of the system. For the Ising model, the magnetization is one-dimensional, always along a given axis; for a fluid, the order parameter is also a one-dimensional quantity, the difference in density between liquid and gas. Therefore the behavior of a fluid near its critical point should be the same as that of a three-dimensional Ising model.]

A   Elements of Quantum Mechanics

You don't need to know any quantum mechanics to understand the basic principles of thermal physics. But to predict the detailed thermal properties of specific physical systems (like a gas of nitrogen molecules or the electrons in a chunk of metal), you do need to know what are the possible "states" and corresponding energies of these systems. The states and their energies are determined by the principles of quantum mechanics.

Even so, you don't need to know much quantum mechanics in order to read this book. At each point in the text where a quantum mechanics result is needed I have summarized that result, in enough detail for the calculation at hand. If you don't care where the result comes from, then you need not read this appendix. At some point, however, you may want to see a more systematic overview of the quantum mechanics that is used in this book. This appendix is intended to provide that overview, whether you choose to read it before or after reading the main text.*

A.1 Evidence for Wave-Particle Duality

The historical roots of quantum mechanics are intimately entwined with the development of statistical mechanics around the turn of the 20th century. Especially important in this history was the breakdown of the equipartition theorem, both for electromagnetic radiation (the "ultraviolet catastrophe" described in Section 7.4) and for the vibrational energy of solid crystals (evidenced by anomalously low heat capacities at low temperature, as investigated in Problems 3.24 and 3.25 and in

---

*There are many good quantum mechanics textbooks that you may wish to consult for a less superficial treatment of the subject. I especially recommend *An Introduction to Quantum Physics* by A. P. French and Edwin F. Taylor (Norton, New York, 1978), and *Introduction to Quantum Mechanics* by David J. Griffiths (Prentice-Hall, Englewood Cliffs, NJ, 1995).

Section 7.5). But there is also plenty of more direct evidence for quantum mechanics, that is, for the idea that neither a wave model nor a particle model is adequate to understand matter and energy at the atomic scale. In this section I will briefly describe some of this evidence.

The Photoelectric Effect

If you shine light on a metal surface, it can knock some electrons out of the metal and send them flying off the surface. This phenomenon is called the **photoelectric effect**; it is the basic mechanism of video cameras and a variety of other electronic light detectors.

To study the photoelectric effect quantitatively, you can put the piece of metal (called a photocathode) into a vacuum tube with another piece of metal (the anode) to catch the ejected electrons. Then you can measure either the voltage that builds up as electrons collect on the anode, or the current produced as these electrons run around the circuit back to the cathode (see Figure A.1).

The *current* is a measure of how *many* electrons (per unit time) are ejected from the cathode and collected on the anode. Not surprisingly, the current increases if the light source is made more intense: Brighter light ejects more electrons.

The *voltage*, on the other hand, is a measure of the *energy* that an electron needs to cross the gap between the cathode and the anode. Initially the voltage is zero, but as electrons collect on the anode they create an electric field that pushes other electrons back toward the cathode. Before long the voltage stabilizes at some final value, indicating that no electrons are ejected with sufficient energy to cross. Voltage is just energy per unit charge, so if the final voltage is V, then the maximum kinetic energy of the ejected electrons (as they leave the cathode) must be $K_{\max} = eV$, where e is the magnitude of the electron's charge.

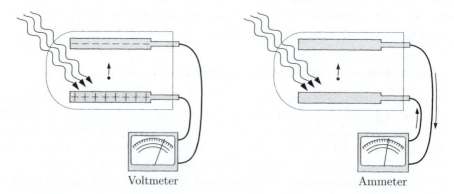

Voltmeter   Ammeter

Figure A.1. Two experiments to study the photoelectric effect. When an ideal voltmeter (with essentially infinite resistance) is connected to the circuit, electrons accumulate on the anode and repel other electrons; the voltmeter measures the *energy* (per unit charge) that an electron needs in order to cross. When an ammeter is connected, it measures the *number* of electrons (per unit time) that collect on the anode and then circulate back to the cathode.

Here's the surprise: The final voltage, and hence the maximum electron kinetic energy, is *independent* of the brightness of the light source. Brighter light ejects *more* electrons, but does not give an individual electron any more energy than faint light. On the other hand, the final voltage *does* depend on the *color* of the light, that is, on the wavelength (λ) or frequency ($f = c/\lambda$). In fact, there is a linear relation between the maximum kinetic energy of the ejected electrons and the frequency of the light:

$$K_{\mathrm{max}} = hf - \phi, \tag{A.1}$$

where h is a universal constant called **Planck's constant** and ϕ is a constant that depends on the metal. This relation was first predicted in 1905 by Einstein, extrapolating from Planck's earlier explanation of blackbody radiation.

Einstein's interpretation of the photoelectric effect is simple: Light comes in tiny bundles or particles, now called **photons**, each with energy equal to Planck's constant times the frequency of the light:

$$E_{\mathrm{photon}} = hf. \tag{A.2}$$

A brighter beam of light contains more photons, but the energy of each photon still depends only on the frequency, not on the brightness. When light hits the photocathode, each electron absorbs the energy of just one photon. The constant ϕ (called the **work function**) is the minimum energy required to get an electron out of the metal; once the electron is free, the maximum energy it can have is the photon's energy (hf) minus ϕ.

We don't normally notice that light comes in discrete bundles, because the bundles are so small: The value of Planck's constant is only 6.63×10^{-34} J·s, so visible-light photons have energies of only about two or three electron-volts. A typical light bulb gives off 10^{20} photons per second. But the technology needed to detect individual photons (from photomultiplier tubes in physics laboratories to CCD cameras for astronomy) is rather commonplace today.

Problem A.1. Photon fundamentals.

(a) Show that $hc = 1240$ eV·nm.

(b) Calculate the energy of a photon with each of the following wavelengths: 650 nm (red light); 450 nm (blue light); 0.1 nm (x-ray); 1 mm (typical for the cosmic background radiation).

(c) Calculate the number of photons emitted in one second by a 1-milliwatt red He-Ne laser ($\lambda = 633$ nm).

Problem A.2. Suppose that, in a photoelectric effect experiment of the type described above, light with a wavelength of 400 nm results in a voltage reading of 0.8 V.

(a) What is the work function for this photocathode?

(b) What voltage reading would you expect to obtain if the wavelength were changed to 300 nm? What if the wavelength were changed to 500 nm? 600 nm?

Electron Diffraction

If light, which everyone thought was a wave, can behave like a stream of particles, then perhaps it's not so surprising that electrons, which everyone thought were particles, can behave like waves.

But let me back up a bit. What do we *mean* when we say that light behaves like a wave? We don't actually see anything waving (as with water waves or waves on a guitar string). What we *can* observe is diffraction and interference effects, when light passes through a narrow opening or around a small obstacle. Perhaps the simplest example is two-slit interference, in which monochromatic light from a single source passes through a pair of closely spaced slits and forms a pattern of alternating light and dark spots on a viewing screen some distance away (see Figure A.2).

Well, electrons do the same thing: Take a beam of electrons (as in a TV picture tube or an electron microscope) and aim it at a pair of very closely spaced slits. On the viewing screen (a TV screen or some other detector) you get an interference pattern, exactly as with light (see Figure A.3).

The wavelength of the electron beam can be determined from the slit spacing and the size of the interference pattern, just as for light. It turns out that the wavelength is inversely proportional to the *momentum* of the electrons, and the constant of proportionality is Planck's constant:

$$\lambda = \frac{h}{p}. \tag{A.3}$$

This famous relation was predicted in 1923 by Louis de Broglie. It holds for photons too, and there it is a direct consequence of the Einstein relation $E = hf$ and the relation $p = E/c$ between the energy and momentum of anything that travels at the speed of light. De Broglie guessed correctly that electrons (and all other "particles") have wavelengths that are related to their momenta in the same way. (The Einstein

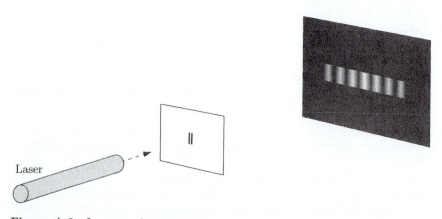

Figure A.2. In a two-slit interference experiment, monochromatic light (often from a laser) is aimed at a pair of slits in a screen. An interference pattern of dark and light bands appears on the viewing screen some distance away.

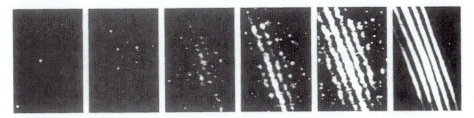

Figure A.3. These images were produced using the beam of an electron microscope. A positively charged wire was placed in the path of the beam, causing the electrons to bend around either side and interfere as if they had passed through a double slit. The current in the electron beam increases from one image to the next, showing that the interference pattern is built up from the statistically distributed light flashes of individual electrons. From P. G. Merli, G. F. Missiroli, and G. Pozzi, *American Journal of Physics* **44**, 306 (1976).

relation $E = hf$ also turns out to apply to electrons and other particles, but this relation is not as useful because the "frequency" of an electron is not directly measurable.)

The fact that both electrons and photons can act like waves and produce interference patterns raises some tricky questions. Each individual particle (electron or photon) can land in only one spot on the viewing screen, so if you send the particles through the apparatus slowly enough, the pattern builds up gradually, dot by dot, as shown in Figure A.3. Apparently, the place where each particle lands is *random*, with the probability varying across the screen as determined by the brightness of the final pattern. This means that *each* photon or electron must somehow pass through *both* slits and then interfere with *itself* to determine the probability distribution for where it will finally land. In other words, the particle behaves like a wave when passing through the slits, and the amplitude of this wave at the location of the screen determines the probabilities that govern the final position. (More precisely, the probability of landing in a particular place is proportional to the *square* of the final wave amplitude, just as the brightness of an electromagnetic wave is proportional to the square of the electric field amplitude.)

Problem A.3. Use the Einstein relation $E = hf$ and the relation $E = pc$ to show that the de Broglie relation A.3 holds for photons.

Problem A.4. Use the relativistic definitions of energy and momentum to show that $E = pc$ for any particle traveling at the speed of light. (For electromagnetic waves this relation can also be derived from Maxwell's equations, but this is much harder.)

Problem A.5. The electrons in a television picture tube are typically accelerated to an energy of 10,000 eV. Calculate the momentum of such an electron, and then use the de Broglie relation to calculate its wavelength.

Problem A.6. In the experiment shown in Figure A.3, the effective slit spacing was 6 μm and the distance from the "slits" to the detection screen was 16 cm. The spacing between the center of one bright line and the next (before magnification) was typically 100 nm. From these parameters, determine the wavelength of the electron beam. What voltage was used to accelerate the electrons?

Problem A.7. The de Broglie relation applies to all "particles," not just electrons and photons.

(a) Calculate the wavelength of a neutron whose kinetic energy is 1 eV.

(b) Estimate the wavelength of a pitched baseball. (Use any reasonable values for the mass and speed.) Explain why you don't see baseballs diffracting around bats.

A.2 Wavefunctions

Given that individual particles can behave like waves, we need a way of describing particles that allows for both particlelike and wavelike properties. For this purpose physicists have invented the quantum-mechanical **wavefunction**. In describing the "state" of a particle, the wavefunction serves the same purpose in quantum mechanics that the position and momentum vectors serve in classical mechanics: It tells us everything there is to know about what the particle is doing at some particular instant. The usual symbol for a particle's wavefunction is Ψ, and it is a function of position, or of the three coordinates x, y, and z. It's simpler, though, to begin with wavefunctions for particles constrained to move in just the x direction. In this case Ψ at any given time is a function only of x.

A particle can have all sorts of wavefunctions. There are narrow, spiky wavefunctions, corresponding to states in which the particle's position is well defined (see Figure A.4). There are also broad, oscillating wavefunctions, corresponding to states in which the particle's momentum is well defined (see Figure A.5). In this

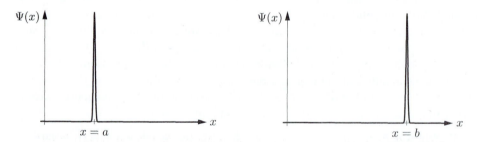

Figure A.4. Wavefunctions for states in which a particle's position is well defined (at $x = a$ and $x = b$, respectively). When a particle is in such a state, its momentum is completely undefined.

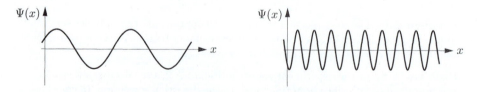

Figure A.5. Wavefunctions for states in which a particle's momentum is well defined (with small and large values, respectively). When a particle is in such a state, its position is completely undefined.

latter case, the momentum p of the particle is related to the wavelength λ by the de Broglie relation, $p = h/\lambda$.

Actually the wavelength of the wavefunction tells us only the *magnitude* of the particle's momentum. Even in one dimension, p_x could be positive or negative, and you can't tell which it is from looking at Figure A.5. In order for Ψ to *completely* determine the state of the particle, we need to make it a two-component object, that is, a *pair* of functions. For a particle with well-defined momentum, the second component has the same wavelength as the first component but is out of phase by $90°$ (see Figure A.6). For the case shown, the momentum is in the $+x$ direction. To give the particle the opposite momentum, we just flip the second component upside down, so it's $90°$ out of phase in the other direction. The two components of Ψ are normally represented by a single complex-valued function, whose "real part" is the first component and whose "imaginary part" (which is no less, or more, real) is the second component. If you want, you can imagine plotting the imaginary part of Ψ along an axis that points up out of the page. Then the three-dimensional graph of a definite-momentum wavefunction is a helix or corkscrew, with a right-handed twist for positive p_x and a left-handed twist for negative p_x.

Besides definite-position wavefunctions and definite-momentum wavefunctions, there are all sorts of others (see Figure A.7). For any wavefunction, though, there is a precise interpretation that's very important. First, take your wavefunction and compute its square modulus:

$$|\Psi(x)|^2 = (\text{Re }\Psi)^2 + (\text{Im }\Psi)^2. \tag{A.4}$$

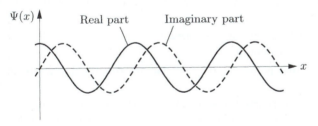

Figure A.6. A more complete illustration of the wavefunction of a particle with well-defined momentum, showing both the "real" and "imaginary" parts of the function.

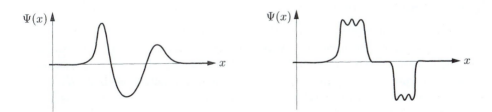

Figure A.7. Other possible wavefunctions for which neither the position nor the momentum of the particle is well defined.

This function, when integrated between any two points x_1 and x_2, gives you the *probability* of finding the particle somewhere between those two points if you were to measure its position at that time. (So $|\Psi|^2$ is a function whose purpose in life is to be integrated.) Qualitatively, you are more likely to find the particle where its wavefunction is large in magnitude, and less likely to find the particle where its wavefunction is small in magnitude. For a narrow spiky wavefunction you're certain to find the particle at the location of the spike, while for a definite-momentum wavefunction you could find the particle absolutely anywhere.

There's also a way to compute the probabilities of getting various outcomes if you were to measure a particle's *momentum*. Unfortunately, the procedure is mathematically intricate: First you have to take the "Fourier transform" of the wavefunction, which is a function of the "wavenumber" $k = 2\pi/\lambda$. Then change variables to $p_x = hk/2\pi$, and square this function to get a function which, when integrated between two values of p_x, gives the probability of getting a momentum within that range.

Qualitatively, you can usually tell just by looking at a wavefunction whether its momentum is reasonably well defined. A perfectly sinusoidal wavefunction (with the proper relation between real and imaginary parts) has a perfectly precise wavelength and therefore a perfectly well-defined momentum, while a definite-position spiky wavefunction has no wavelength at all: If you were to measure the momentum of such a particle, you could get any result whatsoever.

> **Problem A.8.** A definite-momentum wavefunction can be expressed by the formula $\Psi(x) = A(\cos kx + i \sin kx)$, where A and k are constants.
>
> **(a)** How is the constant k related to the particle's momentum? (Justify your answer.)
>
> **(b)** Show that, if a particle has such a wavefunction, you are equally likely to find it at *any* position x.
>
> **(c)** Explain why the constant A must be infinitesimal, if this formula is to be valid for all x.
>
> **(d)** Show that this wavefunction satisfies the differential equation $d\Psi/dx = ik\Psi$.
>
> **(e)** Often the function $\cos\theta + i\sin\theta$ is written instead as $e^{i\theta}$. Treating the i as an ordinary constant, show that the function Ae^{ikx} obeys the same differential equation as in part (d).

The Uncertainty Principle

Another important type of wavefunction is what I sometimes call a "compromise" wavefunction, more often called a **wavepacket**. A wavepacket is approximately sinusoidal within a certain region but then dies out beyond so it's still reasonably localized in space (see Figure A.8). For such a wavefunction both x and p_x are defined approximately, but neither is defined precisely. If we were to measure the position of a particle in such a state, we could get a range of values. If we had a million particles, all in this same state, and we measured their positions, the values would center around some average with a spread that we could quantify by

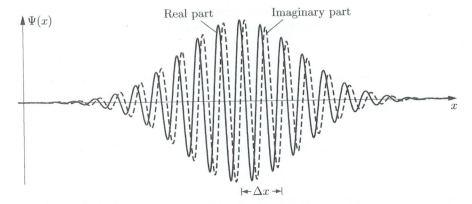

Figure A.8. A **wavepacket**, for which both x and p_x are defined approximately but not precisely. The "width" of the wavepacket is quantified by Δx, technically the standard deviation of the square of the wavefunction. (As you can see, Δx is actually a few times smaller than the "full" width.)

taking the standard deviation of all the values obtained. I'll refer to this standard deviation as Δx; it is a rough measure of the width of the wavepacket.

Similarly, if we had a million particles all in the same state and we measured their momenta, the values would center around some average with a spread that we could quantify by taking the standard deviation. I'll refer to this standard deviation as Δp_x; it is a rough measure of the width of the wavepacket in "momentum space."

We can easily construct a wavepacket with a smaller Δx, just by making the oscillations die out more rapidly on each side. But there is a price: We then get fewer complete oscillations, so the wavelength and momentum of the particle become more poorly defined. By the same token, to construct a wavepacket with a precisely defined momentum we have to include many oscillations, resulting in a large Δx. There is an inverse relation between the width of a wavepacket in position space and its width in momentum space.

To be more precise about this relation, suppose we make a wavepacket so narrow that it includes only one full oscillation before dying out. Then the spread in position is roughly one wavelength, while the spread in momentum is quite large, comparable to the momentum itself:

$$\Delta p_x \sim p_x = \frac{h}{\lambda} \sim \frac{h}{\Delta x}. \tag{A.5}$$

Because a smaller Δp_x implies a larger Δx, the relation

$$(\Delta x)(\Delta p_x) \sim h \tag{A.6}$$

actually applies to any wavepacket, not just a very narrow one. More generally, one can use Fourier analysis to prove that for any wavefunction whatsoever,

$$(\Delta x)(\Delta p_x) \geq \frac{h}{4\pi}. \tag{A.7}$$

This is the famous **Heisenberg uncertainty principle**. It says that if you prepare a million particles with identical wavefunctions, then measure the positions of
half of the particles and the momenta of the other half and compute the standard
deviations, the product of the standard deviations can't be less than $h/4\pi$. Therefore, no matter how you prepare a particle, you can't put it into a state in which
both Δx and Δp_x are arbitrarily small. A properly constructed wavepacket can
attain the best-case limit where the product equals $h/4\pi$; for most wavefunctions,
however, the product is greater.

Problem A.9. The formula for a "properly constructed" wavepacket is

$$\Psi(x) = Ae^{ik_0 x}e^{-ax^2},$$

where A, a, and k_0 are constants. (The exponential of an imaginary number is
defined in Problem A.7. In this problem, just assume that you can manipulate
the i like any other constant.)

(a) Compute and sketch $|\Psi(x)|^2$ for this wavefunction.

(b) Show that the constant A must equal $(2a/\pi)^{1/4}$. (Hint: The probability of
finding the particle somewhere between $x = -\infty$ and $x = \infty$ must equal 1.
See Section 1 of Appendix B for help with the integral.)

(c) The standard deviation Δx can be computed as $\sqrt{\overline{x^2} - \overline{x}^2}$, and the average
value of x^2 is just the sum of all values of x^2, weighted by their probabilities:

$$\overline{x^2} = \int_{-\infty}^{\infty} x^2 |\Psi(x)|^2 \, dx.$$

Use these formulas to show that for this wavepacket, $\Delta x = 1/(2\sqrt{a})$.

(d) The Fourier transform of a function $\Psi(x)$ is defined as

$$\widetilde{\Psi}(k) = \frac{1}{\sqrt{2\pi}} \int_{-\infty}^{\infty} e^{-ikx} \Psi(x) \, dx.$$

Show that $\widetilde{\Psi}(k) = (A/\sqrt{2a}) \exp[-(k-k_0)^2/4a]$ for a properly constructed
wavepacket. Sketch this function.

(e) Using formulas analogous to those in part (c), show that, for this wavefunction, $\Delta k = \sqrt{a}$. (Hint: The standard deviation does not depend on k_0,
so you can simplify the calculation by setting $k_0 = 0$ from the start.)

(f) Compute Δp_x for this wavefunction, and check whether the uncertainty
principle is satisfied.

Problem A.10. Sketch a wavefunction for which the product $(\Delta x)(\Delta p_x)$ is much
greater than $h/4\pi$. Explain how you would estimate Δx and Δp_x for your wavefunction.

Linearly Independent Wavefunctions

As you can tell from the preceding illustrations, the number of possible wavefunctions that a particle can have is enormous. This poses a problem in statistical
mechanics, where we need to *count* how many states are available to a particle.
There is no sensible way to count *all* the wavefunctions; what we need is a way to
count *independent* wavefunctions, in a sense that I will now make precise.

If a wavefunction Ψ can be written in terms of two others Ψ_1 and Ψ_2,

$$\Psi(x) = a\Psi_1(x) + b\Psi_2(x), \tag{A.8}$$

for some (complex) constants a and b, then we say that Ψ is a **linear combination** of Ψ_1 and Ψ_2. On the other hand, if there are no constants a and b for which equation A.8 is true, then we say that Ψ is **linearly independent** of Ψ_1 and Ψ_2. More generally, if we have some set of functions $\Psi_n(x)$, and $\Psi(x)$ cannot be written as a linear combination of the Ψ_n's, then we say that Ψ is linearly independent of the Ψ_n's. And if *none* of the wavefunctions in a collection can be written as a linear combination of the others, then we say they are all linearly independent.

What we want to do in statistical mechanics is count the number of linearly independent wavefunctions available to a particle. If the particle is confined within a finite region and its energy is limited, this number is always finite. Even so, there are many different sets of linearly independent wavefunctions that we can work with. In Section 2.5 I used wavepackets, approximately localized in both position space and momentum space. Usually, though, it is more convenient to use wavefunctions with definite energy, to be discussed in the next section.

> **Problem A.11.** Consider the functions $\Psi_1(x) = \sin(x)$ and $\Psi_2(x) = \sin(2x)$, where x can range from 0 to π. Write down formulas for three different nontrivial linear combinations of Ψ_1 and Ψ_2, and sketch each of your three functions. For simplicity, keep your functions real-valued.

A.3 Definite-Energy Wavefunctions

Among all the possible wavefunctions a particle can have, the most important are the wavefunctions with definite total energy. Total energy is kinetic plus potential; for a nonrelativistic particle in one dimension,

$$E = \frac{p_x^2}{2m} + V(x), \tag{A.9}$$

where the potential energy function $V(x)$ can be practically anything. In the special case where $V(x) = 0$, the total energy is the same as the kinetic energy, which depends only on momentum, and so any definite-momentum wavefunction is also a definite-energy wavefunction. When $V(x)$ is nonzero, however, the potential energy is not well defined for a definite-momentum wavefunction, so the definite-energy wavefunctions will be different.

To find the definite-energy wavefunctions for a given potential energy $V(x)$ you have to solve a differential equation, called the **time-independent Schrödinger equation.*** This equation and methods of solving it are discussed at length in quan-

---

*There's also a time-*dependent* Schrödinger equation, whose purpose is completely different: It tells you how any wavefunction changes with the passage of time. Definite-energy wavefunctions oscillate from real to imaginary and back again with frequency $f = E/h$, while other wavefunctions evolve in more complicated ways. Because definite-energy wavefunctions have the simplest possible time dependence, there is a close mathematical relation between the two Schrödinger equations.

tum mechanics textbooks; here I'll just describe the solutions in a few important special cases.

The Particle in a Box

The simplest nontrivial potential energy function is the "infinite square well,"

$$V(x) = \begin{cases} 0 & \text{for } 0 < x < L, \\ \infty & \text{elsewhere.} \end{cases} \tag{A.10}$$

This idealized potential confines the particle to the region between 0 and L, a one-dimensional "box" (see Figure A.9). Within the box there is no potential energy, while outside the box the particle cannot exist since it would have to have infinite energy.

This potential energy function is so simple that we can find the definite-energy wavefunctions without bothering to solve the time-independent Schrödinger equation. Every allowed wavefunction must be zero outside the box, while inside the box, where there is no potential energy, a definite-energy wavefunction will have definite kinetic energy and therefore definite momentum. Well, almost. The definite-energy wavefunctions need to go continuously to zero at $x = 0$ and at $x = L$, since a discontinuity would introduce an infinite uncertainty in the momentum (that is, zero-wavelength Fourier components). But the definite-momentum wavefunctions don't go to zero anywhere. To make a wavefunction that does have zeros (nodes), we need to add together two wavefunctions with equal and opposite momenta to make a "standing wave." Such a wavefunction will still have definite kinetic energy, since kinetic energy depends only on the *square* of the momentum.

A few of the definite-energy wavefunctions are shown in Figure A.9. In order for the wavefunction to go to zero at *both* ends of the box, only certain wavelengths are permitted: $2L$, $2L/2$, $2L/3$, and so on. For each of these wavelengths we can

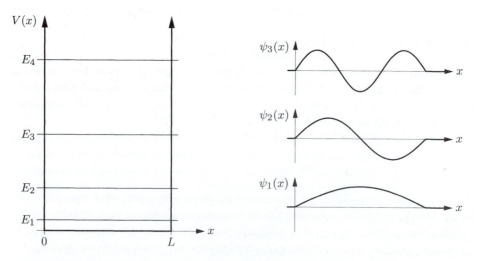

Figure A.9. A few of the lowest energy levels and corresponding definite-energy wavefunctions for a particle in a one-dimensional box.

use the de Broglie relation to find the magnitude of the momentum, then compute the energy as $p^2/2m$. Thus the allowed energies are

$$E_n = \frac{p_n^2}{2m} = \frac{1}{2m}\left(\frac{h}{\lambda_n}\right)^2 = \frac{h^2}{2m}\left(\frac{n}{2L}\right)^2 = \frac{h^2 n^2}{8mL^2}, \tag{A.11}$$

where n is any positive integer. Notice that the energies are quantized: Only certain discretely spaced energies are possible, because the number of half wavelengths that fit within the box must be an integer. More generally, any time a particle is confined within a limited region, its wavefunction must go to zero outside this region and have some whole number of "bumps" inside, so its energy will be quantized.

Definite-energy wavefunctions are important not just because they have definite energy, but also because any *other* wavefunction can be written as a linear combination of definite-energy wavefunctions. (In the case of the particle-in-a-box wavefunctions, this statement is the same as the theorem of Fourier analysis that says that any function within a finite region can be written as a linear combination of sinusoidal functions.) Furthermore, the definite-energy wavefunctions are all linearly independent of each other (at least for a particle in one dimension that is confined to a limited region). So counting the definite-energy wavefunctions gives us a convenient way to count "all" the states available to a particle.

For a particle confined inside a *three*-dimensional box, we can construct a definite-energy wavefunction simply by multiplying together three one-dimensional definite-energy wavefunctions:

$$\psi(x, y, z) = \psi_x(x)\psi_y(y)\psi_z(z), \tag{A.12}$$

where ψ_x, ψ_y, and ψ_z can each be any of the sinusoidal wavefunctions for a one-dimensional box. (For definite-energy wavefunctions it's conventional to use a lower-case ψ.) These products aren't *all* the definite-energy wavefunctions, but the others can be written as linear combinations of these, so counting wavefunctions that decompose in this way suffices for our purposes. The total energy also decomposes nicely into a sum of three terms:

$$E = \frac{|\vec{p}|^2}{2m} = \frac{1}{2m}(p_x^2 + p_y^2 + p_z^2) = \frac{1}{2m}\left[\left(\frac{hn_x}{2L_x}\right)^2 + \left(\frac{hn_y}{2L_y}\right)^2 + \left(\frac{hn_z}{2L_z}\right)^2\right], \tag{A.13}$$

where L_x, L_y, and L_z are the dimensions of the box and n_x, n_y, and n_z are any three positive integers. If the box is a cube, this formula reduces to

$$E = \frac{h^2}{8mL^2}(n_x^2 + n_y^2 + n_z^2). \tag{A.14}$$

Each triplet of n's yields a distinct linearly independent wavefunction, but not every triplet yields a distinct energy: Most of the energy levels are **degenerate**, corresponding to multiple linearly independent states that must be counted separately in statistical mechanics. (The number of linearly independent states that have a given energy is called the **degeneracy** of the level.)

Problem A.12. Make a rough estimate of the minimum energy of a proton confined inside a box of width 10^{-15} m (the size of an atomic nucleus).

Problem A.13. For ultrarelativistic particles such as photons or high-energy electrons, the relation between energy and momentum is not $E = p^2/2m$ but rather $E = pc$. (This formula is valid for massless particles, and also for massive particles in the limit $E \gg mc^2$.)

(a) Find a formula for the allowed energies of an ultrarelativistic particle confined to a one-dimensional box of length L.

(b) Estimate the minimum energy of an electron confined inside a box of width 10^{-15} m. It was once thought that atomic nuclei might contain electrons; explain why this would be very unlikely.

(c) A nucleon (proton or neutron) can be thought of as a bound state of three quarks that are approximately massless, held together by a very strong force that effectively confines them inside a box of width 10^{-15} m. Estimate the minimum energy of three such particles (assuming all three of them to be in the lowest-energy state), and divide by c^2 to obtain an estimate of the nucleon mass.

Problem A.14. Draw an energy level diagram for a nonrelativistic particle confined inside a three-dimensional cube-shaped box, showing all states with energies below $15 \cdot (h^2/8mL^2)$. Be sure to show each linearly independent state separately, to indicate the degeneracy of each energy level. Does the average number of states per unit energy increase or decrease as E increases?

The Harmonic Oscillator

Another important potential energy function is the harmonic oscillator potential,

$$V(x) = \tfrac{1}{2}k_s x^2, \tag{A.15}$$

where k_s is some "spring constant." The definite-energy wavefunctions for a particle subject to this potential are not easy to guess, but can be found by solving the time-independent Schrödinger equation. A few of them are shown in Figure A.10. These are not sinusoidal functions, but they still have an approximate local "wavelength," which is smaller near the middle (where there is less potential energy and hence more kinetic) and larger off to either side (where there is very little kinetic energy).

As with the particle in a box, the definite-energy wavefunctions for a quantum harmonic oscillator must go to zero at each side, with an integer number of "bumps" in between, so the energies are quantized. This time the allowed energies turn out to be

$$E = \tfrac{1}{2}hf, \ \tfrac{3}{2}hf, \ \tfrac{5}{2}hf, \ \dots, \tag{A.16}$$

where $f = \tfrac{1}{2\pi}\sqrt{k_s/m}$ is the natural frequency of the oscillator. The energies are equally spaced, instead of getting farther apart as you go up as they do for a particle in a box. (This is because the harmonically oscillating particle can "travel" farther to either side if its energy is larger, allowing more space for the wavefunction and hence longer wavelengths.) Often it is convenient to measure all energies relative to the ground-state energy, so that the allowed energies become

$$E = 0, \ hf, \ 2hf, \ \dots. \tag{A.17}$$

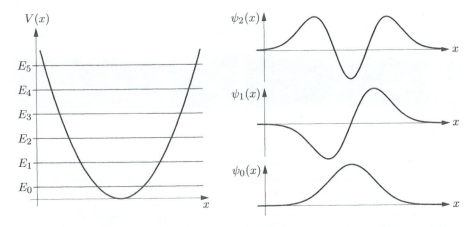

Figure A.10. A few of the lowest energy levels and corresponding wavefunctions for a one-dimensional quantum harmonic oscillator.

Shifting the zero-point in this way has no effect on thermal interactions. See Problem A.24, however, for a situation in which the zero-point energy does matter.

Many real-world systems oscillate harmonically, at least to a first approximation. A good example of a quantum oscillator is the vibrational motion of a diatomic molecule such as N_2 or CO. The vibrational energies can be measured by looking at the light emitted as the molecule makes a transition from one state to another; an example is shown in Figure A.11.

Problem A.15. A CO molecule can vibrate with a natural frequency of $6.4 \times 10^{13} \text{ s}^{-1}$.

    (a) What are the energies (in eV) of the five lowest vibrational states of a CO molecule?

    (b) If a CO molecule is initially in its ground state and you wish to excite it into its first vibrational level, what wavelength of light should you aim at it?

Problem A.16. In this problem you will analyze the spectrum of molecular nitrogen shown in Figure A.11. You may assume that all of the transitions are correctly identified in the energy level diagram.

    (a) What is the approximate difference in energy between the upper and lower electronic states, neglecting any vibrational energy (aside from the zero-point energies $\frac{1}{2}hf$)?

    (b) Determine the approximate spacing in energy between the vibrational levels, for both the lower and upper electronic states.

    (c) Repeat part (b) using a different set of spectral lines, to verify that the diagram is consistent.

    (d) How can you tell from the spectrum that the vibrational levels (for either electronic state) are not quite evenly spaced? (This is an indication that the potential energy function is not exactly quadratic.)

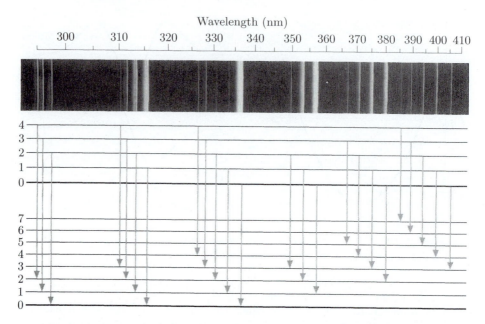

Figure A.11. A portion of the emission spectrum of molecular nitrogen, N_2. The energy level diagram shows the transitions corresponding to the various spectral lines. All of the lines shown are from transitions between the same pair of electronic states. In either electronic state, however, the molecule can also have one or more "units" of vibrational energy; these numbers are labeled at left. The spectral lines are grouped according to the number of units of vibrational energy gained or lost. The splitting within each group of lines occurs because the vibrational levels are spaced farther apart in one electronic state than in the other. From Gordon M. Barrow, *Introduction to Molecular Spectroscopy* (McGraw-Hill, New York, 1962). Photo originally provided by J. A. Marquisee.

(e) For the lower electronic state, what is the effective "spring constant" of the bond that holds the two nitrogen atoms together? (Hint: First determine the spring constant for each *half* of the spring, by considering each atom to be oscillating relative to the fixed center of mass. Then think carefully about how the spring constant (force per amount of stretch) of a whole spring is related to the spring constant of each half.)

Problem A.17. A *two*-dimensional harmonic oscillator can be considered as a system of two independent one-dimensional oscillators. Consider an isotropic two-dimensional oscillator, for which the natural frequency is the same in both directions. Write a formula for the allowed energies of this system, and draw an energy level diagram showing the degeneracy of each level.

Problem A.18. Repeat the previous problem for a *three*-dimensional isotropic oscillator. Find a formula for the number of degenerate states with any given energy.

The Hydrogen Atom

A third important potential energy function is

$$V(r) = -\frac{k_e e^2}{r},$$ (A.18)

the Coulomb potential experienced by the electron in a hydrogen atom. (Here e is the charge of the proton and k_e is the Coulomb constant, 8.99×10^9 N·m^2/C^2.) This is a three-dimensional problem, and solving the time-independent Schrödinger equation is a bit of a chore, but the resulting formula for the energy levels is quite simple:

$$E = -\frac{2\pi^2 m_e e^4 k_e^2}{h^2} \frac{1}{n^2} = -\frac{13.6 \text{ eV}}{n^2},$$ (A.19)

for $n = 1, 2, 3, \ldots$. The number of linearly independent wavefunctions corresponding to level n is n^2: 1 for the ground state, 4 for the first excited state, and so on. In addition to these negative-energy states, there can also be states with *any* positive energy; for these states the electron is ionized, no longer bound to the proton.

An energy level diagram for the hydrogen atom is shown in Figure A.12. The definite-energy wavefunctions are interesting and important, but hard to draw in a small space because they depend on three variables (and there are so many of them). You can find pictures of the wavefunctions (or, more commonly, the squares of the wavefunctions) in most textbooks of modern physics or introductory chemistry.

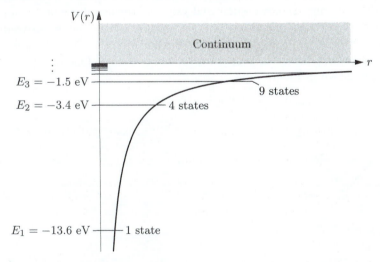

Figure A.12. Energy level diagram for a hydrogen atom. The heavy curve is the potential energy function, proportional to $-1/r$. In addition to the discretely spaced negative-energy states, there is a continuum of positive-energy (ionized) states.

Problem A.19. Suppose that a hydrogen atom makes a transition from a high-n state to a low-n state, emitting a photon in the process. Calculate the energy and wavelength of the photon for each of the following transitions: $2 \rightarrow 1$, $3 \rightarrow 2$, $4 \rightarrow 2$, $5 \rightarrow 2$.

A.4 Angular Momentum

Besides position, momentum, and energy, we might also want to know a particle's *angular* momentum (about some origin). More specifically, we might want to know the magnitude of its angular momentum vector, $|\vec{L}|$, or equivalently, the square of this magnitude, $|\vec{L}|^2$. In addition, we might want to know the three components of the angular momentum, L_x, L_y, and L_z.

Here again, there are special wavefunctions for which any particular variable is well defined. However, there are *no* wavefunctions for which more than one of the components L_x, L_y, and L_z is well defined (except in the trivial case where all three components are zero). The best we can do, with any particular wavefunction, is to specify the value of $|\vec{L}|^2$ and also the value of any *one* component of $\vec{L}$; usually we call it the z component.

The angular momentum of a particle determines only the angular dependence of its wavefunction—in spherical coordinates, the dependence on the angular variables θ and ϕ. Wavefunctions with well-defined angular momentum turn out to be various sinusoidal functions of these variables. For the wavefunction to be single-valued, these sinusoidal functions must go through a whole number of oscillations when you go around any complete circle. Thus only certain "wavelengths" of oscillation are allowed, corresponding to certain quantized values of the angular momentum. The allowed values of $|\vec{L}|^2$ turn out to be $\ell(\ell+1)\hbar^2$, where ℓ is any nonnegative integer and $\hbar$ is an abbreviation for $h/2\pi$. For a given value of ℓ, the allowed values of L_z (or L_x or L_y) are $m\hbar$, where m is any integer from $-\ell$ to ℓ. One way to visualize these states is shown in Figure A.13.

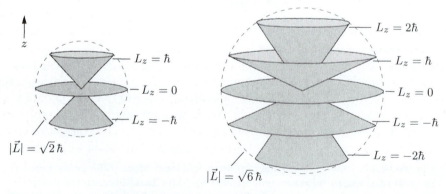

Figure A.13. A particle with well-defined $|\vec{L}|$ and L_z has completely undefined L_x and L_y, so we can visualize its angular momentum "vector" as a cone, smeared over all possible L_x and L_y values. Shown here are the allowed states for $\ell = 1$ and $\ell = 2$.

Angular momentum is most important in problems with rotational symmetry. Then, classically, angular momentum is conserved. In quantum mechanics, rotational symmetry implies that we can find definite-energy wavefunctions that also have definite angular momentum. An important example is the hydrogen atom, for which the ground state must have $|\vec{L}|^2 = 0$, the first excited state ($n = 2$) can have $|\vec{L}|^2 = 0$ or $|\vec{L}|^2 = 2\hbar^2$ (that is, $\ell = 0$ or $\ell = 1$), and so on. (The general rule for the hydrogen atom is that ℓ must be less than n, the integer that determines the energy. By the way, integers like n, ℓ, and m are called **quantum numbers**.)

Problem A.20. A very naive, but partially correct, way to understand quantization of angular momentum is as follows. Imagine that a particle is confined to travel around a circle of radius r. Its angular momentum about the center is then $\pm rp$, where p is the magnitude of its linear momentum at any moment. Let s be a coordinate that labels the position of the particle around the circle, so that s ranges from 0 to $2\pi r$. The wavefunction of this particle is a function of s. Now suppose that the wavefunction is sinusoidal, so that p is well defined. Using the fact that the wavefunction must undergo an integer number of complete oscillations over the entire circle, find the allowed values of p and the allowed values of the angular momentum.

Problem A.21. Enumerate the quantum numbers (n, ℓ, and m) for all the independent states of a hydrogen atom with definite E, $|\vec{L}|^2$, and L_z, up to $n = 3$. Check that the number of independent states for level n is equal to n^2.

Rotating Molecules

An important application of angular momentum in thermal physics is to the rotation of molecules in a gas. The analysis divides conveniently into the three cases of monatomic, diatomic, and polyatomic molecules.

A monatomic molecule (that is, a single atom) doesn't really have any rotational states. It's true that the electrons in the atom can carry angular momentum, and this angular momentum can have various orientations (all with the same energy if the atom is isolated). But to change the *magnitude* of the electrons' angular momentum would require putting them into excited states, which typically requires a few electron-volts of energy, more than is available at ordinary temperatures. In any case, these excited states are classified as electronic states, not molecular rotational states. The nucleus, in addition, can possess an intrinsic angular momentum ("spin"), which can have various orientations, but changing the magnitude of *its* angular momentum requires huge amounts of energy, typically 100,000 eV.

More generally, when we talk about the rotational states of a molecule, we are not interested in the rotation of individual nuclei, nor are we interested in excited electronic states. It's appropriate, therefore, to model the nuclei as point masses, and to neglect the electrons entirely since they're merely "along for the ride." I'll make both of these simplifications throughout this discussion.

In a diatomic molecule, the bond holding the two atoms together is normally quite stiff, so we can picture the two nuclei being held together by a rigid, massless rod. Let's suppose that the center of mass of this "dumbbell" is at rest (that is, we'll neglect any translational motion). Then classically, the configuration of the system in space depends only on two angles, θ and ϕ, specifying the direction toward one of the nuclei in spherical coordinates. (The position of the second nucleus is then completely determined, on the side opposite the first.) The energy of the molecule is determined by its angular momentum vector, conventionally called $\vec{J}$ (instead of $\vec{L}$) in this case. More precisely, the energy is just the usual rotational kinetic energy,

$$E_{\text{rot}} = \frac{|\vec{J}|^2}{2I}, \tag{A.20}$$

where I is the moment of inertia about the center of mass (that is, $I = m_1 r_1^2 + m_2 r_2^2$, where m_i is the mass and r_i is the distance from the rotation axis for the ith nucleus).

Quantum mechanically, the wavefunction of this system is a function only of the angles θ and ϕ. Therefore, specifying the angular momentum state ($|\vec{J}|^2$ and J_z) is sufficient to specify the entire wavefunction, and the number of independent wavefunctions available to the molecule is the same as the number of such angular momentum states. Furthermore, the value of $|\vec{J}|^2$ determines the molecule's rotational energy, according to equation A.20. Quantization of $|\vec{J}|^2$ therefore implies quantization of energy, with the allowed energies being

$$E_{\text{rot}} = \frac{j(j+1)\hbar^2}{2I}. \tag{A.21}$$

Here j is basically the same as the quantum number ℓ used above, with the allowed values 0, 1, 2, ... The degeneracy of each energy level is simply the number of distinct J_z values for that value of j, namely $2j + 1$. An energy level diagram for a rotating diatomic molecule is shown in Figure 6.6.

What I've just said, however, applies only to diatomic molecules made of *distinguishable* atoms: CO, or CN, or even H_2 where the two hydrogen atoms are of different isotopes. If the two atoms are *indistinguishable*, then there are only half as many distinct states, because interchanging the two atoms with each other results in exactly the same configuration. Basically this means that half of the j values in equation A.21 are allowed and half are not; Problem 6.30 explains how to figure out which half is which.

For any diatomic molecule, the spacing between rotational energy levels is proportional to $\hbar^2/2I$. This quantity is largest when the moment of inertia is small, but even for the smallest molecules it turns out to be less than $1/100$ eV. Generally, therefore, the rotational energy levels of a molecule are much more closely spaced than the vibrational levels (see Figure A.14). Because $kT \gg \hbar^2/2I$ for nearly all molecules at room temperature, rotational "degrees of freedom" normally hold a significant amount of thermal energy.

A linear polyatomic molecule, like CO_2, is similar to a diatomic molecule in that its rotational configuration can be specified in terms of only two angles. The rotational energies of such a molecule are therefore again given by equation A.21.

Most polyatomic molecules, however, are more complicated. For example, the orientation of an H_2O molecule is not completely specified by the position of one of the hydrogen atoms relative to the center of mass; even holding the hydrogen atoms

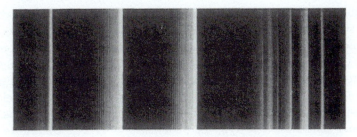

Figure A.14. Enlargement of a portion of the N_2 spectrum shown in Figure A.11, covering approximately the range 370–390 nm. Each of the broad lines is actually split into a "band" of many narrow lines, due to the multiplicity of rotational levels for each vibrational level. From Gordon M. Barrow, *Introduction to Molecular Spectroscopy* (McGraw-Hill, New York, 1962). Photo originally provided by J. A. Marquisee.

fixed, the oxygen atom can still travel around in a little circle. So to specify the orientation of a nonlinear polyatomic molecule we need a third angle. This means that the rotational wavefunctions are now functions of three variables instead of two, and the total number of states available is greater than for a diatomic molecule. The energy level structure is usually quite complex, because the moments of inertia about the three possible rotation axes are usually all different. At reasonably high temperatures, where many rotational states are available, the number of such states is enough to count as three "degrees of freedom." Beyond this important fact, the detailed behavior of polyatomic molecules is beyond the scope of this book.*

Problem A.22. In Section 6.2 I used the symbol ϵ as an abbreviation for the constant $\hbar^2/2I$. This constant is ordinarily measured by microwave spectroscopy: bombarding the molecule with microwaves and looking at what frequencies are absorbed.

(a) For a CO molecule, the constant ϵ is approximately 0.00024 eV. What microwave frequency would induce a transition from the $j = 0$ level to the $j = 1$ level? What frequency would induce a transition from the $j = 1$ level to the $j = 2$ level?

(b) Use the measured value of ϵ to calculate the moment of inertia of a CO molecule.

(c) From the moment of inertia and the known atomic masses, calculate the "bond length," or distance between the nuclei, for a CO molecule.

Spin

In addition to angular momentum due to its motion through space, a quantum-mechanical particle can have an internal or "intrinsic" angular momentum, called **spin**. Sometimes, if you look closely enough, the particle turns out to have internal structure and its spin is merely the result of the motion of its constituents. But in the case of "elementary" particles like electrons and photons, it's best to just

---

*More details on polyatomic molecules can be found in physical chemistry textbooks such as Atkins (1998).

think of spin as an intrinsic angular momentum that can't be visualized in terms of internal structure.

As with other forms of angular momentum, the magnitude of the spin angular momentum can have only certain values: $\sqrt{s(s+1)}\hbar$, where s is a quantum number analogous to ℓ. However, it turns out that s does *not* have to be an integer; it can also be a half-integer, that is, $1/2$ or $3/2$ or $5/2$, etc. Each species of elementary particle has its own value of s, which is fixed once and for all. For electrons, protons, neutrons, and neutrinos, $s = 1/2$; for photons, $s = 1$. For composite particles there are various rules for combining the spins and orbital angular momenta of the constituents to obtain the total spin; a helium-4 atom in its ground state turns out to have $s = 0$, for instance, because the spins of its constituents cancel each other out.

(By the way, when the angular momentum of a system comes from a combination of orbital motion and spin, or when we don't want to commit ourselves as to which it is, we normally call it $\vec{J}$, and use the quantum number j instead of ℓ or s.)

Given the value of s for a particle, the component of its spin angular momentum along the z axis (or any arbitrary axis) can still take on several possible values. Just as with orbital angular momentum, these values range from $s\hbar$ down to $-s\hbar$ in integer steps. So if $s = 1/2$, for instance, the z component of the angular momentum can be $+\hbar/2$ or $-\hbar/2$. If $s = 3/2$, there are four possible values for the z component: $3\hbar/2$, $\hbar/2$, $-\hbar/2$, and $-3\hbar/2$. For a *massless* particle, there's one more twist to the rules: Only the two most extreme values of the z component are allowed, for instance, $\pm\hbar$ for the photon ($s = 1$), and $\pm 2\hbar$ for the graviton ($s = 2$).

A spinning *charged* particle acts like a little bar magnet, whose strength and orientation are characterized by a **magnetic moment vector**, $\vec{\mu}$. For a macroscopic loop of electric current, $|\vec{\mu}|$ is the product of the amount of current and the area enclosed by the loop, while the direction of $\vec{\mu}$ is determined by a right-hand rule. This definition isn't very useful for microscopic magnets, though, so it's best to just define $\vec{\mu}$ in terms of the energy needed to twist the particle when it's in an external magnetic field $\vec{B}$. The energy is lowest when $\vec{\mu}$ is parallel to $\vec{B}$ and highest when it is antiparallel; if we take the energy to be zero when $\vec{\mu}$ and $\vec{B}$ are perpendicular, the general formula is

$$E_{\text{magnetic}} = -\vec{\mu} \cdot \vec{B}. \tag{A.22}$$

It's usually most convenient to call the direction of $\vec{B}$ the z direction, in which case

$$E_{\text{magnetic}} = -\mu_z |\vec{B}|. \tag{A.23}$$

Now since $\vec{\mu}$ is proportional to a particle's angular momentum, a quantum-mechanical particle has quantized values of μ_z: two possible values for a spin-1/2 particle, three for a spin-1 particle, and so on. For the important case of spin-1/2 particles introduced in Section 2.1, I've written

$$\mu_z = \pm\mu, \tag{A.24}$$

so $\mu \equiv |\mu_z|$. But while this notation is convenient, please note that this μ is *not* the same as $|\vec{\mu}|$, any more than $|J_z|$ is the same as $|\vec{J}|$ for a quantum-mechanical angular momentum.

> **Problem A.23.** Draw a cone diagram, as in Figure A.13, showing the spin states of a particle with $s = 1/2$. Repeat for a particle with $s = 3/2$. Draw both diagrams on the same scale, and be as accurate as you can with magnitudes and directions.

A.5 Systems of Many Particles

A system of *two* quantum-mechanical particles has only *one* wavefunction. In one spatial dimension, the wavefunction of a two-particle system is a function of two variables, x_1 and x_2, corresponding to the positions of the two particles. More precisely, if you integrate the square of the wavefunction over some range of x_1 values *and* over some range of x_2 values, you get the probability of finding the first particle within the first range and the second particle within the second range.

Some two-particle wavefunctions can be factored into a product of single-particle wavefunctions:

$$\Psi(x_1, x_2) = \Psi_a(x_1)\Psi_b(x_2). \tag{A.25}$$

This is an enormous simplification, which is valid only for a tiny fraction of all two-particle wavefunctions. Fortunately, though, all *other* two-particle wavefunctions can be written as linear combinations of wavefunctions that factor in this way. So if we're only interested in counting linearly independent wavefunctions, we're free to consider only those wavefunctions that factor. (The preceding statement is true whether or not the two particles interact with each other. But if they do *not* interact, there is a further simplification: The total energy of the system is then the sum of the energies of the two particles, and if we take Ψ_a and Ψ_b to be the appropriate single-particle definite-energy wavefunctions, then their product will be a definite-energy wavefunction for the combined system.)

If the two particles in question are distinguishable from each other, there's not much more to say. But quantum mechanics also allows for particles to be absolutely *in*distinguishable, so that no possible measurement can reveal which is which. In this case, the probability of finding particle 1 at position a and particle 2 at position b must be the same as the probability of finding particle 1 at position b and particle 2 at position a. In other words, the square of the wavefunction must be unchanged under the operation of interchanging its two arguments:

$$|\Psi(x_1, x_2)|^2 = |\Psi(x_2, x_1)|^2. \tag{A.26}$$

This almost implies that Ψ itself is unchanged under this operation, but not quite; another possibility is for Ψ to change sign:

$$\Psi(x_1, x_2) = \pm\Psi(x_2, x_1). \tag{A.27}$$

(Since interchanging the arguments a second time *must* restore Ψ to its original form, these are the only two possibilities; multiplying by i or some other complex number won't do.)

It turns out that nature has taken advantage of both possible signs in equation A.27. For some types of particles, called **bosons**, Ψ is unchanged under interchange of its arguments. For other particles, called **fermions**, Ψ changes sign under this operation:

$$\Psi(x_1, x_2) = \begin{cases} +\Psi(x_2, x_1) & \text{for bosons,} \\ -\Psi(x_2, x_1) & \text{for fermions.} \end{cases} \qquad (A.28)$$

(In the first case we say that Ψ is "symmetric," while in the second case we say that Ψ is "antisymmetric.") Examples of bosons include photons, pions, and many types of atoms and atomic nuclei. Examples of fermions include electrons, protons, neutrons, neutrinos, and many other types of atoms and nuclei. In fact, it turns out that all particles with integer spin (or more precisely, integer values of the quantum number s) are bosons, while all particles with half-integer spin (that is, $s = 1/2$, $3/2$, etc.) are fermions.

The most straightforward application of rule A.28 is to the case where both particles are in the *same* single-particle state:

$$\Psi(x_1, x_2) = \Psi_a(x_1)\Psi_a(x_2). \qquad (A.29)$$

For bosons, this equation guarantees that Ψ will be symmetric under interchange of x_1 and x_2. For fermions, however, such a state isn't possible at all, because such a function cannot be equal to minus itself (unless it is zero, which isn't allowed).

(When we take the spin orientation of a particle into account, the situation is actually a bit more complex. The "state" of a particle includes not just its spatial wavefunction but also its spin state, so for a fermion, the spatial part of the wavefunction can be symmetric as long as the spin part is antisymmetric. For the important case of spin-1/2 particles, the bottom line is that any given spatial wavefunction can be occupied by at most *two* such particles of the same species, provided that they are in what's called an antisymmetric spin configuration.)

All of the statements and formulas of this section generalize in a natural way to systems of three or more particles. The wavefunction of a system of several identical bosons must be unchanged under the interchange of *any* pair of the corresponding arguments, while the wavefunction of a system of several identical fermions must change sign under the interchange of any pair. Any given single-particle state (where "state" means both the spatial wavefunction and the spin configuration) can hold arbitrarily many identical bosons, but at most one fermion.

A.6 Quantum Field Theory

Classical mechanics deals not just with systems of pointlike particles, but also with continuous systems: strings, vibrating solids, and even "fields" such as the electromagnetic field. The usual approach is to first pretend that the continuous object is really a bunch of point particles connected together by little springs, and eventually take the limit where the number of particles goes to infinity and the space between them goes to zero. The result is generally some kind of partial differential equation (for instance, the linear wave equation or Maxwell's equations) that governs the motion as a function of place and time.

When this partial differential equation is linear, it is most easily solved by **Fourier analysis**. Think of the initial shape of the system (say a string) as a superposition of sinusoidal functions of different wavelengths. Each of these "modes" oscillates sinusoidally in time, with its own characteristic frequency. To find the shape of the string at some future time, you first figure out what each mode will look like at that time, then add them back up in the same proportions as initially.

So much for classical continuum mechanics; what if we now want to apply *quantum* mechanics to a continuous system? Here again, the most fruitful approach is usually to work with the Fourier modes of the system. Each mode behaves as a quantum harmonic oscillator, with quantized energy levels determined by the natural frequency of oscillation:

$$E = \tfrac{1}{2}hf, \ \tfrac{3}{2}hf, \ \tfrac{5}{2}hf, \ \dots \tag{A.30}$$

So for any given mode, there is a "zero-point" energy of $\tfrac{1}{2}hf$ (which we usually neglect), plus any integer number of energy units with size hf. Since different modes have different frequencies, the system as a whole can have energy units with lots of different sizes.

In the case of the electromagnetic field, these units of energy are called **photons**. Being discrete entities, they behave very much like particles that are in definite-energy wavefunctions. And since definite-energy states are not the *only* states of the field, we can even mix different modes together to make a "photon" that is localized in space. Thus, we are again confronted with wave-particle duality, this time in an even richer context. We started with a classical field, spread out in space. By applying the principles of quantum mechanics to this system, we see that the field acts in many ways like a collection of discrete particles. But now we have a system in which the number of particles is dependent on the current state, rather than being built in and fixed from the outset. In fact, the field can be in states for which the number of particles is not even well defined. These are just the features required to build an accurate model of the quantum fluctuations of the electromagnetic field, and to describe reactions in which photons are created and destroyed.*

Analogously, the units of vibrational energy in a solid crystal are called **phonons**. Like photons, they can be localized or spread-out, and can be created and destroyed in various reactions. Fundamentally, of course, phonons are not "real" particles: Their wavelengths and energies are limited by the nonzero atomic spacing in the crystal lattice, and they behave simply only when their wavelengths are much larger than this distance. For this reason, phonons are called **quasiparticles**. Still,

---

*For a good, brief treatment of the quantized electromagnetic field, see Ramamurti Shankar, *Principles of Quantum Mechanics*, second edition (Plenum, New York, 1994), Section 18.5. For a more complete introduction to quantum field theory, a good place to start is F. Mandl and G. Shaw, *Quantum Field Theory*, second edition (Wiley, Chichester, 1993).

the phonon picture provides a beautifully accurate description of the low-energy excitations of a crystal. Furthermore, the low-energy excitations of many other materials, from magnetized iron to liquid helium, can be similarly described in terms of various types of quasiparticles.

At a more fundamental level, we can use quantum fields to describe all the other species of "elementary" particles found in nature. Thus there is a "chromodynamic" field, for the force that holds the proton together, manifested as particles called "gluons." There is also an electron field, a muon field, various neutrino fields, quark fields, and so on. Fields corresponding to particles that are fermions need to be set up rather differently, so that each mode can hold only zero units of energy or one, not an unlimited number as with bosons. Fields corresponding to charged particles (electrical or otherwise) turn out to have two types of excitations, one for the particle and another for its "antiparticle," a particle with the same mass but opposite charge. Quite generally, the quantum theory of fields seems to include all the features needed to build an accurate model of elementary particle physics as we now understand it. However, it seems very likely that at sufficiently short wavelengths and high energies, this model will break down, just as the phonon model breaks down when the wavelength becomes comparable to the atomic spacing. Perhaps someday we will discover a new level of structure at some very small length scale, and conclude that all the "particles" of nature are actually quasiparticles.

> **Problem A.24.** According to equation A.30, each mode of a quantum field has a "zero-point" energy of $\frac{1}{2}hf$ even when no further units of energy are present. If the field is really a vibrating string or some other material object, this isn't a problem because the total number of modes is finite: You can't have a mode whose wavelength is shorter than half the atomic spacing (see Section 7.5). But for the electromagnetic field and other fundamental fields corresponding to elementary particles, there is no obvious limit on the number of modes, and the zero-point energies can add up to something embarassing.
>
> **(a)** Consider just the electromagnetic field inside a box of volume L^3. Use the methods of Chapter 7 to write down a formula for the *total* zero-point energy of all the modes of the field inside this box, in terms of a triple integral over the mode numbers in the x, y, and z directions.
>
> **(b)** There are good reasons to believe that most of our current laws of physics, including quantum field theory, break down at the very small length scale where quantum gravity becomes important. By dimensional analysis, you can guess that this length scale is of order $\sqrt{G\hbar/c^3}$, a quantity called the **Planck length**. Show that the Planck length indeed has units of length, and calculate it numerically.
>
> **(c)** Going back to your expression from part (a), cut off the integrals at a mode number corresponding to a wavelength of the Planck length. Then evaluate your expression to obtain an estimate of the energy per unit volume in empty space, due to the zero-point energy of the electromagnetic field. Express your answer in J/m^3, then divide by c^2 to obtain the equivalent *mass* density of empty space (in kg/m^3). Compare to the average mass density of ordinary matter in the universe, which is roughly equivalent to one proton per cubic meter. [Comment: Since most physical effects depend only on *differences* in energy, and since the zero-point energy never

changes, a large energy density in "empty" space would be harmless as far as most of the laws of physics are concerned. The only exception, and it's a big one, is gravity: Energy gravitates, so a large energy density in empty space would affect the expansion rate of the universe. The energy density of empty space is therefore known as the **cosmological constant**. From the observed expansion rate of the universe, cosmologists estimate that the actual cosmological constant cannot be any greater than 10^{-7} J/m^3. The discrepancy between this observational bound and your calculated value is one of the greatest paradoxes in theoretical physics. (The obvious solution would be to find some negative contribution to the energy density coming from some other source. In fact, fermionic fields give a negative contribution to the cosmological constant, but nobody knows how to make this negative contribution cancel the positive contribution from bosonic fields to the required precision.*)]

---

*For more about the cosmological constant paradox and various proposed solutions, see Larry Abbott, *Scientific American* **258**, 106–113 (May, 1988); Ronald J. Adler, Brendan Casey, and Ovid C. Jacob, *American Journal of Physics* **63**, 620–626 (1995); and/or Steven Weinberg, *Reviews of Modern Physics* **61**, 1–82 (1989).

B Mathematical Results

Although no mathematics beyond multivariable calculus is needed to understand the material in this book, in a few places I have quoted mathematical results that are not normally derived in a first course in calculus. The purpose of this appendix is to derive those results. If you're willing to take the results on faith (or better, check them approximately or in special cases), then there's no need to read this appendix. But some of the tools used in the derivations are more broadly applicable in theoretical physics, while all of the derivations themselves are quite lovely. So I hope you'll read on and enjoy this excursion along some of the less-traveled (but very scenic) byways of calculus.

B.1 Gaussian Integrals

The function e^{-x^2} (called a **Gaussian**) has an antiderivative, but there's no way to express that antiderivative in terms of familiar functions (like roots and powers and exponentials and logs). So if you're confronted with an integral of this function you'll probably just want to evaluate it numerically.

However, if the limits of the integral are 0 or $\pm\infty$, you're in luck. It turns out that the integral of e^{-x^2} from $-\infty$ to ∞ is exactly equal to $\sqrt{\pi}$,

$$\int_{-\infty}^{\infty} e^{-x^2}\, dx = \sqrt{\pi}, \tag{B.1}$$

and since the integrand is an even function (see Figure B.1), the integral from 0 to ∞ is just half of this, $\sqrt{\pi}/2$. The proof of this simple result makes use of a *two*-dimensional integral in polar coordinates.

Let me define

$$I = \int_{-\infty}^{\infty} e^{-x^2}\, dx. \tag{B.2}$$

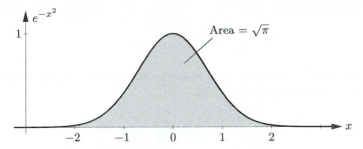

Figure B.1. The Gaussian function e^{-x^2}, whose integral from $-\infty$ to ∞ is $\sqrt{\pi}$.

The trick is to square this quantity:

$$I^2 = \left(\int_{-\infty}^{\infty} e^{-x^2}\, dx \right) \left(\int_{-\infty}^{\infty} e^{-y^2}\, dy \right), \qquad (B.3)$$

where I've carefully renamed the integration variable to y in the second factor so as not to confuse it with the integration variable in the first factor. Now the second factor is just a constant, so I can move it inside the x integral. And the function e^{-x^2} is independent of y, so I can move it inside the y integral:

$$I^2 = \int_{-\infty}^{\infty} e^{-x^2} \left(\int_{-\infty}^{\infty} e^{-y^2}\, dy \right) dx = \int_{-\infty}^{\infty} \int_{-\infty}^{\infty} e^{-x^2} e^{-y^2}\, dy\, dx. \qquad (B.4)$$

What we now have is the integral over all of two-dimensional space of the function $e^{-(x^2+y^2)}$. I'll carry out this integral in polar coordinates, r and ϕ (see Figure B.2). The integrand is simply e^{-r^2}, while the region of integration is r from 0 to ∞ and ϕ from 0 to 2π. Most importantly, the infinitesimal area element $dx\, dy$ becomes $(dr)(r\, d\phi)$ in polar coordinates, as shown in the figure. Therefore our double integral becomes

$$I^2 = \int_0^{\infty} \int_0^{2\pi} e^{-r^2} r\, d\phi\, dr = 2\pi \int_0^{\infty} r\, e^{-r^2}\, dr = (2\pi)\left(-\frac{1}{2}e^{-r^2}\right)\Big|_0^{\infty} = \pi, \qquad (B.5)$$

verifying formula B.1.

Figure B.2. In polar coordinates, the infinitesimal area element is $(dr)(r\, d\phi)$.

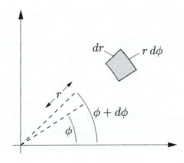

From equation B.1 you can perform a simple substitution to get the more general result

$$\int_0^\infty e^{-ax^2}\,dx = \frac{1}{2}\sqrt{\frac{\pi}{a}}, \tag{B.6}$$

where a is any positive constant. And from *this* equation we can get another useful result by differentiating with respect to a:

$$\frac{d}{da}\int_0^\infty e^{-ax^2}\,dx = \frac{\sqrt{\pi}}{2}\frac{d}{da}a^{-1/2}. \tag{B.7}$$

On the left-hand side we can move the derivative inside the integral, where it hits e^{-ax^2} and brings down a factor of $-x^2$. Evaluating the right-hand side and canceling the minus signs gives

$$\int_0^\infty x^2 e^{-ax^2}\,dx = \frac{1}{4}\sqrt{\frac{\pi}{a^3}}. \tag{B.8}$$

This trick of "differentiating under the integral" is an incredibly handy way to evaluate all sorts of definite integrals of transcendental functions multiplied by powers of x. (The alternative is to integrate by parts, but that's much slower.)

Integrals of Gaussian functions come up all the time in physics and mathematics, so you may want to make yourself a small reference table of the results of this section (including the problems below). In statistical mechanics, Gaussian integrals arise most commonly as integrals of a Boltzmann factor, where the energy is a quadratic function of the integration variable (as in Sections 6.3 and 6.4).

Problem B.1. Sketch an antiderivative of the function e^{-x^2}.

Problem B.2. Take another derivative of equation B.8 to evaluate $\int_0^\infty x^4 e^{-ax^2}\,dx$.

Problem B.3. The integral of $x^n e^{-ax^2}$ is easier to evaluate when n is odd.

(a) Evaluate $\int_{-\infty}^\infty x e^{-ax^2}\,dx$. (No computation allowed!)

(b) Evaluate the *indefinite* integral (i.e., the antiderivative) of xe^{-ax^2}, using a simple substitution.

(c) Evaluate $\int_0^\infty x e^{-ax^2}\,dx$.

(d) Differentiate the previous result to evaluate $\int_0^\infty x^3 e^{-ax^2}\,dx$.

Problem B.4. Sometimes you need to integrate only the "tail" of a Gaussian function, from some large x up to infinity:

$$\int_x^\infty e^{-t^2}\,dt = ? \qquad \text{when } x \gg 1.$$

Evaluate this integral approximately as follows. First, change variables to $s = t^2$, to obtain a simple exponential times something proportional to $s^{-1/2}$. The integral is dominated by the region near its lower limit, so it makes sense to expand $s^{-1/2}$

in a Taylor series about that point, keeping only the first few terms in the series. Do this to obtain a series expansion for the integral. Evaluate the first three terms of the series explicitly to obtain

$$\int_x^\infty e^{-t^2}\, dt = e^{-x^2}\left(\frac{1}{2x} - \frac{1}{4x^3} + \frac{3}{8x^5} - \cdots\right).$$

Note: When x is fairly large, the first few terms of this series will converge very rapidly toward the exact answer. However, if you calculate too many terms, the coefficients in the numerators will eventually start to grow more rapidly than the denominators and the series will diverge. This happens sooner or later no matter how large x is! Series expansions of this type are called **asymptotic expansions**. They're incredibly useful, though they make me rather queasy.

Problem B.5. Use the methods of the previous problem to find an asymptotic expansion for the integral of $t^2 e^{-t^2}$, from x to ∞, when $x \gg 1$.

Problem B.6. The antiderivative of e^{-x^2}, set equal to zero at $x = 0$ and multiplied by $2/\sqrt{\pi}$, is called the **error function**, abbreviated erf x:

$$\operatorname{erf} x \equiv \frac{2}{\sqrt{\pi}} \int_0^x e^{-t^2}\, dt.$$

(a) Show that $\operatorname{erf}(\pm\infty) = \pm 1$.

(b) Evaluate $\int_0^x t^2 e^{-t^2}\, dt$ in terms of erf x.

(c) Use the result of Problem B.4 to find an approximate expression for erf x when $x \gg 1$.

B.2 The Gamma Function

If you start with the integral

$$\int_0^\infty e^{-ax}\, dx = a^{-1} \tag{B.9}$$

and repeatedly differentiate with respect to a, you'll eventually be convinced that

$$\int_0^\infty x^n e^{-ax}\, dx = (n!)a^{-(n+1)}. \tag{B.10}$$

Setting $a = 1$ then gives a formula for $n!$:

$$n! = \int_0^\infty x^n e^{-x}\, dx. \tag{B.11}$$

I'll use this formula in the following section to derive Stirling's approximation for $n!$.

The integral B.11 can be evaluated (not necessarily analytically) even when n isn't an integer, so it gives us a way to generalize the factorial function to nonintegers. The generalization is called the **gamma function**, denoted $\Gamma(n)$, and for some reason it's defined with an offset of 1 in its argument:

$$\Gamma(n+1) \equiv \int_0^\infty x^n e^{-x}\, dx. \tag{B.12}$$

So for integer arguments,

$$\Gamma(n+1) = n!. \tag{B.13}$$

Perhaps the handiest property of the gamma function is the recursion relation

$$\Gamma(n+1) = n\Gamma(n). \tag{B.14}$$

When n is an integer, this formula is essentially the definition of a factorial. But it works for noninteger n too, as you can show from the definition B.12.

From either the definition B.12 or the recursion formula B.14 you can see that $\Gamma(n)$ blows up at $n = 0$. When the argument of the gamma function is negative, the definition B.12 continues to diverge, but we can still *define* the gamma function (for noninteger arguments) by the recursion formula B.14. A plot of the gamma function for both positive and negative arguments is shown in Figure B.3.

The gamma function gives meaning to some ambiguous expressions for factorials that occur in the text of this book, for instance,

$$0! = \Gamma(1) = 1; \qquad (\tfrac{d}{2} - 1)! = \Gamma(\tfrac{d}{2}). \tag{B.15}$$

The gamma function also arises in the evaluation of many definite integrals that occur in theoretical physics. We'll see it again in Section B.4.

Problem B.7. Prove the recursion formula B.14. Do not assume that n is an integer.

Problem B.8. Evaluate $\Gamma(\tfrac{1}{2})$. (Hint: Change variables to convert the integrand to a Gaussian.) Then use the recursion formula to evaluate $\Gamma(\tfrac{3}{2})$ and $\Gamma(-\tfrac{1}{2})$.

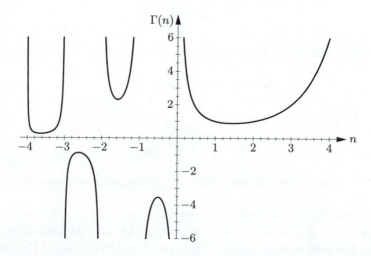

Figure B.3. The gamma function, $\Gamma(n)$. For positive integer arguments, $\Gamma(n) = (n-1)!$. For positive nonintegers, $\Gamma(n)$ can be computed from equation B.12, while for negative nonintegers, $\Gamma(n)$ can be computed from equation B.14.

Problem B.9. Carry out the integral B.12 numerically to evaluate $\Gamma(\frac{1}{3})$ and $\Gamma(\frac{2}{3})$. A useful identity whose proof is beyond the scope of this book is

$$\Gamma(n)\Gamma(1-n) = \frac{\pi}{\sin(n\pi)}.$$

Check this formula numerically for $n = 1/3$.

B.3 Stirling's Approximation

In Section 2.4 I introduced **Stirling's approximation**,

$$n! \approx n^n e^{-n}\sqrt{2\pi n}, \tag{B.16}$$

which is accurate when $n \gg 1$. Since this formula is so important, I'll derive it not once but twice.

The first derivation is easier, but not as accurate. Let's work with the natural log of $n!$:

$$\begin{aligned} \ln n! &= \ln\left[n \cdot (n-1) \cdot (n-2) \cdots 1\right] \\ &= \ln n + \ln(n-1) + \ln(n-2) + \cdots + \ln 1. \end{aligned} \tag{B.17}$$

This sum of logarithms can be represented as the area under a bar graph (see Figure B.4). Now if n is fairly large, the area under the bar graph is approximately equal to the area under the smooth curve of the logarithm function. Therefore,

$$\ln n! \approx \int_0^n \ln x \, dx = \left(x \ln x - x\right)\Big|_0^n = n \ln n - n. \tag{B.18}$$

In other words, $n! \approx (n/e)^n$. This result agrees with equation B.16, aside from the final factor of $\sqrt{2\pi n}$. When n is sufficiently large, as is nearly always the case in statistical mechanics, that factor can be omitted so this result is all we need.

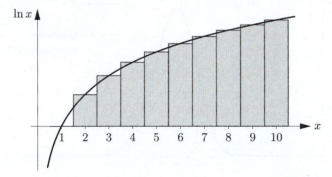

Figure B.4. The area under the bar graph, up to any integer n, equals $\ln n!$. When n is large, this area can be approximated by the area under the smooth curve of the logarithm function.

To derive a more accurate formula for $n!$, you can repeat the previous calculation but choose the limits on the integral more carefully (see Problem B.10). But to really get it right, I'll use a completely different method, starting from the exact formula B.11:

$$n! = \int_0^\infty x^n e^{-x} \, dx. \tag{B.19}$$

Let's think about the integrand, $x^n e^{-x}$, when n is large. The first factor, x^n, rises very rapidly as a function of x, while the second factor, e^{-x}, falls very rapidly to zero. The product is a function that rises and then falls, as shown in Figure B.5. You can easily show that the maximum value is reached precisely at the point $x = n$ (see Problem B.11), and that the height of the peak is $n^n e^{-n}$. What we want is the area under the graph, and to estimate this area we can approximate the function as a Gaussian. To find the Gaussian function that best fits the exact function near $x = n$, let me first write the function as a single exponential:

$$x^n e^{-x} = e^{n \ln x - x}. \tag{B.20}$$

Next, define $y \equiv x - n$, rewrite the exponent in terms of y, and get ready to expand the logarithm:

$$n \ln x - x = n \ln(n + y) - n - y$$
$$= n \ln\left[n\left(1 + \frac{y}{n}\right)\right] - n - y \tag{B.21}$$
$$= n \ln n - n + n \ln\left(1 + \frac{y}{n}\right) - y.$$

Near the peak of the graph, y is much less than n so we can expand the logarithm in a Taylor series:

$$\ln\left(1 + \frac{y}{n}\right) \approx \frac{y}{n} - \frac{1}{2}\left(\frac{y}{n}\right)^2. \tag{B.22}$$

The linear term is canceled by the final $-y$ in equation B.21. Putting everything

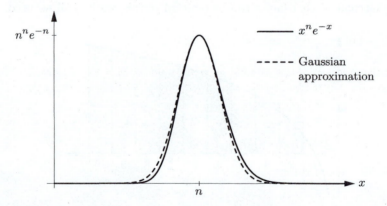

Figure B.5. The function $x^n e^{-x}$ (solid curve), plotted for $n = 50$. The area under this curve is $n!$. The dashed curve shows the best Gaussian fit, whose area gives Stirling's approximation to $n!$.

else together, we obtain the approximation

$$x^n e^{-x} \approx n^n e^{-n} e^{-y^2/2n}, \qquad \text{with } y = x - n. \tag{B.23}$$

This is the best Gaussian approximation to the exact integrand in equation B.19; it is shown as the dashed curve in Figure B.5. To get $n!$, we want to integrate this function from $x = 0$ to $x = \infty$. But we might as well start the integral at $x = -\infty$, since the function is negligible at negative x values anyway. Using the integration formula B.6, we obtain

$$n! \approx n^n e^{-n} \int_{-\infty}^{\infty} e^{-y^2/2n} \, dy = n^n e^{-n} \sqrt{2\pi n}, \tag{B.24}$$

in agreement with equation B.16.

Problem B.10. Choose the limits on the integral in equation B.18 more carefully, to derive a more accurate approximation to $n!$. (Hint: It's the upper limit that is more critical. There's no obvious best choice for the lower limit, but do the best you can.)

Problem B.11. Prove that the function $x^n e^{-x}$ reaches its maximum value at $x = n$.

Problem B.12. Use a computer to plot the function $x^n e^{-x}$, and the Gaussian approximation to this function, for $n = 10$, 20, and 50. Notice how the relative width of the peak (compared to n) decreases as n increases, and how the Gaussian approximation becomes more accurate as n increases. If your computer software permits it, try looking at even higher values of n.

Problem B.13. It is possible to improve Stirling's approximation by keeping more terms in the expansion of the logarithm (B.22). The exponential of the new terms can then be expanded in a Taylor series to yield a polynomial in y multiplied by the same Gaussian as before. Carry out this procedure, consistently keeping all terms that will end up being smaller than the leading term by one power of n. (Since the Gaussian cuts off when y is of order $\sqrt{n}$, you can estimate the sizes of various terms by setting $y = \sqrt{n}$.) When the smoke clears, you should find

$$n! \approx n^n e^{-n} \sqrt{2\pi n} \left(1 + \frac{1}{12n} \right).$$

Check the accuracy of this formula for $n = 1$ and for $n = 10$. (In practice, the correction term is rarely needed. But it does provide a handy way to estimate the error in Stirling's approximation.)

B.4   Area of a d-Dimensional Hypersphere

In Section 2.5 I claimed that the surface "area" of a d-dimensional "hypersphere" of radius r is

$$A_d(r) = \frac{2\pi^{d/2}}{(\frac{d}{2} - 1)!} r^{d-1} = \frac{2\pi^{d/2}}{\Gamma(\frac{d}{2})} r^{d-1}. \tag{B.25}$$

For $d = 2$ this formula gives the circumference of a circle, $A_2(r) = 2\pi r$, while for

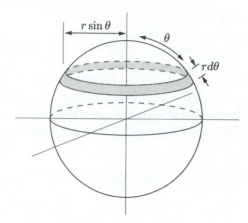

Figure B.6. To calculate the area of a sphere, divide it into loops and integrate. To calculate the area of a *hypersphere*, do the same thing.

$d = 3$ it gives the surface area of a sphere, $A_3(r) = 4\pi r^2$. (For $d = 1$ it gives $A_1(r) = 2$, the number of points bounding a line segment.)

Before proving equation B.25 in general, let's warm up by considering just the case $d = 3$, a true three-dimensional sphere. The surface of the sphere can be built out of loops, as shown in Figure B.6. Each loop has width $r\,d\theta$ and circumference $A_2(r \sin \theta) = 2\pi r \sin \theta$, so the total area of the sphere is

$$A_3(r) = \int_0^\pi A_2(r \sin \theta)\, r\, d\theta = 2\pi r^2 \int_0^\pi \sin \theta\, d\theta = 4\pi r^2. \tag{B.26}$$

By a completely analogous calculation, we can prove equation B.25 for any d, assuming by induction that it holds for $d-1$. Imagine building the surface of a d-dimensional sphere out of $(d-1)$-dimensional "loops," each with width $r\,d\theta$ and with "circumference" $A_{d-1}(r \sin \theta)$. The total "area" is again the integral from 0 to π:

$$
\begin{aligned}
A_d(r) &= \int_0^\pi A_{d-1}(r \sin \theta)\, r\, d\theta \\
&= \int_0^\pi \frac{2\pi^{(d-1)/2}}{\Gamma(\frac{d-1}{2})} (r \sin \theta)^{d-2}\, r\, d\theta \\
&= \frac{2\pi^{(d-1)/2}}{\Gamma(\frac{d-1}{2})} r^{d-1} \int_0^\pi (\sin \theta)^{d-2}\, d\theta.
\end{aligned}
\tag{B.27}
$$

In Problem B.14 you can show that

$$\int_0^\pi (\sin \theta)^n\, d\theta = \frac{\sqrt{\pi}\,\Gamma(\frac{n}{2} + \frac{1}{2})}{\Gamma(\frac{n}{2} + 1)}, \tag{B.28}$$

so that

$$A_d(r) = \frac{2\pi^{(d-1)/2}}{\Gamma(\frac{d-1}{2})} r^{d-1} \cdot \frac{\pi^{1/2}\Gamma(\frac{d}{2} - \frac{1}{2})}{\Gamma(\frac{d}{2})} = \frac{2\pi^{d/2}}{\Gamma(\frac{d}{2})} r^{d-1}, \tag{B.29}$$

as claimed.

Problem B.14. The proof of formula B.28 is by induction.

(a) Check formula B.28 for $n = 0$ and for $n = 1$.

(b) Show that

$$\int_0^\pi (\sin \theta)^n \, d\theta = \left(\frac{n-1}{n}\right) \int_0^\pi (\sin \theta)^{n-2} \, d\theta.$$

(Hint: First write $(\sin \theta)^n$ as $(\sin \theta)^{n-2}(1 - \cos^2 \theta)$. Integrate the second term by parts, differentiating one factor of $\cos \theta$ and integrating everything else.)

(c) Use the results of parts (a) and (b) to prove formula B.28 by induction.

Problem B.15. A cleaner, but much trickier, derivation of formula B.25 is similar to the method used in Section B.1 to evaluate the basic Gaussian integral. The trick is to consider the integral of the function e^{-r^2}, over all space, in d dimensions.

(a) First evaluate this integral in rectangular coordinates. You should obtain $\pi^{d/2}$.

(b) Because the integral has spherical symmetry, you can also evaluate it in d-dimensional spherical coordinates. Explain why the angular integrals must give a factor of $A_d(1)$, the area of a d-dimensional unit hypersphere. Thus, show that the integral is equal to $A_d(1) \cdot \int_0^\infty r^{d-1} e^{-r^2} \, dr$.

(c) Evaluate the integral over r in terms of the gamma function, and thus derive equation B.25.

Problem B.16. Derive a formula for the *volume* of a d-dimensional hypersphere.

B.5 Integrals of Quantum Statistics

In quantum statistics (Chapter 7) we frequently encounter integrals of the form

$$\int_0^\infty \frac{x^n}{e^x \pm 1} \, dx, \tag{B.30}$$

when summing over states for a system of bosons ($-$ in the denominator) or fermions ($+$ in the denominator). These integrals can, of course, be done numerically. When n is an odd integer, however, the answer can be expressed exactly in terms of π.

The first step is to rewrite the integral as an infinite series. Momentarily putting aside the factor of x^n, note that the rest of the integrand can be written as a geometric series:

$$
\begin{aligned}
\frac{1}{e^x \pm 1} &= \frac{e^{-x}}{1 \pm e^{-x}} = e^{-x} \mp (e^{-x})^2 + (e^{-x})^3 \mp \cdots \\
&= e^{-x} \mp e^{-2x} + e^{-3x} \mp e^{-4x} + \cdots .
\end{aligned} \tag{B.31}
$$

Now it's easy to multiply by x^n and integrate term by term. For the case $n = 1$ we obtain

$$
\begin{aligned}
\int_0^\infty \frac{x}{e^x \pm 1} \, dx &= \int_0^\infty \left(xe^{-x} \mp xe^{-2x} + xe^{-3x} \mp \cdots \right) dx \\
&= 1 \mp \frac{1}{2^2} + \frac{1}{3^2} \mp \frac{1}{4^2} + \cdots .
\end{aligned} \tag{B.32}
$$

This type of infinite series comes up a lot in mathematics, so mathematicians have given it a name. The **Riemann zeta function**, $\zeta(n)$, is defined as

$$\zeta(n) \equiv 1 + \frac{1}{2^n} + \frac{1}{3^n} + \cdots = \sum_{k=1}^{\infty} \frac{1}{k^n}. \tag{B.33}$$

Therefore we can write simply

$$\int_0^{\infty} \frac{x}{e^x - 1}\, dx = \zeta(2). \tag{B.34}$$

When the integrand has a plus in the denominator the series alternates, so we need to do a bit of manipulation:

$$\int_0^{\infty} \frac{x}{e^x + 1}\, dx = \left(1 + \frac{1}{2^2} + \frac{1}{3^2} + \cdots\right) - 2\left(\frac{1}{2^2} + \frac{1}{4^2} + \frac{1}{6^2} + \cdots\right)$$
$$= \zeta(2) - \frac{2}{2^2}\left(1 + \frac{1}{2^2} + \frac{1}{3^2} + \cdots\right)$$
$$= \zeta(2) - \frac{1}{2}\zeta(2)$$
$$= \frac{1}{2}\zeta(2). \tag{B.35}$$

It's only a little harder (see Problem B.17) to derive the more general results

$$\int_0^{\infty} \frac{x^n}{e^x - 1}\, dx = \Gamma(n+1)\zeta(n+1);$$
$$\int_0^{\infty} \frac{x^n}{e^x + 1}\, dx = \left(1 - \frac{1}{2^n}\right)\Gamma(n+1)\zeta(n+1). \tag{B.36}$$

(When n is an integer, $\Gamma(n+1) = n!$.)

Now the problem is "simply" to sum the infinite series B.33 that defines the Riemann zeta function. Unfortunately, getting a simple answer is not a simple task at all. I'll do it in a very tricky, roundabout way that uses a Fourier series.*

The trick is to consider a square-wave function, with period 2π and amplitude $\pi/4$ (see Figure B.7). **Fourier's theorem** states that any periodic function can be written as a linear superposition of sines and cosines. For an odd function such as ours only sines are necessary, so we can write

$$f(x) = \sum_{k=1}^{\infty} a_k \sin(kx), \tag{B.37}$$

---

*How anyone ever thought of this method in the first place is beyond me. I learned it from Mandl (1988).

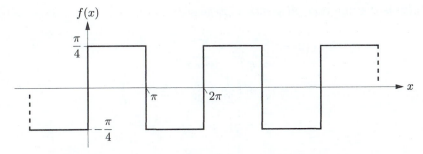

Figure B.7. A square-wave function with period 2π and amplitude $\pi/4$. The Fourier series for this function yields values of $\zeta(n)$ when n is an even integer.

for some set of coefficients a_k. Notice that the first sine wave in the sum has the same period as $f(x)$, while the rest have periods of $1/2$, $1/3$, $1/4$ as much, and so on. To solve for the coefficients we can use "Fourier's trick": Multiply by $\sin(jx)$ (where j is any positive integer) and integrate over one period of the function:

$$\int_0^{2\pi} f(x) \sin(jx)\, dx = \sum_{k=1}^{\infty} a_k \int_0^{2\pi} \sin(kx) \sin(jx)\, dx. \tag{B.38}$$

The integral on the right-hand side is zero except when $k = j$, when it equals π. Keeping only this nonzero term and renaming $j \to k$, we obtain, for any k,

$$a_k = \frac{1}{\pi} \int_0^{2\pi} f(x) \sin(kx)\, dx = \frac{2}{\pi} \int_0^{\pi} f(x) \sin(kx)\, dx. \tag{B.39}$$

This formula gives the Fourier coefficients of any odd function $f(x)$ with period 2π. For our square-wave function, the coefficients are

$$a_k = \frac{2}{\pi} \int_0^{\pi} \frac{\pi}{4} \sin(kx)\, dx = \begin{cases} 1/k & \text{for } k = 1,\, 3,\, 5,\, \ldots, \\ 0 & \text{for } k = 2,\, 4,\, 6,\, \ldots. \end{cases} \tag{B.40}$$

Therefore, for $0 < x < \pi$,

$$\frac{\pi}{4} = \sum_{\text{odd } k} \frac{\sin(kx)}{k}. \tag{B.41}$$

The final trick is to integrate this expression successively with respect to x, then evaluate the result at $\pi/2$. Integrating carefully from $x = 0$ to $x = x'$ gives

$$\frac{\pi x'}{4} = \sum_{\text{odd } k} \frac{1}{k} \int_0^{x'} \sin(kx)\, dx = \sum_{\text{odd } k} \frac{1}{k^2}\left(1 - \cos kx'\right), \tag{B.42}$$

and plugging in $x' = \pi/2$ yields simply

$$\frac{\pi^2}{8} = \sum_{\text{odd } k} \frac{1}{k^2}. \tag{B.43}$$

But $\zeta(2)$ is the sum over *all* positive integers:

$$\zeta(2) = \sum_{\text{odd } k} \frac{1}{k^2} + \sum_{\text{even } k} \frac{1}{k^2}$$

$$= \frac{\pi^2}{8} + \left(\frac{1}{2^2} + \frac{1}{4^2} + \frac{1}{6^2} + \cdots\right)$$

$$= \frac{\pi^2}{8} + \frac{1}{4}\left(\frac{1}{1^2} + \frac{1}{2^2} + \frac{1}{3^2} + \cdots\right)$$

$$= \frac{\pi^2}{8} + \frac{1}{4}\zeta(2). \tag{B.44}$$

In other words,

$$\zeta(2) = \frac{4}{3}\frac{\pi^2}{8} = \frac{\pi^2}{6}. \tag{B.45}$$

This result suffices to evaluate our original integral (B.30) for the case $n = 1$, with either sign in the denominator. For higher odd values of n the procedure is to take more derivatives of equation B.42, then again evaluate the result at $\pi/2$ and manipulate the series slightly (see Problem B.19). Unfortunately, this method does not yield any values of $\zeta(n)$ for odd n; in fact, these cannot be written in terms of π so they must be evaluated numerically.

Problem B.17. Derive the general integration formulas B.36.

Problem B.18. Use a computer to plot the sum of sine waves on the right-hand side of equation B.41, terminating the sum first at $k = 1$, then at $k = 3$, 5, 15, and 25. Notice how the series *does* converge to the square-wave function that we started with, but the convergence is not particularly fast.

Problem B.19. Integrate equation B.42 twice more, then plug in $x = \pi/2$ to obtain a formula for $\sum_{\text{odd}}(1/k^4)$. Use this formula to show that $\zeta(4) = \pi^4/90$, and thus evaluate the integrals B.36 for the case $n = 3$. Explain why this procedure does *not* yield a value for $\zeta(3)$.

Problem B.20. Evaluate equation B.41 at $x = \pi/2$, to obtain a famous series for π. How many terms in this series must you evaluate to obtain π to three significant figures?

Problem B.21. In calculating the heat capacity of a degenerate Fermi gas in Section 7.3, we needed the integral

$$\int_{-\infty}^{\infty} \frac{x^2 e^x}{(e^x + 1)^2}\, dx = \frac{\pi^2}{3}.$$

To derive this result, first show that the integrand is an even function, so it suffices to integrate from 0 to ∞ and then multiply by 2. Then integrate by parts to relate this integral to the one in equation B.35.

Problem B.22. Evaluate $\zeta(3)$ by numerically summing the series. How many terms do you need to keep to get an answer that is accurate to three significant figures?

Suggested Reading

Undergraduate Thermal Physics Texts

Callen, Herbert B., *Thermodynamics and an Introduction to Thermostatistics*, second edition (Wiley, New York, 1985). Develops thermodynamics from an abstract, logically rigorous approach. The application chapters are somewhat easier, and clearly written.

Carrington, Gerald, *Basic Thermodynamics* (Oxford University Press, Oxford, 1994). A nice introduction that sticks to pure classical thermodynamics.

Kittel, Charles, and Herbert Kroemer, *Thermal Physics*, second edition (W. H. Freeman, San Francisco, 1980). An insightful text with a great variety of modern applications.

Mandl, F., *Statistical Physics*, second edition (Wiley, Chichester, 1988). A clearly written text that emphasizes the statistical approach.

Reif, F., *Fundamentals of Statistical and Thermal Physics* (McGraw-Hill, New York, 1965). More advanced than most undergraduate texts. Emphasizes the statistical approach and includes extensive chapters on transport theory.

Stowe, Keith, *Introduction to Statistical Mechanics and Thermodynamics* (Wiley, New York, 1984). Perhaps the easiest book that takes the statistical approach. Very well written, but unfortunately marred by an incorrect treatment of chemical potential.

Zemansky, Mark W., and Richard H. Dittman, *Heat and Thermodynamics*, seventh edition (McGraw-Hill, New York, 1997). A classic text that includes good descriptions of experimental results and techniques. Earlier editions contain a wealth of material that didn't make it into the most recent edition; I especially like the fifth edition (1968, written by Zemansky alone).

Graduate-Level Texts

Chandler, David, *Introduction to Modern Statistical Mechanics* (Oxford University Press, New York, 1987). My favorite advanced text: short and well written, with lots of inviting problems. A partial solution manual is also available.

Landau, L. D., and E. M. Lifshitz, *Statistical Physics*, third edition, Part I, trans. J. B. Sykes and M. J. Kearsley (Pergamon Press, Oxford, 1980). An authoritative classic.

Pathria, R. K., *Statistical Mechanics*, second edition (Butterworth-Heinemann, Oxford, 1996). Good systematic coverage of statistical mechanics.

Pippard, A. B., *The Elements of Classical Thermodynamics* (Cambridge University Press, Cambridge, 1957). A concise summary of the theory as well as several applications.

Reichl, L., *A Modern Course in Statistical Physics*, second edition (Wiley, New York, 1998). Encyclopedic in coverage and very advanced.

Introductory Texts

Ambegaokar, Vinay, *Reasoning About Luck: Probability and its Uses in Physics* (Cambridge University Press, Cambridge, 1996). An elementary text that teaches probability theory and touches on many physical applications.

Fenn, John B., *Engines, Energy, and Entropy: A Thermodynamics Primer* (W. H. Freeman, San Francisco, 1982). A gentle introduction to classical thermodynamics, emphasizing everyday applications and featuring cartoons of Charlie the Caveman.

Feynman, Richard P., Robert B. Leighton, and Matthew Sands, *The Feynman Lectures on Physics* (Addison-Wesley, Reading, MA, 1963). Chapters 1, 3, 4, 6, and 39–46 treat topics in thermal physics, with Feynman's incredibly high density of deep insights per page.

Moore, Thomas A., *Six Ideas that Shaped Physics, Unit T* (McGraw-Hill, New York, 1998). This very clearly written text inspired my approach to the second law in Sections 2.2 and 2.3.

Reif, F., *Statistical Physics: Berkeley Physics Course—Volume 5* (McGraw-Hill, New York, 1967). A rather advanced introduction, but much more leisurely than Reif (1965).

Popularizations

Atkins, P. W., *The Second Law* (Scientific American Books, New York, 1984). A nice coffee-table book with lots of pictures.

Goldstein, Martin, and Inge F. Goldstein, *The Refrigerator and the Universe* (Harvard University Press, Cambridge, MA, 1993). An extensive tour of thermal physics and its diverse applications.

Zemansky, Mark W., *Temperatures Very Low and Very High* (Van Nostrand, Princeton, 1964; reprinted by Dover, New York, 1981). A short paperback that focuses on physics at extreme temperatures. Very enjoyable reading, except when the author slips into textbook mode.

Engines and Refrigerators

Moran, Michael J., and Howard N. Shapiro, *Fundamentals of Engineering Thermodynamics*, third edition (Wiley, New York, 1995). One of several good encyclopedic texts.

Whalley, P. B., *Basic Engineering Thermodynamics* (Oxford University Press, Oxford, 1992). Refreshingly concise.

Chemical Thermodynamics

Atkins, P. W., *Physical Chemistry*, sixth edition (W. H. Freeman, New York, 1998). One of several good physical chemistry texts, packed with information.

Findlay, Alexander, *Phase Rule*, ninth edition, revised by A. N. Campbell and N. O. Smith (Dover, New York, 1951). Everything you ever wanted to know about phase diagrams.

Haasen, Peter, *Physical Metallurgy*, third edition, trans. Janet Mordike (Cambridge University Press, Cambridge, 1996). An authoritative monograph that doesn't shy away from the physics.

Rock, Peter A., *Chemical Thermodynamics* (University Science Books, Mill Valley, CA, 1983). A well-written introduction to chemical thermodynamics with plenty of interesting applications.

Smith, E. Brian, *Basic Chemical Thermodynamics*, fourth edition (Oxford University Press, Oxford, 1990). A nice short book that covers the basics.

Biology

Asimov, Isaac, *Life and Energy* (Doubleday, Garden City, NY, 1962). A popular account of thermodynamics and its applications in biochemistry. Old but still very good.

Stryer, Lubert, *Biochemistry*, fourth edition (W. H. Freeman, New York, 1995). Marvelously detailed, though not as quantitative as one might like.

Tinoco, Ignacio, Jr., Kenneth Sauer, and James C. Wang, *Physical Chemistry: Principles and Applications in Biological Sciences*, third edition (Prentice-Hall, Englewood Cliffs, NJ, 1995). Less comprehensive than a standard physical chemistry text, but with many more biochemical applications.

Earth and Environmental Science

Anderson, G. M., *Thermodynamics of Natural Systems* (Wiley, New York, 1996). A practical introduction to chemical thermodynamics, with a special emphasis on geological applications.

Bohren, Craig F., *Clouds in a Glass of Beer: Simple Experiments in Atmospheric Physics* (Wiley, New York, 1987). Short, elementary, and delightful. Begins by observing that "a glass of beer is a cloud inside out." Bohren has also written a sequel, *What Light Through Yonder Window Breaks?* (Wiley, New York, 1991).

Bohren, Craig F., and Bruce A. Albrecht, *Atmospheric Thermodynamics* (Oxford University Press, New York, 1998). Though intended for meteorology students, this textbook will appeal to anyone who knows basic physics and is curious about the everyday world. Great fun to read and full of food for thought.

Harte, John, *Consider a Spherical Cow: A Course in Environmental Problem Solving* (University Science Books, Sausalito, CA, 1988). A wonderful book that applies undergraduate-level physics and mathematics to dozens of interesting environmental problems.

Kern, Raymond, and Alain Weisbrod, *Thermodynamics for Geologists*, trans. Duncan McKie (Freeman, Cooper and Company, San Francisco, 1967). Features a nice selection of worked examples.

Nordstrom, Darrell Kirk, and James L. Munoz, *Geochemical Thermodynamics*, second edition (Blackwell Scientific Publications, Palo Alto, CA, 1994). A well-written advanced textbook for serious geochemists.

Astrophysics and Cosmology

Carroll, Bradley W., and Dale A. Ostlie, *An Introduction to Modern Astrophysics* (Addison-Wesley, Reading, MA, 1996). A clear, comprehensive introduction to astrophysics at the intermediate undergraduate level.

Peebles, P. J. E., *Principles of Physical Cosmology* (Princeton University Press, Princeton, NJ, 1993). An advanced treatise on cosmology with a detailed discussion of the thermal history of the early universe.

Shu, Frank H., *The Physical Universe: An Introduction to Astronomy* (University Science Books, Mill Valley, CA, 1982). An astrophysics book for physics students, disguised as an introductory astronomy text. Full of physical insight, this book portrays all of astrophysics as a competition between gravity and the second law of thermodynamics.

Weinberg, Steven, *The First Three Minutes* (Basic Books, New York, 1977). A classic account of the history of the early universe. Written for lay readers, yet gives a physicist plenty to think about.

Condensed Matter Physics

Ashcroft, Neil W., and N. David Mermin, *Solid State Physics* (Saunders College, Philadelphia, 1976). An excellent text that is somewhat more advanced than Kittel (below).

Collings, Peter J., *Liquid Crystals: Nature's Delicate Phase of Matter* (Princeton University Press, Princeton, NJ, 1990). A short, elementary overview of both the basic physics and applications.

Goodstein, David L., *States of Matter* (Prentice-Hall, Englewood Cliffs, NJ, 1975; reprinted by Dover, New York, 1985). A well written graduate-level text that surveys the properties of gases, liquids, and solids.

Gopal, E. S. R., *Specific Heats at Low Temperatures* (Plenum, New York, 1966). A nice short monograph that emphasizes comparisons between theory and experiment.

Kittel, Charles, *Introduction to Solid State Physics*, seventh edition (Wiley, New York, 1996). The classic undergraduate text.

Wilks, J., and D. S. Betts, *An Introduction to Liquid Helium*, second edition (Oxford University Press, Oxford, 1987). A concise and reasonably accessible overview.

Yeomans, J. M., *Statistical Mechanics of Phase Transitions* (Oxford University Press, Oxford, 1992). A brief, readable introduction to the theory of critical phenomena.

Computer Simulations

Gould, Harvey, and Jan Tobochnik, *An Introduction to Computer Simulation Methods*, second edition (Addison-Wesley, Reading, MA, 1996). Covers far-ranging applications at a variety of levels, including plenty of statistical mechanics.

Whitney, Charles A., *Random Processes in Physical Systems: An Introduction to Probability-Based Computer Simulations* (Wiley, New York, 1990). A good elementary textbook that takes you from coin flipping to stellar pulsations.

History and Philosophy

Bailyn, Martin, *A Survey of Thermodynamics* (American Institute of Physics, New York, 1994). A textbook that gives a good deal of history on each topic covered.

Brush, Stephen G., *The Kind of Motion We Call Heat: A History of the Kinetic Theory of Gases in the 19th Century* (North-Holland, Amsterdam, 1976). A very scholarly treatment.

Kestin, Joseph (ed.), *The Second Law of Thermodynamics* (Dowden, Hutchinson & Ross, Stroudsburg, PA, 1976). Reprints (in English) of original papers by Carnot, Clausius, Thomson, and others, with helpful editorial comments.

Leff, Harvey S., and Andrew F. Rex (eds.), *Maxwell's Demon: Entropy, Information, Computing* (Princeton University Press, Princeton, NJ, 1990). An anthology of important papers on the meaning of entropy.

Mendelssohn, K., *The Quest for Absolute Zero*, second edition (Taylor & Francis, London, 1977). A popular history of low-temperature physics, from the liquefaction of oxygen to the properties of superfluid helium.

Von Baeyer, Hans Christian, *Maxwell's Demon: Why Warmth Disperses as Time Passes* (Random House, New York, 1998). A brief popular history of thermal physics with an emphasis on the deeper issues. Highly recommended.

Tables of Thermodynamic Data

Keenan, Joseph H., Frederick G. Keyes, Philip G. Hill, and Joan G. Moore, *Steam Tables (S.I. Units)* (Wiley, New York, 1978). Fascinating.

Lide, David R. (ed.), *CRC Handbook of Chemistry and Physics*, 75th edition (Chemical Rubber Company, Boca Raton, FL, 1994). Cumbersome but widely available. Editions published since 1990 are better organized and use more modern units.

National Research Council, *International Critical Tables of Numerical Data* (McGraw-Hill, New York, 1926–33). A seven-volume compendium of a great variety of data.

Reynolds, William C., *Thermodynamic Properties in SI* (Stanford University Dept. of Mechanical Engineering, Stanford, CA, 1979). A handy compilation of properties of 40 important fluids.

Vargaftik, N. B., *Handbook of Physical Properties of Liquids and Gases* (Hemisphere, Washington, DC, 1997). Detailed property tables for a variety of fluids.

Woolley, Harold W., Russell B. Scott, and F. G. Brickwedde, "Compilation of Thermal Properties of Hydrogen in its Various Isotopic and Ortho-Para Modifications," *Journal of Research of the National Bureau of Standards* **41**, 379–475 (1948). Definitive but not very accessible.

*   *   *

An awkward aspect of reading any new textbook is getting used to the notation. Fortunately, many of the notations of thermal physics have become widely accepted and standardized through decades of use. There are several important exceptions, however, including the following:

Quantity	This book	Other symbols
Total energy	U	E
Multiplicity	Ω	W, g
Helmholtz free energy	F	A
Gibbs free energy	G	F
Grand free energy	Φ	Ω
Partition function	Z	Q, q
Maxwell speed distribution	$\mathcal{D}(v)$	$P(v)$
Quantum length	ℓ_Q	λ, λ_T
Quantum volume	v_Q	λ_T^3, $1/n_Q$
Fermi-Dirac distribution	$\overline{n}_{\text{FD}}(\epsilon)$	$f(\epsilon)$
Density of states	$g(\epsilon)$	$D(\epsilon)$

Reference Data

Physical Constants

$$k = 1.381 \times 10^{-23} \text{ J/K}$$
$$= 8.617 \times 10^{-5} \text{ eV/K}$$
$$N_A = 6.022 \times 10^{23}$$
$$R = 8.315 \text{ J/mol·K}$$
$$h = 6.626 \times 10^{-34} \text{ J·s}$$
$$= 4.136 \times 10^{-15} \text{ eV·s}$$
$$c = 2.998 \times 10^8 \text{ m/s}$$
$$G = 6.673 \times 10^{-11} \text{ N·m}^2/\text{kg}^2$$
$$e = 1.602 \times 10^{-19} \text{ C}$$
$$m_e = 9.109 \times 10^{-31} \text{ kg}$$
$$m_p = 1.673 \times 10^{-27} \text{ kg}$$

Unit Conversions

$$1 \text{ atm} = 1.013 \text{ bar} = 1.013 \times 10^5 \text{ N/m}^2$$
$$= 14.7 \text{ lb/in}^2 = 760 \text{ mm Hg}$$
$$(T \text{ in } °C) = (T \text{ in } K) - 273.15$$
$$(T \text{ in } °F) = \tfrac{9}{5}(T \text{ in } °C) + 32$$
$$1 \text{ } °R = \tfrac{5}{9} \text{ K}$$
$$1 \text{ cal} = 4.186 \text{ J}$$
$$1 \text{ Btu} = 1054 \text{ J}$$
$$1 \text{ eV} = 1.602 \times 10^{-19} \text{ J}$$
$$1 \text{ u} = 1.661 \times 10^{-27} \text{ kg}$$

The atomic number (top left) is the number of protons in the nucleus. The atomic mass (bottom) is weighted by isotopic abundances in the earth's surface. Atomic masses are relative to the mass of the carbon-12 isotope, defined to be exactly 12 unified atomic mass units (u). Uncertainties range from 1 to 9 in the last digit quoted. Relative isotopic abundances often vary considerably, both in natural and commercial samples. A number in parentheses is the mass of the longest-lived isotope of that element—no stable isotope exists. However, although Th, Pa, and U have no stable isotopes, they do have characteristic terrestrial compositions, and meaningful weighted masses can be given. For elements 110–112, the mass numbers of known isotopes are given. From the Review of Particle Physics by the Particle Data Group, *The European Physical Journal* **C3**, 73 (1998).

Periodic Table of the Elements

1 IA	2 IIA	3 IIIB	4 IVB	5 VB	6 VIB	7 VIIB	8 VIII	9 VIII	10 VIII	11 IB	12 IIB	13 IIIA	14 IVA	15 VA	16 VIA	17 VIIA	18 VIIIA
1 **H** Hydrogen 1.00794																	2 **He** Helium 4.002602
3 **Li** Lithium 6.941	4 **Be** Beryllium 9.012182											5 **B** Boron 10.811	6 **C** Carbon 12.0107	7 **N** Nitrogen 14.0674	8 **O** Oxygen 15.9994	9 **F** Fluorine 18.9984032	10 **Ne** Neon 20.1797
11 **Na** Sodium 22.989770	12 **Mg** Magnesium 24.3050											13 **Al** Aluminum 26.981538	14 **Si** Silicon 28.0855	15 **P** Phosph. 30.973761	16 **S** Sulfur 32.066	17 **Cl** Chlorine 35.4527	18 **Ar** Argon 39.948
19 **K** Potassium 39.0983	20 **Ca** Calcium 40.078	21 **Sc** Scandium 44.955910	22 **Ti** Titanium 47.867	23 **V** Vanadium 50.9415	24 **Cr** Chromium 51.9961	25 **Mn** Manganese 54.938049	26 **Fe** Iron 55.845	27 **Co** Cobalt 58.933200	28 **Ni** Nickel 58.6934	29 **Cu** Copper 63.546	30 **Zn** Zinc 65.39	31 **Ga** Gallium 69.723	32 **Ge** German. 72.61	33 **As** Arsenic 74.92160	34 **Se** Selenium 78.96	35 **Br** Bromine 79.904	36 **Kr** Krypton 83.80
37 **Rb** Rubidium 85.4678	38 **Sr** Strontium 87.62	39 **Y** Yttrium 88.90585	40 **Zr** Zirconium 91.224	41 **Nb** Niobium 92.90638	42 **Mo** Molybd. 95.94	43 **Tc** Technet. (97.907215)	44 **Ru** Ruthen. 101.07	45 **Rh** Rhodium 102.90550	46 **Pd** Palladium 106.42	47 **Ag** Silver 107.8682	48 **Cd** Cadmium 112.411	49 **In** Indium 114.818	50 **Sn** Tin 118.710	51 **Sb** Antimony 121.760	52 **Te** Tellurium 127.60	53 **I** Iodine 126.90447	54 **Xe** Xenon 131.29
55 **Cs** Cesium 132.90545	56 **Ba** Barium 137.327	57–71 Lantha-nides	72 **Hf** Hafnium 178.49	73 **Ta** Tantalum 180.9479	74 **W** Tungsten 183.84	75 **Re** Rhenium 186.207	76 **Os** Osmium 190.23	77 **Ir** Iridium 192.217	78 **Pt** Platinum 195.078	79 **Au** Gold 196.96655	80 **Hg** Mercury 200.59	81 **Tl** Thallium 204.3833	82 **Pb** Lead 207.2	83 **Bi** Bismuth 208.98038	84 **Po** Polonium (208.982415)	85 **At** Astatine (209.987131)	86 **Rn** Radon (222.017570)
87 **Fr** Francium (223.019731)	88 **Ra** Radium (226.025402)	89–103 Actinides	104 **Rf** Rutherford. (261.1089)	105 **Db** Dubnium (262.1144)	106 **Sg** Seaborg. (263.1186)	107 **Bh** Bohrium (262.1231)	108 **Hs** Hassium (265.1306)	109 **Mt** Meitner. (266.1378)	110 (269, 273)	111 (272)	112 (277)						

Lanthanide series

57 **La** Lanthanum 138.9055	58 **Ce** Cerium 140.116	59 **Pr** Praseodym. 140.90765	60 **Nd** Neodym. 144.24	61 **Pm** Prometh. (144.912745)	62 **Sm** Samarium 150.36	63 **Eu** Europium 151.964	64 **Gd** Gadolin. 157.25	65 **Tb** Terbium 158.92534	66 **Dy** Dyspros. 162.50	67 **Ho** Holmium 164.93032	68 **Er** Erbium 167.26	69 **Tm** Thulium 168.93421	70 **Yb** Ytterbium 173.04	71 **Lu** Lutetium 174.967

Actinide series

89 **Ac** Actinium (227.027747)	90 **Th** Thorium 232.0381	91 **Pa** Protactin. 231.03588	92 **U** Uranium 238.0289	93 **Np** Neptunium (237.048166)	94 **Pu** Plutonium (244.064197)	95 **Am** Americium (243.061372)	96 **Cm** Curium (247.070346)	97 **Bk** Berkelium (247.070298)	98 **Cf** Californ. (251.079579)	99 **Es** Einstein. (252.082997)	100 **Fm** Fermium (257.095096)	101 **Md** Mendelev. (258.098427)	102 **No** Nobelium (259.1011)	103 **Lr** Lawrenc. (262.1098)

Thermodynamic Properties of Selected Substances

All of the values in this table are for one mole of material at 298 K and 1 bar. Following the chemical formula is the form of the substance, either solid (s), liquid (l), gas (g), or aqueous solution (aq). When there is more than one common solid form, the mineral name or crystal structure is indicated. Data for aqueous solutions are at a standard concentration of 1 mole per kilogram water. The enthalpy and Gibbs free energy of formation, $\Delta_f H$ and $\Delta_f G$, represent the changes in H and G upon forming one mole of the material starting with elements in their most stable pure states (e.g., C (graphite), O_2 (g), etc.). To obtain the value of ΔH or ΔG for another reaction, subtract Δ_f of the reactants from Δ_f of the products. For ions in solution there is an ambiguity in dividing thermodynamic quantities between the positive and negative ions; by convention, H^+ is assigned the value zero and all others are chosen to be consistent with this value. Data from Atkins (1998), Lide (1994), and Anderson (1996). Please note that, while these data are sufficiently accurate and consistent for the examples and problems in this textbook, not all of the digits shown are necessarily significant; for research purposes you should always consult original literature to determine experimental uncertainties.

Substance (form)	$\Delta_f H$ (kJ)	$\Delta_f G$ (kJ)	S (J/K)	C_P (J/K)	V (cm^3)
Al (s)	0	0	28.33	24.35	9.99
Al$_2$SiO$_5$ (kyanite)	−2594.29	−2443.88	83.81	121.71	44.09
Al$_2$SiO$_5$ (andalusite)	−2590.27	−2442.66	93.22	122.72	51.53
Al$_2$SiO$_5$ (sillimanite)	−2587.76	−2440.99	96.11	124.52	49.90
Ar (g)	0	0	154.84	20.79	
C (graphite)	0	0	5.74	8.53	5.30
C (diamond)	1.895	2.900	2.38	6.11	3.42
CH$_4$ (g)	−74.81	−50.72	186.26	35.31	
C$_2$H$_6$ (g)	−84.68	−32.82	229.60	52.63	
C$_3$H$_8$ (g)	−103.85	−23.49	269.91	73.5	
C$_2$H$_5$OH (l)	−277.69	−174.78	160.7	111.46	58.4
C$_6$H$_{12}$O$_6$ (glucose)	−1273	−910	212	115	
CO (g)	−110.53	−137.17	197.67	29.14	
CO$_2$ (g)	−393.51	−394.36	213.74	37.11	
H$_2$CO$_3$ (aq)	−699.65	−623.08	187.4		
HCO$_3^-$ (aq)	−691.99	−586.77	91.2		
Ca^{2+} (aq)	−542.83	−553.58	−53.1		
CaCO$_3$ (calcite)	−1206.9	−1128.8	92.9	81.88	36.93
CaCO$_3$ (aragonite)	−1207.1	−1127.8	88.7	81.25	34.15
CaCl$_2$ (s)	−795.8	−748.1	104.6	72.59	51.6
Cl$_2$ (g)	0	0	223.07	33.91	
Cl$^-$ (aq)	−167.16	−131.23	56.5	−136.4	17.3
Cu (s)	0	0	33.150	24.44	7.12
Fe (s)	0	0	27.28	25.10	7.11

Substance (form)	$\Delta_f H$ (kJ)	$\Delta_f G$ (kJ)	S (J/K)	C_P (J/K)	V (cm^3)
H$_2$ (g)	0	0	130.68	28.82	
H (g)	217.97	203.25	114.71	20.78	
H$^+$ (aq)	0	0	0	0	
H$_2$O (l)	−285.83	−237.13	69.91	75.29	18.068
H$_2$O (g)	−241.82	−228.57	188.83	33.58	
He (g)	0	0	126.15	20.79	
Hg (l)	0	0	76.02	27.98	14.81
N$_2$ (g)	0	0	191.61	29.12	
NH$_3$ (g)	−46.11	−16.45	192.45	35.06	
Na$^+$ (aq)	−240.12	−261.91	59.0	46.4	−1.2
NaCl (s)	−411.15	−384.14	72.13	50.50	27.01
NaAlSi$_3$O$_8$ (albite)	−3935.1	−3711.5	207.40	205.10	100.07
NaAlSi$_2$O$_6$ (jadeite)	−3030.9	−2852.1	133.5	160.0	60.40
Ne (g)	0	0	146.33	20.79	
O$_2$ (g)	0	0	205.14	29.38	
O$_2$ (aq)	−11.7	16.4	110.9		
OH$^-$ (aq)	−229.99	−157.24	−10.75	−148.5	
Pb (s)	0	0	64.81	26.44	18.3
PbO$_2$ (s)	−277.4	−217.33	68.6	64.64	
PbSO$_4$ (s)	−920.0	−813.0	148.5	103.2	
SO$_4^{2-}$ (aq)	−909.27	−744.53	20.1	−293	
HSO$_4^-$ (aq)	−887.34	−755.91	131.8	−84	
SiO$_2$ (α quartz)	−910.94	−856.64	41.84	44.43	22.69
H$_4$SiO$_4$ (aq)	−1449.36	−1307.67	215.13	468.98	

Index